MODULAR MATHEMATICS

Module B: Pure Maths 2

L. Bostock, B.Sc.

S. Chandler, B.Sc.

Stanley Thornes (Publishers) Ltd

First published in 1991 by Stanley Thornes (Publishers) Ltd,
Old Station Drive, Leckhampton, CHELTENHAM GL53 0DN

British Library Cataloguing in Publication Data

Bostock, L.
    Modular mathematics: module B: pure maths 2.
    – (Modular mathematics)
    I. Title II. Chandler, S. III. Series
    510

    ISBN 0-7487-1110-4

Typeset by Tech-Set, Gateshead, Tyne & Wear.
Printed and bound in Great Britain at The Bath Press, Avon.

CONTENTS

vi

PREFACE

This is the second book in a series of modular mathematics courses for students wishing to gain academic qualifications beyond GCSE. An AS-level subject in mathematics requires two modules and an A-level needs four modules.

This book, together with Module A, covers the work necessary for AS-level Pure Mathematics and the pure mathematics content of A-level mathematics. Further modules will be produced in pure mathematics, mechanics and statistics to cover a variety of options for both AS-level and A-level mathematics subjects.

The contents of this book assumes knowledge of the topics in Module A; consolidation sections appear at regular intervals throughout the book, each one contains a summary of the work in preceding chapters, a set of multiple choice questions and an exercise containing examination questions. The multiple choice questions are useful for self-testing even if they do not form part of the assessment of the examination to be taken. The examination questions are not intended for use immediately a topic has been covered. They are better used later for revision and practice once confidence has been built up.

There are many computer programmes that aid in the understanding and the use of mathematics. In particular, a good graph drawing package is invaluable. In a few places we have indicated a program that we think is relevant. This is either *Super Graph* or *132 Short Programs for the Mathematics Classroom*. *Super Graph* by David Tall is a flexible graph drawing package and is available from Glentop Press Ltd. *132 Short Programs for the Mathematics Classroom* is produced by The Mathematical Association and is published in book form by Stanley Thornes (Publishers) Ltd.

We are grateful to the following examination boards for permission to reproduce questions from their past examination papers.
(Part questions are indicated by the suffix p):

University of London (U of L)
Joint Matriculation Board (JMB)
University of Cambridge Local Examinations Syndicate (C)
The Associated Examining Board (AEB)
Welsh Joint Education Committee (WJEC)
Cambridge, Oxford and Southern Universities Examination Council (O/C, SU & C)

1991   L. Bostock
                                                             S. Chandler

NOTES ON USE OF THE BOOK

Notation

$=$	is equal to	$\Leftrightarrow$	implies and is implied by
$\equiv$	is identical to	$\in$	is a member of
$\approx$	is approximately equal to*	$:$	is such that
$>$	is greater than	$\mathbb{N}$	the natural numbers
$\geqslant$	is greater than or equal to	$\mathbb{Z}$	the integers
$<$	is less than	$\mathbb{Q}$	the rational numbers
$\leqslant$	is less than or equal to	$\mathbb{R}$	the real numbers
∞	infinity; infinitely large	$\mathbb{R}^+$	the positive real numbers
$\propto$	is proportional to		excluding zero
$\rightarrow$	maps to	$\mathbb{C}$	the complex numbers
	or approaches	$[a, b]$	the interval $\{x : a \leqslant x \leqslant b\}$
$\Rightarrow$	implies		

A stroke through a symbol negates it, e.g. $\neq$ means 'is not equal to'

Abbreviations

$\parallel$	is parallel to	w.r.t.	with respect to
+ve	positive	exp	exponential, e.g. $\exp x$ means e^x
−ve	negative		

Useful Formulae

For a cone with base radius r, height h and slant height l

volume $= \frac{1}{3}\pi r^2 h$ curved surface area $= \pi r l$

For a sphere of radius r

volume $= \frac{4}{3}\pi r^3$ surface area $= 4\pi r^2$

For any pyramid with height h and base area a

volume $= \frac{1}{3}ah$

*Practical problems rarely have exact answers. Where numerical answers are given they are correct to two or three decimal places depending on their context, e.g. π is 3.142 correct to 3 d.p. and although we write $\pi = 3.142$ it is understood that this is not an exact value. We reserve the symbol $\approx$ for those cases where the approximation being made is part of the method used.

Computer Program References

Marginal symbols indicate a computer program which is helpful, programs being identified in the following manner,

47 Program No. 47 from *132 Short Programs for the Mathematics Classroom*

SG *Super Graph*

Instructions for Answering Multiple Choice Exercises

These exercises are included in each consolidation section. The questions are set in groups; the answering techniques are different for each group and are classified as follows:

TYPE I
These questions consist of a problem followed by several alternative answers, only *one* of which is correct.

Write down the letter corresponding to the correct answer.

TYPE II
In this type of question some information is given and is followed by a number of responses. *One or more* of these follow(s) directly and necessarily from the information given.

Write down the letter(s) corresponding to the correct response(s).
e.g. PQR is a triangle

    A   PQ + QR is less than PR
    B   if $\angle P$ is obtuse, $\angle Q$ and $\angle R$ must both be acute
    C $\angle P = 90°$, $\angle Q = 45°$, $\angle R = 45°$

The correct response is **B**.
A is definitely incorrect and **C** may or may not be true of triangle PQR, i.e. it does not follow directly and necessarily from the information given. Responses of this kind should not be regarded as correct.

TYPE III
A single statement is made. Write T if it is true and F if it is false.

CHAPTER 1

FUNCTIONS

POLYNOMIAL FUNCTIONS

The general form of a polynomial function is

$$f(x) = a_n x^n + a_{n-1} x^{n-1} + \ldots + a_2 x^2 + a_1 x + a_0$$

where $a_n, a_{n-1}, \ldots, a_0$ are constants, n is a positive integer and $a_n \neq 0$

Examples of polynomials are

$$f(x) = 3x^4 - 2x^3 + 5, \quad f(x) = x^5 - 2x^3 + x, \quad f(x) = x^2$$

The *order* of a polynomial is the highest power of x in the function. Thus, the order of $x^4 - 7$ is 4, and the order of $2x - 1$ is 1

We have already investigated the graphs of polynomials of order 1,

e.g. $\quad f(x) = 2x - 1$

which gives a straight line,

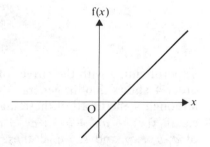

and of order 2,

e.g. $\quad f(x) = x^2 - 4$

which gives a parabola,

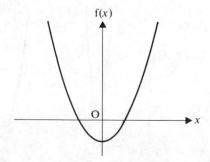

and of order 3,

e.g. $f(x) = x^3 - 2x + 1$

which gives a cubic curve.

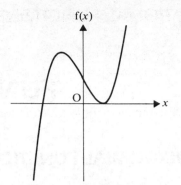

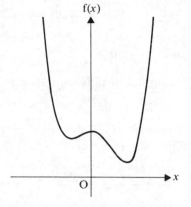

The shape of the curve
$f(x) = x^4 - 3x^3 + 2x^2 + 1$
looks like this.

Experimenting with the curves of other polynomial functions of order 4 shows that, in general, the curve has three turning points although some or all of these may merge,
e.g. if $f(x) = ax^4 + bx^3 + cx^2 + dx + e,$ then for various values of a, b, c, d and e, we get these curves.

$a > 0$

$a > 0$

$a < 0$

RATIONAL FUNCTIONS

> A rational function is one in which both numerator and denominator are polynomials.

Examples of rational functions of x are

$$\frac{1}{x}, \quad \frac{x}{x^2 - 1}, \quad \frac{3x^2 + 2x}{x - 1}$$

The graph of the simplest rational function, $f(x) = \dfrac{1}{x}$, is familiar.

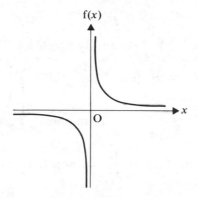

As $x \to \pm\infty$, $\dfrac{1}{x}$ tends to zero, so the curve approaches the x-axis but does not cross it.

Also, as $x \to 0$, $\dfrac{1}{x} \to \pm\infty$, so the curve approaches the line $x = 0$ but does not cross it, i.e. the line $x = 0$ is an asymptote. Further, when $x = 0$, $\dfrac{1}{x}$ is undefined, so $x = 0$ must be excluded from the domain of f.

If, for any rational function f, c is a value of x for which the denominator is zero then

1) $f(x)$ is undefined when $x = c$, so c must be excluded from the domain of f,

2) as $x \to c$, $f(x) \to \pm\infty$, so $x = c$ is a vertical asymptote to the curve $y = f(x)$.

Example 1a

Sketch the graph of the function given by $f(x) = \dfrac{1}{2-x}$

$f(x) = \dfrac{1}{2-x}$ does not exist when $x = 2$, so the curve $y = f(x)$
does not cross the line $x = 2$

As $x \to 2$ from above, $2 - x$ is negative and approaches zero,

$$\text{so} \qquad \frac{1}{2-x} \to -\infty$$

As $x \to 2$ from below, $2 - x$ is positive and approaches zero,

$$\text{so} \qquad \frac{1}{2-x} \to \infty$$

Therefore the line $x = 2$ is an asymptote.

As $x \to \infty$, $\dfrac{1}{2-x} \to 0$ from below

and as $x \to -\infty$, $\dfrac{1}{2-x} \to 0$ from above

$\left. \right\}$ $\therefore$ the x-axis is an asymptote.

As this is similar to $f(x) = \dfrac{1}{x}$, we now have enough information to
sketch the graph.

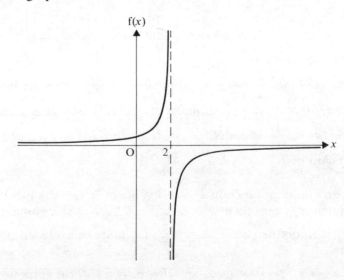

EXPONENTIAL FUNCTIONS

Exponent is another word for index or power.

An exponential function is one where the variable is in the index.

For example, 2^x, 3^{-x}, 10^{x+1} are exponential functions of x

Consider the function $f(x) = 2^x$ for which a table of corresponding values of x and $f(x)$ and a graph are given below.

x	$-\infty \leftarrow \ldots$	-10	-1	$-\frac{1}{10}$	0	$\frac{1}{10}$	1	10	$\ldots \rightarrow \infty$
$f(x)$	$0 \leftarrow \ldots$	$\frac{1}{1024}$	$\frac{1}{2}$	0.93	1	1.07	2	1024	$\ldots \rightarrow \infty$

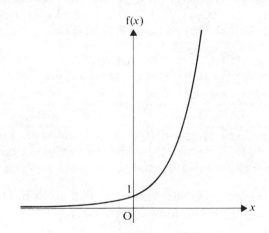

From these we see that

1) 2^x has a real value for all real values of x and 2^x is positive for all values of x, i.e. the range of f is $f(x) > 0$

2) As $x \rightarrow -\infty$, $f(x) \rightarrow 0$, i.e. the x-axis is an asymptote.

3) As x increases, $f(x)$ increases at a rapidly accelerating rate.

Note also that the curve crosses the y-axis at $(0, 1)$, i.e. $f(0) = 1$. In fact for any function of the form $f(x) = a^x$, where a is a constant and greater than 1, $f(0) = 1$, and the curve representing it is similar in shape to that for 2^x.

Exponential functions can be used as mathematical models for many growth and decay situations. For example, if inflation is running at a steady 10% p.a., then what £1 buys at the beginning of a year will cost £(1.1) after one year, £(1.1)^2 after two years, £(1.1)^3 after three years, and so on. If after x years the cost is £y, then the relationship between x and y is

$$y = (1.1)^x$$

EXERCISE 1a

1. Draw sketch graphs of the following functions.

   (a) 3^x (b) $\dfrac{1}{2x}$ (c) 4^{2x} (d) $\dfrac{1}{x-3}$

2. If $f(x) = (\frac{1}{2})^x$ write down the values of $f(x)$ corresponding to $x = -4, -3, -2, -1, 0, 1, 2, 3$ and 4. From these values deduce the behaviour of $f(x)$ as $x \to \pm\infty$ and hence sketch the graph of the function.

3. What value of x must be excluded from the domain of

   $$f(x) = \frac{1}{x+2}?$$

   Describe the behaviour of $f(x)$ as x approaches this value from above and from below. Describe also the behaviour of $f(x)$ as $x \to \pm\infty$. Using this information to sketch the graph of $f(x)$.

4. By following a procedure similar to that given in Question 3 draw sketch graphs of the following functions.

   (a) $-\dfrac{1}{x}$ (b) $\dfrac{1}{1-2x}$ (c) $\dfrac{2}{x+1}$ (d) $1 + \dfrac{1}{x}$

5. Find the values of x where the curve $y = f(x)$ cuts the x-axis and sketch the curve when

   (a) $f(x) = x(x-1)(x+1)$ (b) $f(x) = x(x-1)(x+1)(x-2)$

   (c) $f(x) = (x^2-1)(2-x)$ (d) $f(x) = (x^2-1)(4-x^2)$

6. The radioactivity of a certain substance is found to decrease at a rate such that the reading on a Geiger counter is $\frac{9}{10}$ of the reading 24 hours earlier. If the initial reading is 3, what is the reading

   (a) one day later   (b) two days later   (c) three days later?

   If the reading x days later is y, what is the relationship between x and y?

PERIODIC FUNCTIONS

A function whose graph consists of a basic pattern which repeats at regular intervals is called a *periodic* function. The width of the basic pattern is the *period* of the function.

$f(x) = \sin x$, for example, is periodic and its period is 2π

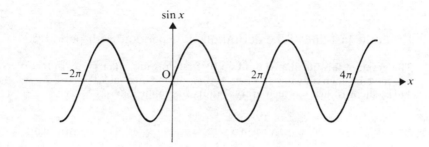

If $f(x)$ is periodic with period a, then it follows that

$$f(x + a) = f(x) \quad \text{for all values of } x$$

Therefore the definition of $f(x)$ within one period (i.e. for $0 < x \leqslant a$) together with the definition $f(x + a) = f(x)$ for $x \in \mathbb{R}$, defines a periodic function.

For example, if $\begin{cases} f(x) = 2x - 1 & \text{for } 0 < x \leqslant 1 \\ f(x + 1) = f(x) & \text{for all values of } x \end{cases}$

then we know that the function is periodic with a period of 1

The graph of this function can be sketched by drawing $f(x) = 2x - 1$ for $0 < x \leqslant 1$ and repeating the pattern at unit intervals in either direction.

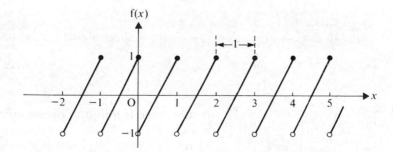

The basic pattern in the graph of a periodic function can be made up of two or more different definitions. The next worked example illustrates such a compound periodic function.

Example 1b

Sketch the graph of the function f defined by

$$f(x) = x \qquad \text{for} \quad 0 \leqslant x < 1$$
$$f(x) = 4 - x^2 \quad \text{for} \quad 1 \leqslant x < 2$$
$$f(x + 2) = f(x) \qquad \text{for all real values of } x$$

From the last line of the definition, f is periodic with period 2.

The graph of this function is built up by first drawing $f(x) = x$ in the interval $0 \leqslant x < 1$, then drawing $f(x) = 4 - x^2$ in the interval $1 \leqslant x < 2$. This pattern is then repeated every 2 units along the x-axis.

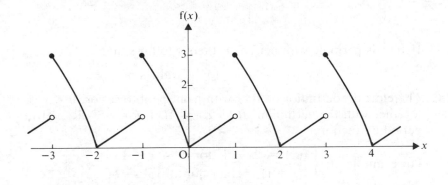

CONTINUOUS FUNCTIONS

A function f is continuous at $x = a$ if $f(a)$ is defined *and* if $f(x) \rightarrow f(a)$ as $x \rightarrow a$ from above and from below.

For example, $f(x) = x^2$ is continuous at $x = 1$ because $f(1)$ is defined and is equal to 1 and $f(x) \rightarrow f(1)$ as $x \rightarrow 1$ from above and below.

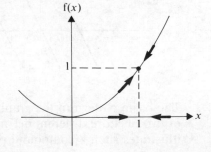

Now consider the function f
defined by

$$f(x) = x \quad \text{for} \ 0 \leqslant x < 1$$

and $\ f(x) = \tfrac{1}{2}x \quad \text{for} \ 1 \leqslant x < 2$

When $x = 1$, $f(1)$ is defined
and is equal to $\tfrac{1}{2}$

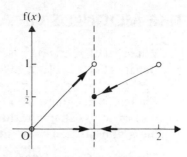

but

$$\left. \begin{array}{l} \text{as} \ x \to 1 \ \text{from below,} \ \ f(x) \to 1 \\ \text{as} \ x \to 1 \ \text{from above,} \ \ f(x) \to \tfrac{1}{2} \end{array} \right\} \ \text{so} \ f(x) \ \text{is not continuous at} \ x = 1$$

> **A *continuous function* satisfies the conditions for continuity
> at all values of x in its domain.**

EXERCISE 1b

1. Sketch each function and state whether or not it is periodic, and/or
   continuous.

   (a) $\cos x$ (b) $\tan x$ (c) 3^x (d) $x^2 - 4$

   (e) $\cot x$ (f) $(x - 1)^2$ (g) $x - 1$ (h) $1/x$

2. Sketch the graph of $f(x)$ within the interval $-4 < x \leqslant 6$ if

   $$f(x) = 4 - x^2 \quad \text{for} \ 0 < x \leqslant 2$$

   and $\qquad\qquad f(x) = f(x - 2) \quad \text{for all values of} \ x$

3. If $\qquad\qquad\qquad f(\theta) = \sin \theta \quad \text{for} \ 0 < x \leqslant \tfrac{1}{2}\pi$

   $$f(\theta) = \cos \theta \quad \text{for} \ \tfrac{1}{2}\pi < x \leqslant \pi$$

   and $\qquad\qquad f(\theta + \pi) = f(\theta) \quad \text{for all values of} \ \theta$

   sketch the function $f(\theta)$ for the range $-2\pi < \theta \leqslant 2\pi$

4. A function $f(x)$ is periodic with a period of 4. Sketch the graph of
   the function for $-6 \leqslant x \leqslant 6$, given that

   $$f(x) = -x \qquad \text{for} \ 0 < x \leqslant 3$$

   $$f(x) = 3x - 12 \quad \text{for} \ 3 < x \leqslant 4$$

THE MODULUS OF A FUNCTION

SG

When $f(x) = x$, $f(x)$ is negative when x is negative.
But if $g(x) = |x|$, g takes the *positive* numerical value of x,
e.g. when $x = -3$, $|x| = 3$, so $g(x)$ is always positive.
Therefore the graph of $g(x) = |x|$ can be obtained from the graph of
$f(x) = x$ by changing, to the equivalent positive values, the part of
the graph of $f(x) = x$ for which $f(x)$ is negative.

Thus for negative values of $f(x)$, the graph of $g(x) = |x|$ is the
reflection of $f(x) = x$ in the y-axis.

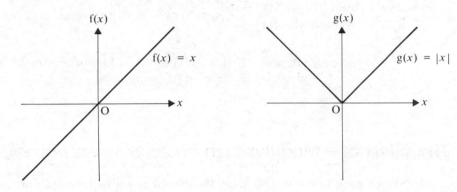

In general, the curve C_1 whose equation is $y = |f(x)|$ is obtained
from the curve C_2 with equation $y = f(x)$, by reflecting in the x-axis
the parts of C_2 for which $f(x)$ is negative. The remaining sections of
C_1 are not changed.

For example, to sketch $y = |(x - 1)(x - 2)|$ we start by sketching the
curve $y = (x - 1)(x - 2)$. We then reflect in the x-axis the part of
this curve which is below the x-axis.

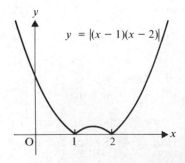

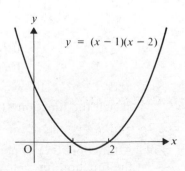

Note that for any function f, the mapping $x \rightarrow |f(x)|$ is also a
function.

EXERCISE 1c

Sketch the following graphs.

1. $y = |2x - 1|$ 2. $y = |x(x - 1)(x - 2)|$

3. $y = |x^2 - 1|$ 4. $y = |x^2 + 1|$

5. $y = |\sin x|$ 6. $y = |\cos x|$

7. $y = |x^2 - x - 20|$ 8. $y = |\tan x|$

9. $y = \left| \dfrac{1}{x + 1} \right|$ 10. $y = |x^3|$

The Effect of a Modulus Sign on a Cartesian Equation

When a section of the curve $y = f(x)$ is reflected in Oy, the equation of that part of the curve becomes $y = -f(x)$,

e.g., if $y = |x|$ for $x \in \mathbb{R}$
we can write this equation as

$$\begin{cases} y = x & \text{for } x \geqslant 0 \\ y = -x & \text{for } x < 0 \end{cases}$$

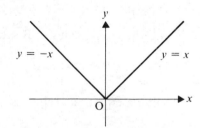

Intersection

To find the points of intersection between two graphs whose equations involve a modulus, we first sketch the graphs to locate the points roughly. Then we identify the equations in non-modulus form for each part of the graph. If these equations are written on the sketch then the correct pair of equations for solving simultaneously can be identified.

For example, the points common to $y = x - 1$ and $y = |x^2 - 3|$ can be seen from the sketch.

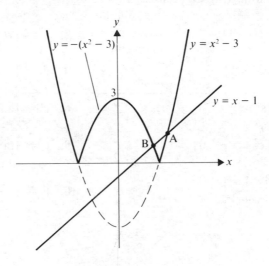

We can also see from the sketch that

the coordinates of A satisfy the equations

$$y = x - 1 \quad \text{and} \quad y = x^2 - 3 \qquad\qquad [1]$$

and the coordinates of B satisfy the equations

$$y = x - 1 \quad \text{and} \quad y = 3 - x^2 \qquad\qquad [2]$$

Solving equations [1] gives $x^2 - x - 2 = 0 \quad \Rightarrow \quad x = -1 \text{ or } 2$

It is clear from the diagram that $x \neq 1$, so A is the point $(2, 1)$

Similarly, solving equations [2] gives $x^2 + x - 4 = 0$

$$\Rightarrow \qquad\qquad x = -\tfrac{1}{2}(1 \pm \sqrt{17})$$

Again from the diagram, it is clear that the x-coordinate of B is positive, so at B, $x = -\tfrac{1}{2}(1 - \sqrt{17})$
Then using $y = x - 1$ gives $y = \tfrac{1}{2}(-3 + \sqrt{17})$

This example illustrates the importance of checking solutions to see if they are relevant to the given problem.

Solution of Equations Involving a Modulus Sign

An equation such as $1 - |x| = 3x - 1$ can be solved by finding the values of x at the points of intersection of two graphs. As we are dealing with an equation, we can rearrange it if necessary to give graphs that are as simple as possible.

Example 1d _____

Solve the equation $1 - |x| = 3x - 1$

$$1 - |x| = 3x - 1 \quad \Rightarrow \quad |x| = 2 - 3x$$

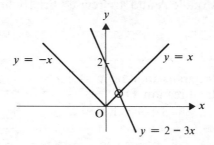

A sketch of the graphs $y = |x|$ and $y = 2 - 3x$ shows that $|x| = 2 - 3x$ where $x = 2 - 3x$

$$x = 2 - 3x \quad \Rightarrow \quad x = \tfrac{1}{2}$$

$\therefore \quad 1 - |x| = 3x - 1$ when $x = \tfrac{1}{2}$

_____

EXERCISE 1d

Find the points of intersection of the graphs.

1. $y = |x|$ and $y = x^2$

2. $y = x$ and $y = |x^2 - 1|$

3. $y = |1/x|$ and $y = |x|$

4. $y = 9 - x^2$ and $y = |3x - 1|$

5. $y = |x^2 - 4|$ and $y = 2x + 1$

6. $y = |x^2 - 1|$ and $y = |x - 2|$

Solve the equations

7. $2|x| = 2x + 1$

8. $|x + 1| = |x - 1|$

9. $|2x - 1| + 2 = 4x$

10. $|x^2 - 1| - 1 = 3x - 2$

11. $|2 - x^2| + 2x + 1 = 0$

12. $4 - |x^2 - 4| = 5x$

CHAPTER 2

CURVE SKETCHING

TRANSFORMATIONS

The graphs of many functions can be obtained from transformations of the curves representing basic functions. These transformations were introduced in Module A and are revised briefly here.

$y = f(x) + c$ is a translation of $y = f(x)$ by c units in the direction Oy

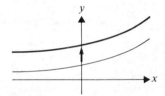

$y = f(x + c)$ is a translation of $y = f(x)$ by $-c$ units in the direction Ox

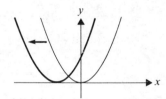

$y = -f(x)$ is the reflection of $y = f(x)$ in the x-axis.

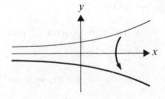

$y = f(-x)$ is the reflection of $y = f(x)$ in the y-axis.

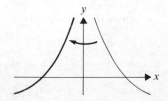

$y = af(x)$ is a one-way stretch of
$y = f(x)$ by a factor a parallel to the
y-axis.

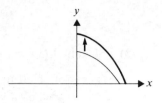

$y = f(ax)$ is a one-way stretch of
$y = f(x)$ by a factor $\frac{1}{a}$ parallel to the
x-axis.

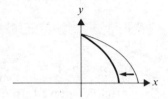

Now consider the curve $y = \dfrac{x}{x + 1}$

It looks as if $\dfrac{x}{x + 1}$ is related to $f(x) = \dfrac{1}{x}$ and we see that we can

write $y = \dfrac{x}{x + 1} = 1 - \dfrac{1}{x + 1}$, i.e. $y = 1 - f(x + 1)$

We can now build up a picture of the curve $y = \dfrac{x}{x + 1}$ in stages,
with rough sketches.

1) Sketch $f(x) = \dfrac{1}{x}$

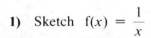
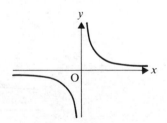

2) Sketch $g(x) = \dfrac{1}{x + 1} = f(x + 1)$
   ($f(x)$ moved one unit to the left.)

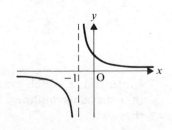

3)   Sketch $h(x) = -\dfrac{1}{x + 1} = -g(x)$

   ($g(x)$ reflected in the x-axis.)

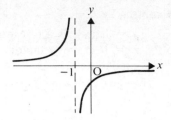

4)   Sketch $y = 1 - \dfrac{1}{x + 1} = 1 + h(x)$

   ($h(x)$ moved up one unit.)

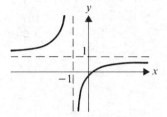

From the last sketch, we see that the asymptotes to the curve are
$y = 1$ and $x = -1$

From the equation of the curve, $y = 0$ when $x = 0$, so the curve
goes through the origin and there are no other intercepts on the axes.
We can now draw a more accurate sketch.

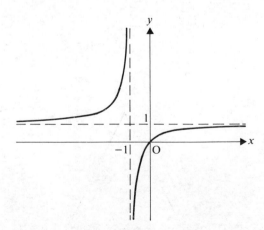

In general, a sketch graph should clearly show the following features
of a curve: asymptotes, intercepts on the axes and turning points when
they exist.

Example 2a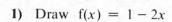

Sketch the graph of $y = 3 - |1 - 2x|$

We can use transformations to build up the picture in stages.

1) Draw $f(x) = 1 - 2x$

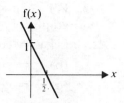

2) Draw $g(x) = |1 - 2x|$

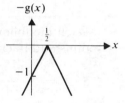

3) Draw $-|1 - 2x|$
($-g(x)$ is the reflection of $g(x)$ in Ox.)

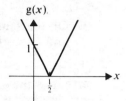

4) $y = 3 - |1 - 2x|$ is the graph of $-g(x)$ translated 3 units upwards.

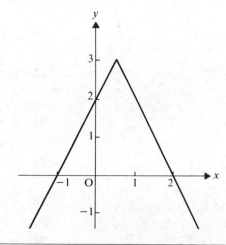

EXERCISE 2a

1. Find the values of a and b such that $x^2 - 4x + 1 \equiv (x - a)^2 + b$. On the same set of axes sketch the curves $y = x^2$ and $y = x^2 - 4x + 1$

2. For $-2\pi < x < 2\pi$ and on the same set of axes sketch the graphs of
   (a) $y = \cos x$ and $y = 3\cos x$
   (b) $y = \sin x$ and $y = \sin 2x$
   (c) $y = \cos x$ and $y = \cos(x - \frac{1}{6}\pi)$
   (d) $y = \sin x$ and $y = 2\sin(\frac{1}{6}\pi - x)$

3. On the same set of axes sketch the graphs of $y = 1/x$, $y = 3/x$ and $y = 1/3x$

4. Show that $\dfrac{x - 2}{x - 3} \equiv 1 + \dfrac{1}{x - 3}$. Sketch the curve $y = \dfrac{x - 2}{x - 3}$ clearly showing the asymptotes and the intercepts on the axes.

Sketch the following curves clearly showing any asymptotes, turning points and intercepts on the axes.

5. $y = \dfrac{1}{x - 1}$ 6. $y = 1 - 2\sin x$

7. $y = 1 - (x - 2)^2$ 8. $y = \dfrac{1}{1 - x}$

9. $y = |x + 2| - 3$ 10. $y = 2 - |x - 1|$

11. $y = \dfrac{1 - x}{x}$ 12. $y = 1 - x^3$

13. $y = \dfrac{2x + 1}{x}$ 14. $y = \dfrac{x + 1}{x - 1}$

15. $y = 3 - (x - 2)^2$ 16. $y = 2\sin(x - \frac{1}{3}\pi)$

17. $y = 3\cos(x + \frac{1}{6}\pi)$ 18. $y = \dfrac{1 + 2x}{1 - x}$

19. $y = |x^2 - 1| - 1$ 20. $y = 1 - |x^2 - 2|$

21. $y = 3x^3 - 4$

22. $y = 3 - 2x^4$

23. $y = 3 - (x + 2)^3$

24. $y = (x - 1)^4$

RECIPROCAL CURVES

SG

Consider the curve $y = \dfrac{1}{f(x)}$ when the graph of $f(x)$ is known.

The following simple properties of reciprocals enable the graph of $1/f(x)$ to be deduced from the graph of $f(x)$.

1) For a given value of x, $f(x)$ and $1/f(x)$ both have the same sign.

2) When the value of $f(x)$ is increasing the value of $1/f(x)$ is decreasing, i.e. when $\dfrac{d}{dx} f(x)$ is positive, $\dfrac{d}{dx}\left(\dfrac{1}{f(x)}\right)$ is negative and conversely.

3) If $f(x) = 1$ then $\dfrac{1}{f(x)} = 1$ also.

Similarly when $f(x) = -1$, $\dfrac{1}{f(x)} = -1$

4) If $f(x) \to \infty$ then $\dfrac{1}{f(x)} \to 0$ from above,

and if $f(x) \to -\infty$ then $\dfrac{1}{f(x)} \to 0$ from below.

5) If $f(x) \to 0$ from above then $\dfrac{1}{f(x)} \to \infty$,

and if $f(x) \to 0$ from below then $\dfrac{1}{f(x)} \to -\infty$

6) If $f(x)$ has a maximum value then $\dfrac{1}{f(x)}$ has a minimum value, and conversely.

Examples 2b

1. Use the curve $y = x^2$ to sketch the curve whose equation is
   $y = 1/x^2$

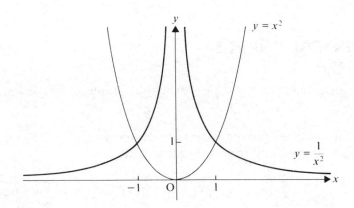

From the sketch of $y = x^2$ we see that

(a) $x^2 > 0$ for all values of x, so $1/x^2 > 0$ for all values of x;

(b) when $x > 0$, x^2 is increasing so $1/x^2$ is decreasing, and when
$x < 0$, x^2 is decreasing so $1/x^2$ is increasing;

(c) when $x \rightarrow \pm\infty$, $x^2 \rightarrow \infty$ so $1/x^2 \rightarrow 0$
and when $x \rightarrow 0$, $x^2 \rightarrow 0$ so $1/x^2 \rightarrow \infty$

Notice that when $x = 0$, $1/x^2$ is undefined so the fact that
$f(x) = x^2$ has a minimum value when $x = 0$ has no relevance in
this case.

From this information the graph of $y = 1/x^2$ can be drawn.

_____

2. Sketch the curve $y = \dfrac{1}{x^2 + 2}$

Sketching the graph of $f(x) = x^2 + 2$ and then making the following

observations enables the curve $y = \dfrac{1}{f(x)}$ to be drawn.

(a) $f(x) > 0$ for all values of x, therefore so is $1/f(x)$;

(b) for $x < 0$, $f'(x)$ is negative so the gradient of $1/f(x)$ is positive; for $x > 0$, $f'(x)$ is positive so the gradient of $1/f(x)$ is negative;

(c) when $x \to \pm\infty$, $f(x) \to \infty$ so $1/f(x) \to 0$;

(d) when $x = 0$, $f(x)$ has a minimum value of 2 so $1/f(x)$ has a maximum value of $\frac{1}{2}$

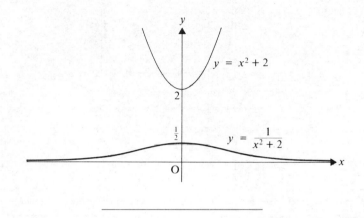

---

3. Sketch the curve $y = \dfrac{1}{(1+x)(4-x)}$ clearly showing asymptotes and any turning points.

First we sketch the graph of $f(x) = (1+x)(4-x)$

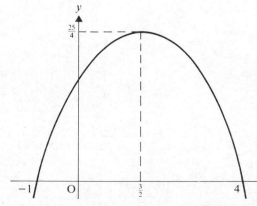

When $x = -1$ and 4, $f(x) = 0 \Rightarrow 1/f(x) \to \infty$

i.e. $x = -1$ and $x = 4$ are asymptotes to $y = 1/f(x)$

Where $x = \frac{3}{2}$, $f(x)$ has a maximum value of $\frac{25}{4}$

therefore $1/f(x)$ has a minimum value of $\frac{4}{25}$

Remembering that when $f(x)$ is +ve, $1/f(x)$ is positive also, and conversely, and noting that as $x \to \pm\infty$, $f(x) \to -\infty$, so $1/f(x) \to 0$ from below, we can now sketch the curve $y = \dfrac{1}{(1+x)(4-x)}$

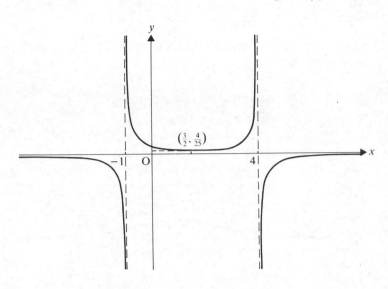

EXERCISE 2b

Sketch each curve, clearly showing any asymptotes, turning points and intercepts on the axes.

1. $y = \dfrac{1}{x^2 - 3x + 2}$

2. $y = \dfrac{1}{x^2 - 4}$

3. $y = \dfrac{1}{\sin x}$

4. $y = \dfrac{1}{1 - x^2}$

5. $y = \dfrac{1}{(x-1)(x-2)}$

6. $y = \dfrac{1}{x^3}$

7. $y = \dfrac{1}{x^3 - 1}$

8. $y = \dfrac{1}{x(x-1)(x-2)}$

9. $y = 1 - \dfrac{1}{x^2}$

10. $y = \dfrac{1}{x^2 - 1}$

MIXED EXERCISE 2

1. Sketch the curve whose equation is $y = 1 - \dfrac{1}{x - 3}$

   Give the equations of the asymptotes and the coordinates of the intercepts on the axes.

2. A function f is defined by $f(x) = \dfrac{1}{(1 - x)}, \; x \neq 1$

   (a) Why is 1 excluded from the domain of f?

   (b) Sketch the curve $y = f(x)$

   (c) Sketch the curve $y = f(x + 1)$

   (d) Express $\dfrac{x + 1}{x - 1}$ in terms of $f(x)$ and hence sketch the curve
   $$y = \frac{x + 1}{x - 1}$$

3. Draw sketches of the following curves.

   (a) $y = (x^2 - 1)(x^2 - 2)$ (b) $y = \dfrac{1}{3 - x}$

   (c) $y = 2^{4 - x}$ (d) $y = 1 + |2 + x|$

4. The function f is defined by
   $$f(x) = \sin x \qquad \text{for } 0 \leqslant x < \tfrac{1}{2}\pi$$
   $$f(x) = \pi - x \qquad \text{for } \tfrac{1}{2}\pi \leqslant x < \pi$$
   $$f(x) = f(x + \pi) \quad \text{for } x \in \mathbb{R}$$
   Sketch the graph of $f(x)$ for $0 \leqslant x < 4\pi$

5. Draw a sketch of the curve whose equation is $y = x^2 - 1$.
   Hence superimpose a sketch of the curve whose equation is
   $$y = \frac{1}{(x^2 - 1)}$$

6. Sketch the graph of $g(x) = \dfrac{x + 2}{x + 1}$

   On the same set of axes sketch the graph of the function g^{-1} and hence state the range of g^{-1}

7. Find the values of x for which $|x - 1| = 1 + |1 + x|$

8. The function f is periodic with period 2 and
$$f(x) = x^2 \qquad \text{for} \quad 0 \leqslant x < 1$$
$$f(x) = 3 - 2x \quad \text{for} \quad 1 \leqslant x < 2$$
Sketch the graph of f for $-2 \leqslant x \leqslant 4$
For which values of x in this range is $f(x)$ not continuous?

9. State whether $f(x)$ is odd, even, periodic (in which case give the period) or none of these.
(a) $f(x) = \tan x$ (b) $f(x) = (x + 1)(x)(x - 1)$
(c) $f(x) = x^4$ (d) $f(x) = \sin(x - \frac{1}{2}\pi)$

10. Sketch the graph of $y = |\sin x|$
Which of the following statements apply to $f(x) = |\sin x|$?
(a) $f(x)$ is even
(b) $f(x)$ is odd
(c) $f(x)$ is periodic
(d) $f(x)$ is continuous.

11. Find the minimum value of $f(x)$ and the values of x for which $f(x) = 0$ where $f(x) = 2x^2 + x - 6$
On the same set of axes sketch the curves $y = f(x)$ and $y = 1/f(x)$. Use your sketch to deduce the number of distinct roots of the equation $[f(x)]^2 = 1$

CHAPTER 3

ALGEBRA

PARTIAL FRACTIONS

Two separate fractions such as $\dfrac{3}{x+4} + \dfrac{2}{x-5}$ can be expressed as a single fraction with a common denominator,

i.e. $\dfrac{3}{x+4} + \dfrac{2}{x-5} = \dfrac{3(x-5) + 2(x+4)}{(x+4)(x-5)} = \dfrac{5x-7}{(x+4)(x-5)}$

Later on in the course it is necessary to reverse this operation, i.e. to take an expression such as $\dfrac{x-2}{(x+3)(x-4)}$ and express it as the sum of two separate fractions.

This process is called splitting up, or decomposing, into *partial fractions*.

Consider again $\dfrac{x-2}{(x+3)(x-4)}$

This fraction is a *proper fraction* because the highest power of x in the numerator (1 in this case) is less than the highest power of x in the denominator (2 in this case when the brackets are expanded).

Therefore its separate (or partial) fractions also will be proper,

i.e. $\dfrac{x-2}{(x+3)(x-4)}$ can be expressed as $\dfrac{A}{x+3} + \dfrac{B}{x-4}$

where A and B are numbers. The worked example which follows shows how the values of A and B can be found.

25

Example 3a

Express $\dfrac{x - 2}{(x + 3)(x - 4)}$ in partial fractions.

$$\frac{x - 2}{(x + 3)(x - 4)} = \frac{A}{x + 3} + \frac{B}{x - 4}$$

Expressing the separate fractions on the RHS as a single fraction over a common denominator gives

$$\frac{x - 2}{(x + 3)(x - 4)} = \frac{A(x - 4) + B(x + 3)}{(x + 3)(x - 4)}$$

This is not an equation because the RHS is just another way of expressing the LHS. It follows that, as the denominators are identical the numerators also are identical.

i.e. $\qquad\qquad x - 2 = A(x - 4) + B(x + 3)$

Remembering that this is *not* an equation but two ways of writing the same expression, it follows that LHS = RHS for any value that we choose to give to x.

Choosing to substitute 4 for x (to eliminate A) gives

$$2 = A(0) + B(7)$$

$$\Rightarrow \qquad\qquad B = \frac{2}{7}$$

Choosing to substitute -3 for x (to eliminate B) gives

$$-5 = A(-7) + B(0)$$

$$\Rightarrow \qquad\qquad A = \frac{5}{7}$$

Therefore $\qquad \dfrac{x - 2}{(x + 3)(x - 4)} = \dfrac{5/7}{x + 3} + \dfrac{2/7}{x - 4}$

$$= \frac{5}{7(x + 3)} + \frac{2}{7(x - 4)}$$

EXERCISE 3a

Express the following fractions in partial fractions.

1. $\dfrac{x - 2}{(x + 1)(x - 1)}$

2. $\dfrac{2x - 1}{(x - 1)(x - 7)}$

3. $\dfrac{4}{(x+3)(x-2)}$

4. $\dfrac{7x}{(2x-1)(x+4)}$

5. $\dfrac{2}{x(x-2)}$

6. $\dfrac{2x-1}{x^2-3x+2}$

7. $\dfrac{3}{x^2-9}$

8. $\dfrac{6x+7}{3x(x+1)}$

9. $\dfrac{9}{2x^2+x}$

10. $\dfrac{x+1}{6x^2-x-2}$

The Cover-up Method

This is a quicker method for finding the numerators of the separate fractions.

Consider $f(x) = \dfrac{x}{(x-2)(x-3)} = \dfrac{A}{x-2} + \dfrac{B}{x-3}$

$$= \dfrac{A(x-3) + B(x-2)}{(x-2)(x-3)}$$

$\Rightarrow \qquad\qquad\qquad x \equiv A(x-3) + B(x-2)$

When $x = 2$, $A = \dfrac{2}{2-3} = -2$,

which is the value of $\dfrac{x}{(x-3)}$ when $x = 2$

i.e. $A = f(2)$ with the factor $(x-2)$ 'covered up'.

Similarly when $x = 3$, $B = \dfrac{3}{3-2} = 3$

which is the value of $\dfrac{x}{(x-2)}$ when $x = 3$

i.e. $B = f(3)$ with the factor $(x-3)$ covered up.

Hence $\qquad \dfrac{x}{(x-2)(x-3)} \equiv \dfrac{-2}{x-2} + \dfrac{3}{x-3}$

Note that this method can be used only for linear factors.

Example 3b

Express $\dfrac{1}{(2x-1)(x+3)}$ in partial fractions.

$$\frac{1}{(2x-1)(x+3)} = \frac{\text{f}(\frac{1}{2}) \text{ with } (2x-1) \text{ covered up}}{(2x-1)} + \frac{\text{f}(-3) \text{ with } (x+3) \text{ covered up}}{(x+3)}$$

$$\frac{1}{(2x-1)(x+3)} = \frac{2}{7(2x-1)} - \frac{1}{7(x+3)}$$

Note that the intermediate step does not need to be written down.

EXERCISE 3b

Express these fractions in partial fractions.

1. $\dfrac{x+4}{(x+3)(x-5)}$ 2. $\dfrac{3x+1}{(2x-1)(x-1)}$

3. $\dfrac{2x-3}{(x-2)(4x-3)}$ 4. $\dfrac{3-x}{(x+1)(2x-1)}$

5. $\dfrac{3}{x(2x+1)}$ 6. $\dfrac{4}{x^2-7x-8}$

7. $\dfrac{4x}{4x^2-9}$ 8. $\dfrac{4x-2}{x^2+2x}$

9. $\dfrac{3x}{2x^2-2x-4}$ 10. $\dfrac{3x+2}{2x^2-4x}$

Quadratic Factors in the Denominator

It is also possible to decompose fractions with quadratic or higher degree factors.

Consider $\dfrac{x^2+1}{(x^2+2)(x-1)}$

This is a proper fraction, so its partial fractions are also proper, i.e.

$\dfrac{x^2+1}{(x^2+2)(x-1)}$ can be expressed in the form $\dfrac{Ax+B}{x^2+2} + \dfrac{C}{x-1}$

Using the cover-up method gives $C = \frac{2}{3}$, but to find A and B the partial fraction form must be expressed as a single fraction giving

$$x^2 + 1 \equiv (Ax + B)(x - 1) + \tfrac{2}{3}(x^2 + 2)$$

The values of A and B can then be found by substituting any suitable values for x

We will choose $x = 0$ and $x = -1$ as these are simple values to handle. (We do not choose $x = 1$ as it eliminates A and B and it was used to find C.)

$x = 0$ gives $\qquad 1 = B(-1) + \tfrac{2}{3}(2) \qquad\qquad \Rightarrow \qquad B = \tfrac{1}{3}$

$x = -1$ gives $\qquad 2 = (-A + \tfrac{1}{3})(-2) + \tfrac{2}{3}(3) \qquad \Rightarrow \qquad A = \tfrac{2}{3}$

$\therefore \qquad\qquad \dfrac{x^2 + 1}{(x^2 + 2)(x - 1)} \equiv \dfrac{2x + 1}{3(x^2 + 2)} + \dfrac{2}{3(x - 1)}$

A Repeated Factor in the Denominator

Consider the fraction $\dfrac{2x - 1}{(x - 2)^2}$

This is a proper fraction, and it is possible to express this as two fractions with numerical numerators as we can see if we adjust numerator,

i.e. $\qquad \dfrac{2x - 1}{(x - 2)^2} \equiv \dfrac{2(x - 2) - 1 + 4}{(x - 2)^2} \equiv \dfrac{2}{x - 2} + \dfrac{3}{(x - 2)^2}$

Any fraction whose denominator is a repeated linear factor can be expressed as separate fractions with numerical numerators, for example,

$\dfrac{2x^2 - 3x + 4}{(x - 1)^3} \qquad$ can be expressed as $\qquad \dfrac{A}{x - 1} + \dfrac{B}{(x - 1)^2} + \dfrac{C}{(x - 1)^3}$

In the general case the values of the numerators can be found using the method in the next worked example.

To summarise, a proper fraction can be decomposed into partial fractions and the form of the partial fractions depends on the form of the factors in the denominator where

a linear factor gives a partial fraction of the form $\dfrac{A}{ax + b}$

a quadratic factor gives a partial fraction of the form $\dfrac{Ax + B}{ax^2 + bx + c}$

a repeated factor gives two partial fractions of the form

$$\frac{A}{ax + b} + \frac{B}{(ax + b)^2}$$

Examples 3c

1. Express $\dfrac{x - 1}{(x + 1)(x - 2)^2}$ in partial fractions.

$$\frac{x - 1}{(x + 1)(x - 2)^2} \equiv \frac{-\frac{2}{9}}{x + 1} + \frac{B}{(x - 2)} + \frac{C}{(x - 2)^2}$$

$\Rightarrow \qquad x - 1 \equiv (-\tfrac{2}{9})(x - 2)^2 + B(x + 1)(x - 2) + C(x + 1)$

$x = 2$ gives $C = \tfrac{1}{3}$

Comparing coefficients of x^2 gives $0 = -\tfrac{2}{9} + B \Rightarrow B = \tfrac{2}{9}$

$\therefore \qquad \dfrac{x - 1}{(x + 1)(x - 2)^2} \equiv -\dfrac{2}{9(x + 1)} + \dfrac{2}{9(x - 2)} + \dfrac{1}{3(x - 2)^2}$

Note that C can be found by the cover-up method, but B cannot.

2. Express $\dfrac{x^3}{(x+1)(x-3)}$ in partial fractions.

This fraction is improper and it must be divided out to obtain a mixed fraction before it can be expressed in partial fractions.

$$
\begin{array}{r}
x + 2 \\
x^2 - 2x - 3 \overline{\smash{)}x^3 } \\
\underline{x^3 - 2x^2 - 3x} \\
2x^2 + 3x \\
\underline{2x^2 - 4x - 6} \\
7x + 6
\end{array}
$$

$\therefore \qquad \dfrac{x^3}{(x+1)(x-3)} \equiv x + 2 + \dfrac{7x+6}{(x+1)(x-3)}$

$\qquad\qquad\qquad\qquad \equiv x + 2 + \dfrac{1}{4(x+1)} + \dfrac{27}{4(x-3)}$

EXERCISE 3c

Express in partial fractions,

1. $\dfrac{2}{(x-1)(x^2+1)}$

2. $\dfrac{x^2+1}{x(2x^2+1)}$

3. $\dfrac{x^2+3}{x(x^2+2)}$

4. $\dfrac{2x^2+x+1}{(x-3)(2x^2+1)}$

5. $\dfrac{x^3-1}{(x+2)(2x+1)(x^2+1)}$

6. $\dfrac{x^2+1}{x(2x^2-1)(x-1)}$

7. $\dfrac{x}{(x-1)(x-2)^2}$

8. $\dfrac{x^2-1}{x^2(2x+1)}$

9. $\dfrac{3}{x(3x-1)^2}$

10. $\dfrac{x^2}{(x+1)(x-1)}$

11. $\dfrac{x^2-2}{(x+3)(x-1)}$

12. $\dfrac{x^3+3}{(x-1)(x+1)}$

LOGARITHMS

Consider the statement $10^2 = 100$

If this is expressed in words we have

   the base 10   raised to the power 2   gives 100

Now this relationship can be rearranged to give the same information, but with a different emphasis, i.e.

   the power to which the base 10 must be raised   to give 100   is 2

> **In this form the power is called a logarithm (log)**

The whole relationship can then be abbreviated to read

   the logarithm to the base 10   of 100   is 2

or $\log_{10} 100 = 2$

In the same way, $2^3 = 8$ $\Rightarrow$ $\log_2 8 = 3$

and $3^4 = 81$ $\Rightarrow$ $\log_3 81 = 4$

Similarly $\log_5 25 = 2$ $\Rightarrow$ $5^2 = 25$

and $\log_9 3 = \frac{1}{2}$ $\Rightarrow$ $9^{1/2} = 3$

Although we have so far used only 10, 2 and 3, the base of a logarithm can be any positive number, or even an unspecified number represented by a letter, for example

$$b = a^c \quad \Leftrightarrow \quad \log_a b = c$$

Note that the symbol $\Leftrightarrow$ means that each of these facts implies the other.

Example 3d _____

(a) Write $\log_2 64 = 6$ in index form.
(b) Write $5^3 = 125$ in logarithmic form.
(c) Complete the statement $2^{-3} = ?$ and then write it in logarithmic form.

(a) If $\log_2 64 = 6$ then the base is 2, the number is 64 and the power (i.e. the log) is 6

$$\log_2 64 = 6 \quad \Rightarrow \quad 2^6 = 64$$

(b) If $5^3 = 125$ then the base is 5, the log (i.e. the power) is 3 and the number is 125

$$5^3 = 125 \quad \Rightarrow \quad \log_5 125 = 3$$

(c) $2^{-3} = \frac{1}{8}$

The base is 2, the power (log) is -3 and the number is $\frac{1}{8}$

$$2^{-3} = \frac{1}{8} \quad \Rightarrow \quad \log_2\left(\frac{1}{8}\right) = -3$$

EXERCISE 3d

Convert each of the following facts to logarithmic form.

1. $10^3 = 1000$ **2.** $2^4 = 16$ **3.** $10^4 = 10\,000$

4. $3^2 = 9$ **5.** $4^2 = 16$ **6.** $5^2 = 25$

7. $10^{-2} = 0.01$ **8.** $9^{1/2} = 3$ **9.** $5^0 = 1$

10. $4^{1/2} = 2$ **11.** $12^0 = 1$ **12.** $8^{1/3} = 2$

13. $p = q^2$ **14.** $x^y = 2$ **15.** $p^q = r$

Convert each of the following facts to index form.

16. $\log_{10} 100\,000 = 5$ **17.** $\log_4 64 = 3$ **18.** $\log_{10} 10 = 1$

19. $\log_2 4 = 2$ **20.** $\log_2 32 = 5$ **21.** $\log_{10} 1000 = 3$

22. $\log_5 1 = 0$ **23.** $\log_3 9 = 2$ **24.** $\log_4 16 = 2$

25. $\log_3 27 = 3$ **26.** $\log_{36} 6 = \frac{1}{2}$ **27.** $\log_a 1 = 0$

28. $\log_x y = z$ **29.** $\log_a 5 = b$ **30.** $\log_p q = r$

Evaluating Logarithms

It is generally easier to solve a simple equation in index form than in log form so we often use an index equation in order to evaluate a logarithm. For example to evaluate $\log_{49} 7$ we can say

if $\qquad\qquad x = \log_{49} 7$ then $49^x = 7 \quad \Rightarrow \quad x = \tfrac{1}{2}$

therefore $\qquad\qquad\qquad\qquad \log_{49} 7 = \tfrac{1}{2}$

In particular, for any base b,

if $\qquad\qquad x = \log_b 1$ then $b^x = 1 \quad \Rightarrow \quad x = 0$

i.e. the logarithm to any base of 1 is zero.

EXERCISE 3e

Evaluate

1. $\log_2 4$
2. $\log_{10} 1\,000\,000$
3. $\log_2 64$
4. $\log_3 81$
5. $\log_8 64$
6. $\log_4 64$
7. $\log_9 3$
8. $\log_{1/2} 4$
9. $\log_{10} 0.1$
10. $\log_{121} 11$
11. $\log_5 1$
12. $\log_2 2$
13. $\log_{64} 4$
14. $\log_{99} 1$
15. $\log_{27} 3$
16. $\log_a a^3$

THE LAWS OF LOGARITHMS

When working with indices, we found certain rules that powers obey in the multiplication and division of numbers. Because logarithm is just another word for index or power, it is to be expected that logarithms too obey certain laws and these we are now going to investigate.

Consider $\qquad x = \log_a b$ and $\quad y = \log_a c$

$\Rightarrow \qquad a^x = b \qquad$ and $\quad a^y = c$

$\qquad\qquad$ Now $\qquad bc = (a^x)(a^y)$

$\Rightarrow \qquad\qquad\qquad bc = a^{x+y}$

Therefore $\qquad\qquad \log_a bc = x + y$

i.e. $\qquad\qquad\qquad \log_a bc = \log_a b + \log_a c$

This is the first law of logarithms and, as a can represent *any* base, this law applies to the log of *any* product *provided that the same base is used for all the logarithms in the formula.*

Using x and y again, a law for the log of a fraction can be found.

$$\frac{b}{c} = \frac{a^x}{a^y} \quad \Rightarrow \quad \frac{b}{c} = a^{x-y}$$

Therefore $\log_a (b/c) = x - y$

i.e. $\log_a (b/c) = \log_a b - \log_a c$

A third law allows us to deal with an expression of the type $\log_a b^n$

Using $x = \log_a b^n \quad \Rightarrow \quad a^x = b^n$

i.e. $a^{x/n} = b$

Therefore $x/n = \log_a b \quad \Rightarrow \quad x = n \log_a b$

i.e. $\log_a b^n \quad = \quad n \log_a b$

So we now have the three most important laws of logarithms. Because they are true for *any* base it is unnecessary to include a base in the formula but

in each of these laws every logarithm must be to the same base

$$\log bc = \log b + \log c$$
$$\log b/c = \log b - \log c$$
$$\log b^n = n \log b$$

Examples 3f

1. Express $\log pq^2\sqrt{r}$ in terms of $\log p$, $\log q$ and $\log r$

$$\log pq^2\sqrt{r} = \log p + \log q^2 + \log \sqrt{r}$$
$$= \log p + 2 \log q + \tfrac{1}{2}\log r$$

2. Simplify $3 \log p + n \log q - 4 \log r$

$$3 \log p + n \log q - 4 \log r = \log p^3 + \log q^n - \log r^4$$
$$= \log \frac{p^3 q^n}{r^4}$$

EXERCISE 3f

Express in terms of $\log p$, $\log q$, and $\log r$

1. $\log pq$ **2.** $\log pqr$ **3.** $\log p/q$ **4.** $\log pq/r$

5. $\log p/qr$ **6.** $\log p^2 q$ **7.** $\log q/r^2$ **8.** $\log p\sqrt{q}$

9. $\log p^2 q^3/r$ **10.** $\log \sqrt{(q/r)}$ **11.** $\log q^n$ **12.** $\log p^n q^m$

Simplify

13. $\log p + \log q$ **14.** $2 \log p + \log q$

15. $\log q - \log r$ **16.** $3 \log q + 4 \log p$

17. $n \log p - \log q$ **18.** $\log p + 2 \log q - 3 \log r$

Changing the Base of a Logarithm

Suppose that $x = \log_a c$ and that we wish to express x as a logarithm to the base b

$$\log_a c = x \quad \Rightarrow \quad c = a^x$$

Now taking logs to the base b gives

$$\log_b c = x \log_b a \quad \Rightarrow \quad x = \frac{\log_b c}{\log_b a}$$

i.e.
$$\log_a c = \frac{\log_b c}{\log_b a}$$

In the special case when $c = b$, i.e. when $\log_b c = 1$, this relationship becomes

$$\log_a b = \frac{1}{\log_b a}$$

Evaluating Logarithms

Scientific calculators usually have two sets of logarithms stored in them. These are logs to the base 10 (common logarithms) and logs to the base e (natural logarithms) and the respective keys are marked log and ln. (The number e is special and is introduced later in this book.) At this stage we will use only logs to the base 10.

Example 3g _____

Evaluate $\log_2 3$

First we change the base to 10, then we use a calculator.

$$\log_2 3 = \frac{\log_{10} 3}{\log_{10} 2}$$

$$= 1.58 \text{ correct to 3 s.f.}$$

EXERCISE 3g

1. Change the base of each logarithm to 10
   (a) $\log_3 8$ (b) $\log_5 10$ (c) $\log_{100} 5$

2. Change the base of each logarithm to a.
   (a) $\log_x y$ (b) $\log_x a$ (c) $\log_y 8$

3. Find, correct to three significant figures, the value of
   (a) $\log_8 12$ (b) $\log_5 100$ (c) $\log_3 15$

EXPONENTIAL AND LOGARITHMIC EQUATIONS

When the unknown quantity forms part of an index, taking logs will often transform the index into a factor.

For example, if $5^x = 10$ then taking logs of both sides gives

$$x \log 5 = \log 10 \quad \Rightarrow \quad x = \frac{1}{\log 5}$$

Taking logs can work when the terms containing a power involving x can be expressed as a single term, but this is not always possible.

Consider for example the equation

$$2^{2x} + 3(2^x) - 4 = 0$$

Now $2^{2x} + 3(2x)$ cannot be simplified into a single term. However, recognising that 2^{2x} is $(2^x)^2$, the substitution $y = 2^x$ can be made,

i.e. $y^2 + 3y - 4 = 0$

This is a quadratic equation which can now be solved.

When an equation contains logarithms involving the unknown, first make sure that all logs are to the same base, then check to see if a simple substitution will reduce the equation to a recognisable form. Sometimes the best policy is to remove the logarithms. Equations of each of these types are considered in the following worked examples.

Examples 3h _____

1. Solve the equation $\log_3 x - 4 \log_x 3 + 3 = 0$

The two log terms have different bases so we begin by changing the base of the log in the second term to 3. The given equation then becomes

$$\log_3 x - \frac{4}{\log_3 x} + 3 = 0$$

$\Rightarrow$ $(\log_3 x)^2 - 4 + 3 \log_3 x = 0$

Substituting y for $\log_3 x$ gives $y^2 - 4 + 3y = 0 \Rightarrow y^2 + 3y - 4 = 0$

$\therefore$ $(y + 4)(y - 1) = 0 \Rightarrow y = 1$ or -4

i.e. $\log_3 x = 1$ or $\log_3 x = -4 \Rightarrow x = 3$ or $x = 3^{-4} = \frac{1}{81}$

_____

2. Solve for x and y the equations

$$yx = 16 \text{ and } \log_2 x - 2 \log_2 y = 1$$

$$yx = 16 \qquad\qquad\qquad [1]$$

$$\log_2 x - 2 \log_2 y = 1 \qquad\qquad [2]$$

Using the laws of logs, equation [2] can be written as

$$\log_2 \frac{x}{y^2} = 1 \quad \Rightarrow \quad \frac{x}{y^2} = 2, \quad \text{i.e.} \quad x = 2y^2$$

Substituting $2y^2$ for x in equation [1] gives

$$2y^3 = 16 \quad \Rightarrow \quad y^3 = 8$$

$$\therefore \quad y = 2 \quad \text{and, from [1],} \quad x = 8$$

EXERCISE 3h

Solve the equations.

1. $3^x = 6$

2. $5^x = 4$

3. $2^{2x} = 5$

4. $3^{x-1} = 7$

5. $4^{2x+1} = 3$

6. $5^x(5^{x-1}) = 10$

7. $2(2^{2x}) - 5(2^x) + 2 = 0$

8. $3^{2x+1} - 26(3^x) - 9 = 0$

9. $4^x - 6(2^x) - 16 = 0$

10. $\log_2 x + \log_x 2 = 2$

11. $\log_2 x = \log_4(x + 6)$

12. $4\log_3 x = \log_x 3$

Solve the equations simultaneously.

13. $2\lg y = \lg 2 + \lg x$ and $2^y = 4^x$ (lg means $\log_{10}$)

14. $\log_x y = 2$ and $xy = 8$

15. $\log_3 x = y = \log_9(2x - 1)$

16. $\lg (x + y) = 0$ and $2\lg x = \lg (y + 1)$

MIXED EXERCISE 3

1. Express in partial fractions

(a) $\dfrac{2}{(x + 1)(x - 1)}$

(b) $\dfrac{3}{(x - 2)(x + 1)}$

(c) $\dfrac{1}{x(x - 3)}$

(d) $\dfrac{4}{(x - 1)(x + 3)}$

(e) $\dfrac{1}{x^2 - 1}$

(f) $\dfrac{2}{(2x + 1)(2x - 1)}$

2. Express in partial fractions

(a) $\dfrac{2}{x^2(x-1)}$ (b) $\dfrac{1}{x(x^2+1)}$ (c) $\dfrac{x^2}{x-1}$

(d) $\dfrac{x-1}{(x^2-4)(x+1)}$ (e) $\dfrac{2x-1}{(x-1)^2(2x+1)}$ (f) $\dfrac{x^2+3x}{(x^2+1)(x+1)^2}$

3. Given that $y = \dfrac{1}{x(x-1)}$, express y in partial fractions and hence find $\dfrac{dy}{dx}$.

4. Express $f(x) = \dfrac{2x}{(x-1)(x+1)}$ in partial fractions.

Hence find $f'(x)$ and show that $f(x)$ has no stationary values.

5. Solve the equations

(a) $3^{2x} = 10$ (b) $4^x + 2^x - 6 = 0$

(c) $(2^x)(3^x) = \dfrac{1}{36}$ (d) $2\log_2 x = \log_x 2$

(e) $3\log_4 x - \log_4 6 = 16$ (f) $\log_x 1.2 = 2.1$

CHAPTER 4

STRAIGHT LINES AND CIRCLES

THE ANGLE BETWEEN TWO LINES

Consider two lines with gradients m_1 and m_2, where $m_1 = \tan\theta_1$ and $m_2 = \tan\theta_2$

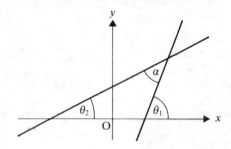

The angle, α, between the lines is given by $\alpha = \theta_1 - \theta_2$

Therefore $\tan\alpha = \tan(\theta_1 - \theta_2) = \dfrac{\tan\theta_1 - \tan\theta_2}{1 + \tan\theta_1 \tan\theta_2}$

i.e. $$\tan\alpha = \dfrac{m_1 - m_2}{1 + m_1 m_2}$$

Examples 4a _____

1. Find the tangent of the angle between the lines $3x - 2y = 5$ and $4x + 5y = 1$

If m_1 is the gradient of $3x - 2y = 5$, then $m_1 = \frac{3}{2}$

If m_2 is the gradient of $4x + 5y = 1$, then $m_2 = -\frac{4}{5}$

Then the angle, α, between these lines is given by

$$\tan \alpha = \frac{\frac{3}{2} - (-\frac{4}{5})}{1 + (\frac{3}{2})(-\frac{4}{5})} = \left(\frac{23}{10}\right)\Big/\left(\frac{-2}{10}\right) = -\frac{23}{2}$$

As $\tan \alpha$ is negative, α is the obtuse angle between the lines. The acute angle between them is $\arctan 23/2$

THE DISTANCE OF A POINT FROM A STRAIGHT LINE

The 'distance of a point from a line' is understood to mean the perpendicular distance.

Consider the line with equation $ax + by + c = 0$ and any point $A(p, q)$ distant d from the line.

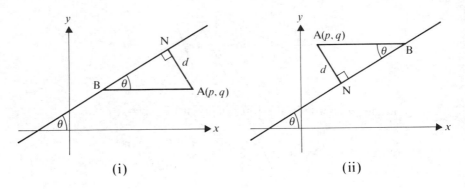

(i)   (ii)

In both diagrams, $d = AN = AB \sin \theta$

and AB is horizontal so, at B, $y = q$ and $x = -(bq + c)/a$

Hence in diagram (i) $AB = p - \left(\dfrac{-(bq + c)}{a}\right) = \dfrac{ap + bq + c}{a}$

and in diagram (ii) $AB = \left(\dfrac{-(bq + c)}{a}\right) - p = -\dfrac{ap + bq + c}{a}$

i.e. for any point A, $AB = \pm \dfrac{ap + bq + c}{a}$

The gradient of the line, $\tan \theta$, is $-\dfrac{a}{b} \implies \sin \theta = \dfrac{a}{\pm\sqrt{(a^2 + b^2)}}$

Now $AN = AB \sin \theta$

hence $AN = \pm \left(\dfrac{ap + bq + c}{a} \right) \left(\dfrac{a}{\sqrt{(a^2 + b^2)}} \right) = \pm \left(\dfrac{ap + bq + c}{\sqrt{(a^2 + b^2)}} \right)$

The two signs ($\pm$) in this formula arise from points that are on opposite sides of the line. If only the *length*, *d*, of AN is required, we take the positive value of this formula, i.e.

$$d = \left| \frac{ap + bq + c}{\sqrt{(a^2 + b^2)}} \right|$$

Examples 4a (continued) _____

2. The vertices of a triangle are the points $A(1, 2)$, $B(3, 1)$, $C(-1, -2)$. Find the length of the altitude through A and hence find the area of the triangle.

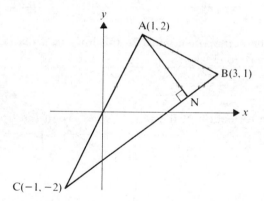

The equation of the line BC is $y = \frac{3}{4}x - \frac{5}{4}$

Writing this equation in the form $3x - 4y - 5 = 0$, then the length of the altitude AN through $A(1, 2)$ is given by

$$d = \left| \frac{3(1) - 4(2) - 5}{\sqrt{(3^2 + \{-4\}^2)}} \right| = \left| -\frac{10}{5} \right| = 2$$

The length of BC is $\sqrt{\{(3)^2 + (4)^2\}} = 5$

Therefore the area of $\triangle ABC$ is $\frac{1}{2}(CB)(AN) = 5$ sq. units.

3. Determine, without the aid of a diagram, whether the points
A$(-3, 4)$ and B$(-2, 3)$ are on the same side of the line
$y + 3x + 4 = 0$. Find the acute angle between this line and AB,
giving your answer in degrees correct to three significant figures.

We need to find the *sign* of the expression used to give the distance of each point
from the line.

Writing the equation of the given line as $3x + y + 4 = 0$, and using

the formula $\dfrac{ap + bq + c}{\sqrt{(a^2 + b^2)}}$ gives

for A, $\quad \dfrac{3(-3) + 4 + 1}{\sqrt{(9 + 1)}} < 0 \quad$ and for B, $\quad \dfrac{3(-2) + 3 + 4}{\sqrt{(9 + 1)}} > 0$

The signs are opposite so A and B are on opposite sides of the line.

The gradient of the given line is -3, and the gradient of AB is -1.
If α is an angle between the given line and AB then

$$\tan \alpha = \frac{m_1 - m_2}{1 + m_1 m_2} = \frac{-3 + 1}{1 + 3} = -\frac{1}{2}$$

Therefore the acute angle between the lines is $\arctan \frac{1}{2}$
which is $26.6°$ to 3 s.f.

_____

4. Find the reflection of the point A$(-1, 5)$ in the line $2y = x + 1$

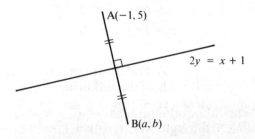

If B(a, b) is the reflection of A in the given line, l, then

AB is bisected by l and AB is at right angles to l

Let C(x, y) be the point where AB and l cut.

As we have two facts about AB, we can use these to give two relationships between
a and b. These can then be solved simultaneously to find a and b. First we can
use the fact that l bisects AB.

$C(x, y)$ is the midpoint of AB, so $x = \frac{1}{2}(a - 1)$ and $y = \frac{1}{2}(b + 5)$
but $C(x, y)$ is on the line $2y = x + 1$, i.e.

$$b + 5 = \tfrac{1}{2}(a - 1) + 1 \quad \Rightarrow \quad a - 2b = 9 \qquad [1]$$

Then we can use the fact that AB and l are perpendicular, so the product of their gradients is -1

AB and l are perpendicular, so

$$\left(\frac{b - 5}{a + 1}\right)\left(\frac{1}{2}\right) = -1 \quad \Rightarrow \quad 2a + b = 3 \qquad [2]$$

Solving equations [1] and [2] simultaneously gives

$$a = 3 \quad \text{and} \quad b = -3$$

EXERCISE 4a

1. Find the distance from A to the given line in each of the following cases.
    (a) A$(3, 4)$; $2x - y = 3$
    (b) A$(-1, -2)$; $3y = 4x - 1$
    (c) A(a, b); $y = mx + c$
    (d) A$(4, -1)$; $x + y = 6$
    (e) A(x, y); $ax + by + c = 0$
    (f) A$(0, 0)$; $ax + by + c = 0$

2. Determine whether A and B are on the same or opposite sides of the given line in each of the following cases.
    (a) A$(1, 2)$, B$(4, -3)$; $3x + y = 7$
    (b) A$(0, 3)$, B$(7, 6)$; $x - 4y + 1 = 0$
    (c) A$(-5, 1)$, B$(-2, 3)$; $7x + y - 6 = 0$

3. Find the tangents of the acute angles between the following pairs of lines.
    (a) $2x + 3y = 7$, $x - 6y = 5$
    (b) $x + 4y - 1 = 0$, $3x + 7y = 2$
    (c) $a_1x + b_1y + c_1 = 0$, $a_2x + b_2y + c_2 = 0$

4. A$(0, 1)$, B$(3, 7)$ and C$(-4, -4)$ are the vertices of a triangle. Find the tangent of each of the three angles in the triangle and the length of each altitude.

5. Find the equations of the two lines through the origin which are inclined at $45°$ to the line $2x + 3y - 4 = 0$

6. Find the image of the point $(5, 6)$ in the line
    (a) $3x - y + 1 = 0$
    (b) $y = 4x + 20$
    (c) $2x + 5y + 18 = 0$

7. A point P(X, Y) is equidistant from the line $x + 2y = 3$ and from the point $(2, 0)$. Find an equation relating X and Y.

8. Show that A$(4, 1)$ and B$(2, -3)$ are equidistant from the line $2x + 5y = 1$. Is A the reflection of B in this line?

9. Write down the distance of the point P(X, Y) from each of the lines $5x - 12y + 3 = 0$ and $3x + 4y - 6 = 0$. By equating these distances find the equations of two lines that bisect the angles between the two given lines [i.e. the equations of the set of points P(X, Y)].

10. A$(4, 4)$ and B$(7, 0)$ are two vertices of a triangle OAB. Find the equation of the line that bisects the angle OBA. If this line meets OA at C show that C divides OA in the ratio OB:BA.

LOCI

In general, within a plane a point P can be anywhere and, further, if (x, y) are the coordinates of P, then x and y can take any values independently of each other.

However, when the possible positions of P are restricted by some condition to a line (curved or straight), the set of points satisfying this condition is called the *locus* of P

Further, the relationship between x and y which applies only to the locus of P defines that locus and is called the *Cartesian equation* of P.

Examples 4b

1. A point, P, is restricted so that it is equidistant from the points
   A(1, 2) and B(−2, −1). Find the Cartesian equation of P.

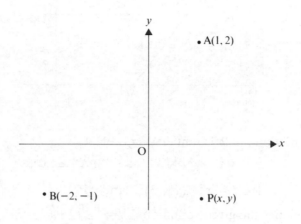

P is restricted to those positions where PA = PB.

Translating this condition into a relationship between x and y gives the equation
of the locus of P.

$$PA^2 = (x - 1)^2 + (y - 2)^2 \quad \text{and} \quad PB^2 = (x + 2)^2 + (y + 1)^2$$

If PA = PB then $PA^2 = PB^2$.

Using the given condition in this form avoids introducing square roots.

$$\therefore \quad PA = PB \quad \Rightarrow \quad (x - 1)^2 + (y - 2)^2 = (x + 2)^2 + (y + 1)^2$$

$$\Rightarrow \quad 0 = 6x + 6y$$

Therefore $x + y = 0$ is the equation of the locus of P.

Note that the line $x + y = 0$ is the perpendicular bisector of AB.

2. A point $P(x, y)$ is twice as far from the point $A(3, 0)$ as it is from the line $x = 5$. Find the equation of the locus of P.

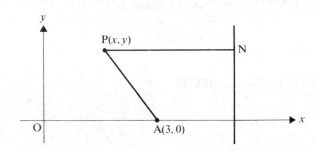

The restriction on P is that $PA = 2PN$.

Now $PA = 2PN \quad \Rightarrow \quad PA^2 = 4PN^2$

But $PA^2 = y^2 + (x - 3)^2$ and $PN^2 = (5 - x)^2$

$\therefore$ P satisfies the given condition $\iff y^2 + (x - 3)^2 = 4(5 - x)^2$

i.e. the equation of the locus of P is $y^2 - 3x^2 + 34x = 91$

EXERCISE 4b

Find the Cartesian equation of the locus of the set of points P in each of the following cases.

1. P is equidistant from the point $(4, 1)$ and the line $x = -2$

2. P is equidistant from $(3, 5)$ and $(-1, 1)$

3. P is three times as far from the line $x = 8$ as from the point $(2, 0)$

4. P is equidistant from the lines $3x + 4y + 5 = 0$ and $12x - 5y + 13 = 0$

5. P is at a constant distance of two units from the point $(3, 5)$

6. P is at a constant distance of five units from the line $4x - 3y = 1$

7. A is the point $(-1, 0)$, B is the point $(1, 0)$ and angle APB is a right angle.

CIRCLES

If a point P is at a constant distance, r, from a fixed point C then the locus of P is a circle whose centre is C and whose radius is r. In this section we look at a variety of methods for dealing with coordinate geometry problems involving circles.

The Equation of a Circle

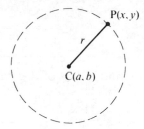

Consider a circle with radius r and centre $C(a, b)$.

A point $P(x, y)$ is on the circle if and only if $CP = r$, i.e. $CP^2 = r^2$

Now $$CP^2 = (x - a)^2 + (y - b)^2$$

∴ $P(x, y)$ is on the circle $\iff$ $(x - a)^2 + (y - b)^2 = r^2$

$$(x - a)^2 + (y - b)^2 = r^2$$

is the equation of a circle with centre (a, b) and radius r

For example, the equation of a circle with centre $(-2, 3)$ and radius 1 is

$$[x - (-2)]^2 + [y - 3]^2 = 1$$

$\Rightarrow$ $$x^2 + y^2 + 4x - 6y + 12 = 0$$

As well as being able to write down the equation of a circle, given its centre and radius, it is equally important to be able to recognise an equation as that of a circle. Expanding and simplifying the equation of a circle with centre (a, b) and radius r gives

$$x^2 + y^2 - 2ax - 2ay + (a^2 + b^2 - r^2) = 0$$

which can be expressed as

$$x^2 + y^2 + 2gx + 2fy + c = 0$$

where g, f and c are constants.

Comparing coefficients gives

$$g = -a, \quad f = -b, \quad c = a^2 + b^2 - r^2 \quad \Rightarrow \quad r^2 = f^2 + g^2 - c$$

So

$x^2 + y^2 + 2gx + 2fy + c = 0$ is the general equation of a circle
provided that $g^2 + f^2 - c > 0$

The centre of the circle is $(-g, -f)$ and the radius is $\sqrt{(g^2 + f^2 - c)}$

Note that the coefficients of x^2 and y^2 are equal and that no xy term is present.

Examples 4c

1. Find the centre and radius of the circle whose equation is

$$x^2 + y^2 + 8x - 2y + 13 = 0$$

There are two ways of finding the centre and radius of this circle. The first method involves forming perfect squares so that we can compare the given equation with $(x - a)^2 + (y - b)^2 = r^2$

$$x^2 + 8x + 16 + y^2 - 2y + 1 = 16 + 1 - 13$$

$\Rightarrow$ $\qquad\qquad\qquad (x + 4)^2 + (y - 1)^2 = 4$

$\therefore$ the centre is $(-4, 1)$ and the radius is 2

Alternatively we can compare the given equation with the general equation of a circle giving

$$2g = 8 \Rightarrow g = 4, \quad 2f = -2 \Rightarrow f = -1 \quad \text{and} \quad c = 13$$

The centre, $(-g, -f)$, is $(-4, 1)$ and the radius, $\sqrt{(g^2 + f^2 - c)}$, is 2

---

2. Show that $2x^2 + 2y^2 - 6x + 10y = 1$ is the equation of a circle and find its centre and radius.

Before we can compare this equation with the general form for the equation of a circle, we must divide the given equation by 2

$$2x^2 + 2y^2 - 6x + 10y - 1 = 0 \quad \Rightarrow \quad x^2 + y^2 - 3x + 5y - \tfrac{1}{2} = 0$$

Comparing with $\qquad\qquad\qquad\qquad x^2 + y^2 + 2gx + 2fy + c = 0$

shows that $2g = -3, \ 2f = 5, \ c = -\tfrac{1}{2}$

$\Rightarrow (g^2 + f^2 - c) = 9$ which is greater than 0

Therefore the equation does represent a circle.

The centre is $(-\frac{3}{2}, \frac{5}{2})$ and the radius is 3

EXERCISE 4c

1. Write down the equation of the circle with

   (a) centre $(1, 2)$, radius 3   (b) centre $(0, 4)$, radius 1

   (c) centre $(-3, -7)$, radius 2   (d) centre $(4, 5)$, radius 3

2. Find the centre and radius of the circle whose equation is

   (a) $x^2 + y^2 + 8x - 2y - 8 = 0$

   (b) $x^2 + y^2 + x + 3y - 2 = 0$

   (c) $x^2 + y^2 + 6x - 5 = 0$

   (d) $2x^2 + 2y^2 - 3x + 2y + 1 = 0$

   (e) $x^2 + y^2 = 4$

   (f) $(x - 2)^2 + (y + 3)^2 = 9$

   (g) $2x + 6y - x^2 - y^2 = 1$

   (h) $3x^2 + 3y^2 + 6x - 3y - 2 = 0$

3. Determine which of the following equations represent circles.

   (a) $x^2 + y^2 = 8$ (b) $2x^2 + y^2 + 3x - 4 = 0$

   (c) $x^2 - y^2 = 8$ (d) $x^2 + y^2 + 4x - 2y + 20 = 0$

   (e) $x^2 + y^2 + 8 = 0$ (f) $x^2 + y^2 + 4x - 2y - 20 = 0$

Tangents to Circles and Other Problems

The following worked examples illustrate how a variety of problems concerning circles can be solved easily with the aid of a diagram and the use of the simple geometric properties of a circle.

It is unnecessary to use calculus methods to find the equation of a tangent to a circle.

Examples 4d

1. Find the equation of the tangent at the point $(3, 1)$ on the circle $x^2 + y^2 - 4x + 10y - 8 = 0$. What is the angle between this tangent and the positive direction of the x-axis?

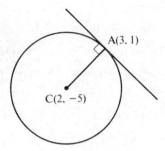

The centre of the circle is $C(2, -5)$

The tangent at A is perpendicular to the radius CA

The gradient of CA is $\dfrac{1 - (-5)}{3 - 2} = 6$

Therefore the gradient of the tangent at A is $-\frac{1}{6}$ and the tangent goes through $A(3, 1)$

So its equation is $\qquad\qquad y - 1 = -\frac{1}{6}(x - 3)$

i.e. $\qquad\qquad\qquad\qquad 6y + x = 9$

If α is the angle between the tangent and the positive direction of the x-axis,

then $\qquad \tan \alpha = -\frac{1}{6}$

$\Rightarrow \qquad\qquad \alpha = 170.5°$

2. Determine whether the lines $5y = 12x - 33$ and $3x + 4y = 9$ are tangents to the circle $x^2 + y^2 + 2x - 8y = 8$

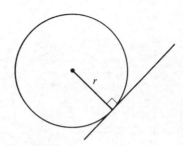

A line is a tangent to a circle if and only if the distance from the centre of the circle to the line is equal to the radius.

Writing the equation of the circle as

$$(x + 1)^2 + (y - 4)^2 = 8 + 1 + 16 = 25$$

shows that $C(-1, 4)$ is the centre and the radius is 5

For the line $5y = 12x - 33$, i.e. $12x - 5y - 33 = 0$, the distance d_1 from the centre $(-1, 4)$ is given by

$$d_1 = \left| \frac{12(-1) - 5(4) - 33}{\sqrt{[12^2 + (-5)^2]}} \right| = \left| -\frac{65}{13} \right| = 5$$

i.e. $d_1 = r$.

Thus $12x - 5y - 33 = 0$ *is* a tangent.

For the line $3x + 4y = 9$, i.e. $3x + 4y - 9 = 0$, the distance d_2 from $(-1, 4)$ is given by

$$d_2 = \left| \frac{3(-1) + 4(4) - 9}{\sqrt{[3^2 + 4^2]}} \right| = \frac{4}{5}$$

i.e. $d_2 \neq 5$.

Thus $3x + 4y - 9 = 0$ *is not* a tangent.

3. Find the equations of the tangents from the origin to the circle
    $x^2 + y^2 - 5x - 5y + 10 = 0$

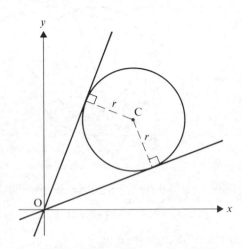

The given circle has centre $(\frac{5}{2}, \frac{5}{2})$

and radius $\sqrt{\left[\left(\dfrac{5}{2}\right)^2 + \left(\dfrac{5}{2}\right)^2 - 10\right]} = \dfrac{\sqrt{10}}{2}$

Any line through the origin has equation $y = mx$, i.e. $mx - y = 0$

If the line is a tangent to the circle, the distance from the centre, $(\frac{5}{2}, \frac{5}{2})$, to the line is equal to the radius

i.e. $\left| \dfrac{m(\frac{5}{2}) - (\frac{5}{2})}{\sqrt{[(m)^2 + (-1)^2]}} \right| = \dfrac{\sqrt{10}}{2}$

$\Rightarrow$ $\left(\dfrac{5}{2}m - \dfrac{5}{2}\right)^2 = \dfrac{10}{4}(m^2 + 1)$

$\Rightarrow$ $3m^2 - 10m + 3 = 0$

$\Rightarrow$ $(3m - 1)(m - 3) = 0$

$\Rightarrow$ $m = \frac{1}{3}$ or 3

So the two tangents from the origin to the given circle are

$$y = 3x \quad \text{and} \quad 3y = x$$

4. Find the equation of the circle whose diameter is the line joining
    the points $A(1, 5)$ and $B(-2, 3)$

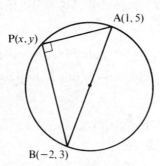

We can use the fact that the angle in a semicircle is $90°$

$P(x, y)$ is a point on the circle if and only if

$$(\text{gradient } AP) \times (\text{gradient } BP) = -1$$

The gradient of AP is $\dfrac{y - 5}{x - 1}$ and the gradient of PB is $\dfrac{y - 3}{x + 2}$

$\therefore \qquad P(x, y)$ is on the circle $\quad \Longleftrightarrow \quad \left(\dfrac{y - 5}{x - 1}\right)\left(\dfrac{y - 3}{x + 2}\right) = -1$

$\therefore$ the equation of the circle is $x^2 + y^2 + x - 8y + 13 = 0$

---

5. Find the equation of the circle that goes through the points $A(0, 1)$,
    $B(4, 7)$ and $C(4, -1)$

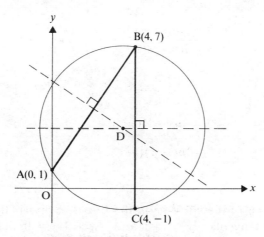

We can use the fact that the centre of a circle lies on the perpendicular bisector of
a chord.

The midpoint of AB is the point $(2, 4)$ and the gradient of AB is $\frac{3}{2}$

∴ the perpendicular bisector of AB is the line

$$2x + 3y = 16 \tag{1}$$

The midpoint of BC is the point $(4, 3)$ and BC is vertical

∴ the perpendicular bisector of BC is horizontal and its equation is

$$y = 3 \tag{2}$$

Solving equations [1] and [2] gives $x = \frac{7}{2}$ and $y = 3$

∴ D is the point $(\frac{7}{2}, 3)$

The radius, r, is the length of DA (or DC or DB), i.e.

$$r^2 = (3 - 1)^2 + (\tfrac{7}{2} - 0)^2 = \tfrac{65}{4}$$

Therefore the equation of the circle is

$$(x - \tfrac{7}{2})^2 + (y - 3)^2 = \tfrac{65}{4} \implies x^2 + y^2 - 7x - 6y + 5 = 0$$

---

6. The lines $3y = 4x$, $4x + 3y = 0$ and $y = 8$ are tangents to a circle. Find the equation of the circle.

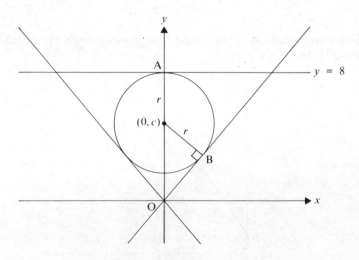

The centre, C, of the circle is on the line that bisects the angle between the tangents $3y = 4x$ and $4x + 3y = 0$. These two tangents are equally inclined to the y-axis so, from symmetry, C lies on the y-axis and its x-coordinate is 0

There is no information from which the y-coordinate of C can be found directly, so we will call it c. We can then use the fact that the distance from C to each tangent is equal to the radius, r, to form an equation in c.

$$r = \text{CA} = |8 - c| \quad \text{and} \quad r = \text{CB} = \pm(0 - 3c)/5$$

$$\therefore \qquad 5(8 - c) = \pm(-3c) \qquad \Rightarrow \qquad c = 20 \quad \text{or} \quad c = 5$$

Therefore there are two circles that satisfy the given conditions.

From $r = |8 - c|$, the corresponding values of the radii are 12 and 3

The equation of the circle is either

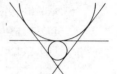

$$x^2 + (y - 20)^2 = 144 \quad \text{or} \quad x^2 + (y - 5)^2 = 9$$

Touching Circles

If two circles touch externally then the distance between their centres is equal to the sum of their radii.

If the circles touch internally then the distance between their centres is equal to the difference of their radii.

7. Show that the circles $(x - 8)^2 + (y - 6)^2 = 25$ and $(5x - 16)^2 + (5y - 12)^2 = 25$ touch each other.

The circle $(x - 8)^2 + (y - 6)^2 = 25$ has centre $(8, 6)$ and radius 5

The equation $(5x - 16)^2 + (5y - 12)^2 = 25$ can be divided by 25, giving

$(x - \frac{16}{5})^2 + (y - \frac{12}{5})^2 = 1$ so the centre is $(\frac{16}{5}, \frac{12}{5})$ and the radius is 1

The distance between the centres is $\sqrt{[(8 - \frac{16}{5})^2 + (6 - \frac{12}{5})^2]} = 6$

The sum of the radii of the circles is $5 + 1 = 6$

Therefore the circles touch externally.

EXERCISE 4d

1. Determine whether the given line is a tangent to the given circle in each of the following cases.

   (a) $3x - 4y + 14 = 0$; $x^2 + y^2 + 4x + 6y - 3 = 0$

   (b) $5x + 12y = 4$; $x^2 + y^2 - 2x - 2y + 1 = 0$

   (c) $x + 2y + 6 = 0$; $x^2 + y^2 - 6x - 4y + 8 = 0$

   (d) $x + 2y + 6 = 0$; $x^2 + y^2 - 6x + 4y + 8 = 0$

2. Write down the equation of the tangent to the given circle at the given point.

   (a) $x^2 + y^2 - 2x + 4y - 20 = 0$; $(5, 1)$

   (b) $x^2 + y^2 - 10x - 22y + 129 = 0$; $(6, 7)$

   (c) $x^2 + y^2 - 8y + 3 = 0$; $(-2, 7)$

Find the equations of the following circles (in some cases more than one circle is possible).

3. A circle passes through the points $(1, 4)$, $(7, 5)$ and $(1, 8)$

4. A circle has its centre on the line $x + y = 1$ and passes through the origin and the point $(4, 2)$

5. The line joining $(2, 1)$ to $(6, 5)$ is a diameter of a circle.

6. A circle with centre $(2, 7)$ passes through the point $(-3, -5)$

7. A circle intersects the y-axis at the origin and at the point $(0, 6)$ and also touches the x-axis.

8. A circle touches the negative x and y axes and also the line $7x + 24y + 12 = 0$

Find the equations of the tangents specified in Questions 9 to 11.

9. Tangents *from* the origin to the circle $x^2 + y^2 - 10x - 6y + 25 = 0$

10. The tangent *at* the origin to the circle $x^2 + y^2 + 2x + 4y = 0$

11. Tangents to the circle $x^2 + y^2 - 4x + 6y - 7 = 0$ which are parallel to the line $2x + y = 3$

12. Show that if the point P is twice as far from the point $(4, -2)$ as it is from the origin then P lies on a circle. Find the centre and radius of this circle.

13. Determine which of the following pairs of circles touch.

(a) $x^2 + y^2 + 2x - 4y + 1 = 0$; $x^2 + y^2 - 6x - 10y + 25 = 0$

(b) $x^2 + y^2 + 8x + 2y - 8 = 0$; $x^2 + y^2 - 16x - 8y = 64$

(c) $x^2 + y^2 + 6x = 0$; $x^2 + y^2 + 6x - 4y + 12 = 0$

(d) $x^2 + y^2 + 2x - 8y + 1 = 0$; $x^2 + y^2 - 6y = 0$

(e) $x^2 + y^2 + 2x = 3$; $x^2 + y^2 - 6x - 3 = 0$

14. If $y = 2x + c$ is a tangent to the circle
$x^2 + y^2 + 4x - 10y - 7 = 0$ find the value(s) of c

15. Find the condition that m and c satisfy if the line $y = mx + c$
touches the circle $x^2 + y^2 - 2ax = 0$

CHAPTER 5

TRIGONOMETRIC IDENTITIES AND EQUATIONS

THE TRIGONOMETRIC FUNCTIONS

The graphs and main properties of the basic trigonometric functions are as follows.

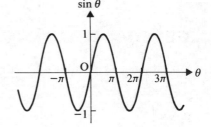

The sine function, $f: \theta \rightarrow \sin \theta$
   f is periodic with period 2π,
   f is odd, i.e. $-\sin \theta = \sin(-\theta)$,
   the range of f is $-1 \leqslant \sin \theta \leqslant 1$

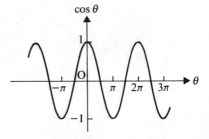

The cosine function, $f: \theta \rightarrow \cos \theta$
   f is periodic with period 2π,
   f is even, i.e. $\cos \theta = \cos(-\theta)$,
   the range of f is $-1 \leqslant \cos \theta \leqslant 1$

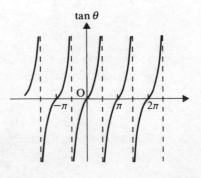

The tangent function, $f: \theta \rightarrow \tan \theta$
   f is periodic with period π,
   f is odd, i.e. $-\tan \theta = \tan(-\theta)$,
   the range of f is all real values.

The reciprocal trig ratios are

$$\sec\theta\left(\equiv\frac{1}{\cos\theta}\right) \qquad \operatorname{cosec}\theta\left(\equiv\frac{1}{\sin\theta}\right) \qquad \cot\theta\left(\equiv\frac{1}{\tan\theta}\right)$$

Some relationships between trig ratios are

$$\tan\theta\equiv\frac{\sin\theta}{\cos\theta}, \qquad\qquad \sin^2\theta+\cos^2\theta\equiv 1$$

$$\tan^2\theta+1\equiv\sec^2\theta$$

$$1+\cot^2\theta\equiv\operatorname{cosec}^2\theta$$

GENERAL SOLUTIONS OF TRIG EQUATIONS

Consider the equation $\sin\theta = 1$

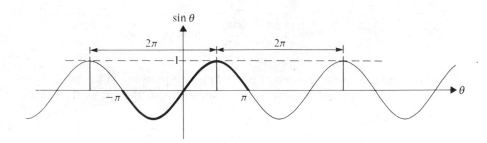

We see that, in the range $-\pi\leqslant\theta\leqslant\pi$, there is one solution, $\theta=\tfrac{1}{2}\pi$
We can also see that there are many more values of θ that satisfy
$\sin\theta = 1$

The general solution is an expression which represents all angles which satisfy the given equation. From the graph we can see that the general solution is an infinite set of angles.

In looking for a general solution, use can be made of the graphs of trig
functions and their periodic nature.

Now consider the equation $\sin \theta = \frac{1}{2}$

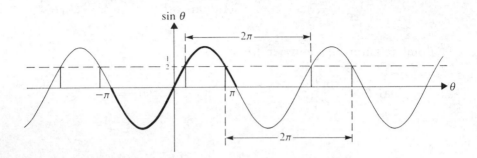

The period of the sine function is covered by the interval $[-\pi, \pi]$ and repeats every 2π

In the interval $[-\pi, \pi]$, $\theta = \frac{1}{6}\pi$ and $\theta = \frac{5}{6}\pi$ and we can get further solutions by adding (or subtracting) any multiple of 2π to $\frac{1}{6}\pi$ or to $\frac{5}{6}\pi$

Therefore the general solution of $\sin \theta = \frac{1}{2}$ can be expressed as

$$\theta = \begin{cases} \frac{1}{6}\pi + 2n\pi \\ \frac{5}{6}\pi + 2n\pi \end{cases} \quad \text{where } n \in \mathbb{Z}$$

The same argument can be applied to any equation $\sin \theta = s$, where $-1 \leqslant s \leqslant 1$, which usually has two solutions in the interval $[-\pi, \pi]$, showing that

the general solution of $\sin \theta = s$, where $-1 \leqslant s \leqslant 1$, is

$$\theta = \begin{cases} \theta_1 + 2n\pi \\ \theta_2 + 2n\pi \end{cases} \quad \text{or} \quad \theta = \begin{cases} \theta_1 + 360n° \\ \theta_2 + 360n° \end{cases}$$

where $n \in \mathbb{Z}$, and θ_1 and θ_2 are the solutions in the range $-\pi \leqslant \theta \leqslant \pi$

A similar situation occurs when the equation $\cos \theta = c$ is considered. The cosine function is periodic with a period of 2π, and one period is covered by the interval $[-\pi, \pi]$.

Within this interval there are usually two solutions, θ_1 and θ_2, and adding or subtracting any multiple of 2π gives another angle with the same cosine ratio. But, because the graph is symmetrical about $\theta = 0$, $\theta_2 = -\theta_1$, the general solution can be expressed in terms of θ_1 alone.

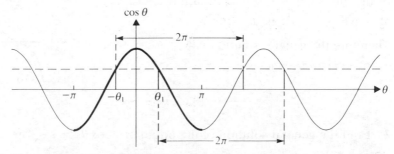

The general solution of the equation $\cos \theta = c$, where $-1 \leqslant c \leqslant 1$ is

$$\theta = \pm\theta_1 + 2n\pi$$

where $n \in \mathbb{Z}$ and θ_1 is a solution in the range $-\pi \leqslant \theta \leqslant \pi$

The situation with the equation $\tan \theta = t$ is different because the tangent function has a period of π which is covered by the interval $[-\frac{1}{2}\pi, \frac{1}{2}\pi]$. Within this interval there is only one solution of the equation $\tan \theta = t$ and adding or subtracting any multiple of π gives another angle with the same tangent ratio.

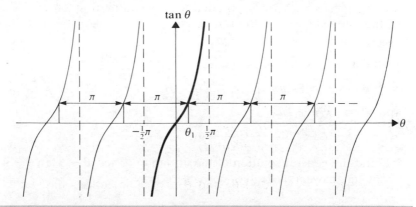

The general solution of the equation $\tan \theta = t$ is

$$\theta = \theta_1 + n\pi$$

where $n \in \mathbb{Z}$ and θ_1 is the solution in the range $-\frac{1}{2}\pi \leqslant \theta \leqslant \frac{1}{2}\pi$

Examples 5a

1. Find the general solution of the equation $\tan\theta = -\sqrt{3}$

In the interval $[-\frac{1}{2}\pi, \frac{1}{2}\pi]$ the solution
of $\tan\theta = -\sqrt{3}$ is $\theta = -\frac{1}{3}\pi$

Therefore the general solution is
$\theta = -\frac{1}{3}\pi + n\pi$ where $n \leqslant \mathbb{Z}$

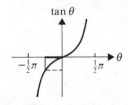

2. Find the general solution of the equation $\cos\theta = \dfrac{1}{\sqrt{2}}$

In the interval $[-\pi, \pi]$, the solutions
of $\cos\theta = \dfrac{1}{\sqrt{2}}$ are $\pm\frac{1}{4}\pi$

So the general solution is
$\theta = \pm\frac{1}{4}\pi + 2n\pi$ where $n \leqslant \mathbb{Z}$

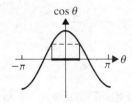

3. Find the general solution of the equation $\sin\theta = -\frac{1}{2}$

In the interval $[-\pi, \pi]$, the solutions
are $\theta = -\frac{1}{6}\pi$ and $\theta = -\frac{5}{6}\pi$

So the general solution is
$$\left.\begin{array}{l}\theta = -\frac{1}{6}\pi + 2n\pi \\ \text{and } \theta = -\frac{5}{6}\pi + 2n\pi\end{array}\right\} \text{ where } n \leqslant \mathbb{Z}$$

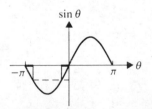

4. Find the general solution of the equation $3\sec^2\theta - 5\tan\theta - 4 = 0$
    giving answers in degrees correct to 1 d.p.

Using $1 + \tan^2\theta \equiv \sec^2\theta$ we have

$$3(1 + \tan^2\theta) - 5\tan\theta - 4 = 0$$
$$\Rightarrow \qquad 3\tan^2\theta - 5\tan\theta - 1 = 0$$

This equation does not have any simple factors so we solve it by the formula.

$$\Rightarrow \qquad \tan\theta = \frac{5 \pm \sqrt{(25+12)}}{6}$$

$$= 1.8471 \text{ or } -0.1805$$

In the interval $[-90°, 90°]$

$\tan\theta = 1.8471 \quad \Rightarrow \quad \theta = 61.6°$

$\tan\theta = -0.1805 \quad \Rightarrow \quad \theta = -10.2°$

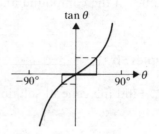

The general solution is therefore $\theta = \begin{cases} 61.6° + 180n° \\ -10.2° + 180n° \end{cases}$ where $n \in \mathbf{Z}$

EXERCISE 5a

Find the general solutions of each of the following equations.
Give answers in radians when they are exact; otherwise give answers
in degrees to 1 d.p.

1. $\sin\theta = \frac{\sqrt{3}}{2}$

2. $\cos\theta = 0$

3. $\tan\theta = -\sqrt{3}$

4. $\sin\theta = -\frac{1}{4}$

5. $\cos\theta = -\frac{1}{2}$

6. $\tan\theta = 1$

7. $\sec\theta = 1$

8. $\operatorname{cosec}\theta = 2$

9. $\sin^2\theta = \frac{1}{4}$

10. $\cos^2\theta = 1$

11. $5\cos\theta - 4\sin^2\theta = 2$

12. $4\cot^2\theta + 12\operatorname{cosec}\theta + 1 = 0$

13. $4\sec^2\theta - 3\tan\theta = 5$

14. $2\cos\theta - 4\sin^2\theta + 2 = 0$

15. $2\sin\theta\cos\theta + \sin\theta = 0$

16. $4\cos\theta = \cos\theta\operatorname{cosec}\theta$

17. $\sqrt{3}\tan\theta = 2\sin\theta$

18. $\cot\theta = \sin\theta$

19. $3\sin^2\theta + \cos\theta = 1$

20. $\tan\theta = \cos\theta$

21. $\cos\theta\sin\theta = 2\cos\theta$

22. $\tan\theta = 2$

23. $\cos^2\theta = 2 - 2\sin^2\theta$

24. $\tan\theta\cos^2\theta = \sin\theta$

EQUATIONS INVOLVING COMPOUND ANGLES

Many trig equations involve ratios of a compound angle, for example.

$$\cos 2\theta = \tfrac{1}{2} \qquad \cot(\tfrac{1}{3}\theta - 90°) = 1$$

Simple equations of this type can be solved by finding first the general values of the compound angle and then the corresponding values of θ.

Examples 5b

1. Find the general solution of the equation $\cos 2\theta = \tfrac{1}{2}$

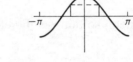

Using $2\theta = \phi$ gives $\cos \phi = \tfrac{1}{2}$

In the interval $[-\pi, \pi]$, the solutions
of $\cos \phi = \tfrac{1}{2}$ are $\phi = \pm\tfrac{1}{3}\pi$

So the general solution for ϕ is $\qquad \phi = \pm\tfrac{1}{3}\pi + 2n\pi$

but $\phi = 2\theta$, therefore $\qquad 2\theta = \pm\tfrac{1}{3}\pi + 2n\pi$

Hence $\qquad\qquad\qquad\qquad\qquad \theta = \pm\tfrac{1}{6}\pi + n\pi$

2. Find the general solution of the equation $\cot(\tfrac{1}{3}\theta - 90°) = 1$, giving the answer in degrees.

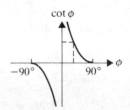

Using $\tfrac{1}{3}\theta - 90° = \phi$ gives

$$\cot(\tfrac{1}{3}\theta - 90°) = \cot \phi$$

The general solution of the equation $\cot \phi = 1$ is

$$\phi = 45° + 360n°$$

But $\phi = \tfrac{1}{3}\theta - 90°$, so $\qquad \tfrac{1}{3}\theta - 90° = 45° + 360n°$

$\Rightarrow \qquad\qquad\qquad\qquad\qquad \theta = 135° + 270° + 1080n°$

i.e. $\qquad\qquad\qquad\qquad\qquad\qquad \theta = 405° + 1080n°$

EXERCISE 5b

Find the general solutions of the following equations, giving your answers in radians.

1. $\tan 3\theta = 1$

2. $\cos 4\theta = -\frac{1}{2}$

3. $\sin \frac{1}{3}\theta = \frac{1}{2}$

4. $\sec 3\theta = 2$

5. $\cos \frac{1}{2}\theta = \frac{1}{2}\sqrt{3}$

6. $\cot 2\theta = -1$

7. $\cos (2\theta - \frac{1}{4}\pi) = 0$

8. $\sin (\frac{1}{4}\theta + \frac{1}{6}\pi) = -1$

9. $\cot (\frac{1}{3}\pi - \theta) = \sqrt{3}$

10. $\cos (2\theta + \frac{1}{6}\pi) = \frac{1}{2}$

Using the General Solution for a Specified Range

Sometimes an equation involving multiple angles requires a solution in a specified range and this can be found from the general solution.

Example 5c _____

Find the angles in the interval $[-180°, 180°]$ which satisfy the equation $\tan 3\theta = -2$

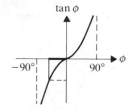

Using $\phi = 3\theta$, gives $\tan \phi = -2$

In the interval $[-90°, 90°]$, $\phi = -63.4°$,

$\therefore$ the general solution is $\quad 3\theta = -63.4° + 180n°$

$\Rightarrow \qquad\qquad\qquad\qquad \theta = -21.1° + 60n°$

Now to get values of θ from $-180°$ to $180°$, we see that we need values of n from -2 to 3

When $n = -2, -1, 0, 1, 2, 3$

$\theta = -141.1°, -81.1°, -21.1°, 38.9°, 98.9°, 158.9°$

Note that when a multiple angle is involved, it is the general solution of the multiple angle that must be found.

EXERCISE 5c

Solve these equations for values of θ in the range $-180° \leqslant \theta \leqslant 180°$.

1. $\sin 2\theta = \frac{1}{2}$ **2.** $\tan \frac{1}{2}\theta = 0.5$

3. $\cos(2\theta - 30°) = -\frac{1}{2}$ **4.** $\sin(\frac{1}{2}\theta + 60°) = 1$

Solve these equations for values of θ in the range $0 \leqslant \theta \leqslant \pi$

5. $\tan(\frac{1}{2}\theta - \frac{1}{4}\pi) = 1$ **6.** $\sec 3\theta = 2$

7. $\cos(2\theta + \frac{1}{2}\pi) = \frac{1}{2}\sqrt{2}$ **8.** $\cot 4\theta = 0$

COMPOUND ANGLE IDENTITIES

It is often useful to be able to express a trig ratio of an angle $A + B$ in terms of trig ratios of A and of B.

It is dangerously easy to think, for instance, that $\sin(A + B)$ is $\sin A + \sin B$. However, this is *false* as can be seen by considering

$$\sin(45° + 45°) = \sin 90° \qquad = 1$$

whereas $\sin 45° + \sin 45° = \frac{1}{2}\sqrt{2} + \frac{1}{2}\sqrt{2} \neq 1$

Thus the sine function is *not distributive* and neither are the other trig functions.

The correct identity is $\sin(A + B) = \sin A \cos B + \cos A \sin B$.

This is proved geometrically when A and B are both acute, from the diagram below.

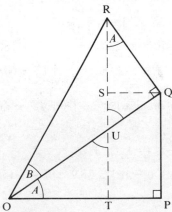

The right-angled triangles OPQ and OQR contain angles A and B are shown.
From the diagram, $\angle URQ = A$

$$\sin(A + B) = \frac{TR}{OR} = \frac{TS + SR}{OR} = \frac{PQ + SR}{OR}$$

$$= \frac{PQ}{OQ} \times \frac{OQ}{OR} + \frac{SR}{QR} \times \frac{QR}{OR}$$

$$\therefore \qquad \sin(A + B) \equiv \sin A \cos B + \cos A \sin B$$

This identity is in fact valid for all angles and it can be adapted to give the full set of compound angle formulae. The reader is left to do this in the following exercise.

EXERCISE 5d

1. In the identity $\sin(A + B) \equiv \sin A \cos B + \cos A \sin B$, replace B by $-B$ to show that $\sin(A - B) \equiv \sin A \cos B - \cos A \sin B$

2. In the identity $\sin(A - B) \equiv \sin A \cos B - \cos A \sin B$, replace A by $(\frac{1}{2}\pi - A)$ to show that $\cos(A + B) \equiv \cos A \cos B - \sin A \sin B$

3. In the identity $\cos(A + B) \equiv \cos A \cos B - \sin A \sin B$, replace B by $-B$ to show that $\cos(A - B) \equiv \cos A \cos B + \sin A \sin B$

4. Use $\dfrac{\sin(A + B)}{\cos(A + B)}$ to show that $\tan(A + B) \equiv \dfrac{\tan A + \tan B}{1 - \tan A \tan B}$

5. Replace B by $-B$ in the formula for $\tan(A + B)$ to show that

$$\tan(A - B) \equiv \frac{\tan A - \tan B}{1 + \tan A \tan B}$$

Collecting these results we have:

$$\sin(A + B) \equiv \sin A \cos B + \cos A \sin B$$

$$\sin(A - B) \equiv \sin A \cos B - \cos A \sin B$$

$$\cos(A + B) \equiv \cos A \cos B - \sin A \sin B$$

$$\cos(A - B) \equiv \cos A \cos B + \sin A \sin B$$

$$\tan(A + B) \equiv \frac{\tan A + \tan B}{1 - \tan A \tan B}$$

$$\tan(A - B) \equiv \frac{\tan A - \tan B}{1 + \tan A \tan B}$$

Examples 5e

1. Find exact values for   (a) $\sin 75°$ (b) $\cos 105°$

To find exact values, we need to express the given angle in terms of angles whose trig ratios are known as exact values, e.g. $30°$, $60°$, $45°$, $90°$, $120°$, ...
Now $75° = 45° + 30°$ (or $120° - 45°$ or other alternative compound angles).

(a) $\sin 75° = \sin(45° + 30°) = \sin 45° \cos 30° + \cos 45° \sin 30°$

$$= \left(\frac{\sqrt{2}}{2}\right)\left(\frac{\sqrt{3}}{2}\right) + \left(\frac{\sqrt{2}}{2}\right)\left(\frac{1}{2}\right)$$

$$= \frac{\sqrt{2}}{4}(\sqrt{3} + 1)$$

(b) $\cos 105° = \cos(60° + 45°) = \cos 60° \cos 45° - \sin 60° \sin 45°$

$$= \left(\frac{1}{2}\right)\left(\frac{\sqrt{2}}{2}\right) - \left(\frac{\sqrt{3}}{2}\right)\left(\frac{\sqrt{2}}{2}\right)$$

$$= \frac{\sqrt{2}}{4}(1 - \sqrt{3})$$

---

2. A is obtuse and $\sin A = \frac{3}{5}$,   B is acute and $\sin B = \frac{12}{13}$.   Find the exact value of $\cos(A + B)$.

$$\cos(A + B) \equiv \cos A \cos B - \sin A \sin B$$

In order to use this formula, we need values for $\cos A$ and $\cos B$.   These can be found using Pythagoras' theorem in the appropriate right-angled triangle.

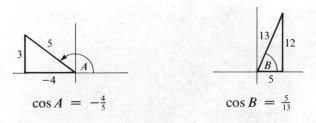

$$\cos A = -\frac{4}{5} \qquad\qquad \cos B = \frac{5}{13}$$

$$\therefore \qquad \cos(A + B) = \left(-\frac{4}{5}\right)\left(\frac{5}{13}\right) - \left(\frac{3}{5}\right)\left(\frac{12}{13}\right) = -\frac{56}{65}$$

---

3. Simplify $\sin\theta\cos\frac{1}{3}\pi - \cos\theta\sin\frac{1}{3}\pi$ and hence find the smallest positive value of θ for which the expression has a minimum value.

$\sin\theta\cos\frac{1}{3}\pi - \cos\theta\sin\frac{1}{3}\pi$ is the expansion of $\sin(A - B)$ with $A = \theta$ and $B = \frac{1}{3}\pi$

$$\sin\theta\cos\tfrac{1}{3}\pi - \cos\theta\sin\tfrac{1}{3}\pi = \sin(\theta - \tfrac{1}{3}\pi)$$

Let $f(\theta) = \sin(\theta - \frac{1}{3}\pi)$

Now the graph $f(\theta)$ is a sine wave, but translated $\frac{1}{3}\pi$ in the direction of the positive θ-axis.

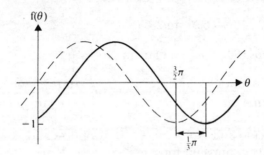

Therefore $f(\theta)$ has a minimum value of -1 and the smallest +ve value of θ at which this occurs is $\frac{3}{2}\pi + \frac{1}{3}\pi = \frac{11}{6}\pi$

---

4. Find the general solution of the equation $2\cos\theta = \sin(\theta + \frac{1}{6}\pi)$

$$2\cos\theta = \sin(\theta + \tfrac{1}{6}\pi)$$
$$= \sin\theta\cos\tfrac{1}{6}\pi + \cos\theta\sin\tfrac{1}{6}\pi = \tfrac{\sqrt{3}}{2}\sin\theta + \tfrac{1}{2}\cos\theta$$
$$\therefore \quad \tfrac{3}{2}\cos\theta = \tfrac{\sqrt{3}}{2}\sin\theta$$
$$\Rightarrow \quad \frac{3}{\sqrt{3}} = \frac{\sin\theta}{\cos\theta} \quad \Rightarrow \quad \tan\theta = \sqrt{3}$$

$\tan\frac{1}{3}\pi = \sqrt{3}$

Therefore the general solution is $\theta = \frac{1}{3}\pi + n\pi$

EXERCISE 5e

Find the exact value of each expression, leaving your answer in surd form where necessary.

1. $\cos 40° \cos 50° - \sin 40° \sin 50°$

2. $\sin 37° \cos 7° - \cos 37° \sin 7°$

3. $\cos 75°$ 4. $\tan 105°$

5. $\sin 165°$ 6. $\cos 15°$

Simplify each of the following expressions.

7. $\sin \theta \cos 2\theta + \cos \theta \sin 2\theta$

8. $\cos \alpha \cos (90° - \alpha) - \sin \alpha \sin (90° - \alpha)$

9. $\dfrac{\tan A + \tan 2A}{1 - \tan A \tan 2A}$ 10. $\dfrac{\tan 3\beta - \tan 2\beta}{1 + \tan 3\beta \tan 2\beta}$

11. A is acute and $\sin A = \frac{7}{25}$, B is obtuse and $\sin B = \frac{4}{5}$. Find an exact expression for
    (a) $\sin (A + B)$ (b) $\cos (A + B)$ (c) $\tan (A - B)$

12. Find the greatest value of each expression and the value of θ between 0 and 360° at which it occurs.
    (a) $\sin \theta \cos 25° - \cos \theta \sin 25°$ (b) $\sin \theta \sin 30° + \cos \theta \cos 30°$
    (c) $\cos \theta \cos 50° - \sin \theta \sin 50°$ (d) $\sin 60° \cos \theta - \cos 60° \sin \theta$

Prove the following identities.

13. $(\sin A + \cos A)(\sin B + \cos B) \equiv \sin (A + B) + \cos (A - B)$

14. $\sin (A + B) + \sin (A - B) \equiv 2 \sin A \cos B$

15. $\sin (\frac{1}{4}\pi + A) + \sin (\frac{1}{4}\pi - A) \equiv \sqrt{2} \cos A$

16. $\cos (A + B) + \cos (A - B) \equiv 2 \cos A \cos B$

17. $\dfrac{\sin (A + B)}{\cos A \cos B} \equiv \tan A + \tan B$

18. $\sin (\theta + 60°) \equiv \sin (120° - \theta)$

Solve the following equations for values of θ in the range $0 \leqslant \theta \leqslant 360°$

19. $\cos(45° - \theta) = \sin\theta$

20. $3\sin\theta = \cos(\theta + 60°)$

21. $\tan(A - \theta) = \frac{2}{3}$ and $\tan A = 3$

22. $\sin(\theta + 60°) = \cos\theta$

Find the general solution of the following equations.

23. $\sin(x + \frac{1}{3}\pi) = \cos x$

24. $\sin x = \cos(x - \frac{2}{3}\pi)$

THE DOUBLE ANGLE IDENTITIES

The compound angle formulae deal with any two angles A and B and can therefore be used for two equal angles, i.e. when B = A

Replacing B by A in the trig identities for (A + B) gives the following set of double angle identities.

$$\sin 2A \equiv 2\sin A \cos A$$
$$\cos 2A \equiv \cos^2 A - \sin^2 A$$
$$\tan 2A \equiv \frac{2\tan A}{1 - \tan^2 A}$$

The second of these identities can be expressed in several forms because

$$\cos^2 A - \sin^2 A \equiv \begin{cases} (1 - \sin^2 A) - \sin^2 A \equiv 1 - 2\sin^2 A \\ \cos^2 A - (1 - \cos^2 A) \equiv 2\cos^2 A - 1 \end{cases}$$

i.e.

$$\cos 2A \equiv \begin{cases} \cos^2 A - \sin^2 A \\ 1 - 2\sin^2 A \\ 2\cos^2 A - 1 \end{cases}$$

Examples 5f _____

1. If $\tan \theta = \frac{3}{4}$, find the values of $\tan 2\theta$ and $\tan 4\theta$

Using $\tan 2A \equiv \dfrac{2 \tan A}{1 - \tan^2 A}$ with $A = \theta$ and $\tan \theta = \frac{3}{4}$ gives

$$\tan 2\theta = \frac{2(\frac{3}{4})}{1 - (\frac{3}{4})^2} = \frac{24}{7}$$

Using the identity for $\tan 2A$ again, but this time with $A = 2\theta$, gives

$$\tan 4\theta = \frac{2 \tan 2\theta}{1 - \tan^2 2\theta} = \frac{2(\frac{24}{7})}{1 - (\frac{24}{7})^2} = -\frac{336}{527}$$

_____

2. Eliminate θ from the equations $x = \cos 2\theta, \quad y = \sec \theta$

Using $\cos 2\theta \equiv 2 \cos^2 \theta - 1$ gives

$$x = 2 \cos^2 \theta - 1 \quad \text{and} \quad y = \frac{1}{\cos \theta}$$

$$\therefore \qquad\qquad x = 2\left(\frac{1}{y}\right)^2 - 1$$

$$\Rightarrow \qquad\qquad (x + 1)y^2 = 2$$

Note that this is a Cartesian equation which has been obtained by _eliminating the parameter_ θ from a _pair of parametric equations_.

_____

3. Prove that $\sin 3A \equiv 3 \sin A - 4 \sin^3 A$

$$\sin 3A \equiv \sin (2A + A)$$

$$\equiv \sin 2A \cos A + \cos 2A \sin A$$

$$\equiv (2 \sin A \cos A)\cos A + (1 - 2 \sin^2 A)\sin A$$

$$\equiv 2 \sin A \cos^2 A + \sin A - 2 \sin^3 A$$

$$\equiv 2 \sin A(1 - \sin^2 A) + \sin A - 2 \sin^3 A$$

$$\equiv 3 \sin A - 4 \sin^3 A$$

_____

4. Find the general solution of the equation $\cos 2x + 3 \sin x = 2$

When a trig equation involves different multiples of an angle, it is usually sensible to express the equation in a form where the trig ratios are all of the same angle and, when possible, only one trig ratio is included.

Using $\cos 2x \equiv 1 - 2 \sin^2 x$ gives

$$1 - 2 \sin^2 x + 3 \sin x = 2$$

$$\Rightarrow \qquad 2 \sin^2 x - 3 \sin x + 1 = 0$$

$$\Rightarrow \qquad (2 \sin x - 1)(\sin x - 1) = 0$$

$$\therefore \qquad \sin x = \tfrac{1}{2} \quad \text{or} \quad \sin x = 1$$

When $\sin x = \tfrac{1}{2}$,

$$x = \begin{cases} \frac{1}{6}\pi + 2n\pi \\ \frac{5}{6}\pi + 2n\pi \end{cases}$$

When $\sin x = 1$,

$$x = \tfrac{1}{2}\pi + 2n\pi$$

The general solution is therefore $x = \frac{1}{6}\pi + 2n\pi, \frac{5}{6}\pi + 2n\pi, \frac{1}{2}\pi + 2n\pi$

EXERCISE 5f

Simplify, giving an exact value where this is possible.

1. $2 \sin 15° \cos 15°$

2. $\cos^2 \frac{1}{8}\pi - \sin^2 \frac{1}{8}\pi$

3. $\sin \theta \cos \theta$

4. $1 - 2 \sin^2 4\theta$

5. $\dfrac{2 \tan 75°}{1 - \tan^2 75°}$

6. $\dfrac{2 \tan 3\theta}{1 - \tan^2 3\theta}$

7. $\sqrt{(1 + \cos 6\theta)}$

8. $2 \cos^2 \frac{3}{8}\pi - 1$

9. $\dfrac{1 + \tan x}{1 - \tan x}$ (Hint. $\tan 45° = 1$)

10. $1 - 2 \sin^2 \frac{1}{8}\pi$

11. Find the value of $\cos 2\theta$ and $\sin 2\theta$ when θ is acute and when
   (a) $\cos \theta = \frac{3}{5}$ (b) $\sin \theta = \frac{7}{25}$ (c) $\tan \theta = \frac{12}{5}$

12. If $\tan\theta = -\frac{7}{24}$ and θ is obtuse, find

    (a) $\tan 2\theta$ (b) $\cos 2\theta$ (c) $\sin 2\theta$ (d) $\cos 4\theta$

13. Eliminate θ from the following pairs of equations.

    (a) $x = \tan 2\theta,\ y = \tan\theta$ (b) $x = \cos 2\theta,\ y = \cos\theta$

    (c) $x = \cos 2\theta,\ y = \operatorname{cosec}\theta$ (d) $x = \sin 2\theta,\ y = \sec 4\theta$

14. Prove the following identities.

    (a) $\dfrac{1 - \cos 2A}{\sin 2A} \equiv \tan A$

    (b) $\sec 2A + \tan 2A \equiv \dfrac{\cos A + \sin A}{\cos A - \sin A}$

    (c) $\tan\theta + \cot\theta \equiv 2\operatorname{cosec} 2\theta$

    (d) $\cos 4A \equiv 8\cos^4 A - 8\cos^2 A + 1$

    (e) $\sin 2\theta = \dfrac{2\tan\theta}{1 + \tan^2\theta}$

    (f) $\dfrac{1 - \cos 2A + \sin 2A}{1 + \cos 2A + \sin 2A} \equiv \tan A$

15. Find the general solutions of the following equations.
Give answers in radians if they are exact: otherwise give them in degrees to 1 d.p.

    (a) $\cos 2x = \sin x$ (b) $\sin 2x + \cos x = 0$

    (c) $4 - 5\cos\theta = 2\sin^2\theta$ (d) $\tan 2\theta \tan\theta = 2$

    (e) $\sin 2\theta - 1 = \cos 2\theta$ (f) $5\cos x \sin 2x + 4\sin^2 x = 4$

THE HALF-ANGLE IDENTITIES

If we replace A by $\frac{1}{2}\theta$ in the double angle formula for $\tan 2A$, we get

$$\tan\theta \equiv \frac{2\tan\frac{1}{2}\theta}{1 - \tan^2\frac{1}{2}\theta} \qquad [1]$$

If we now make the substitution $t = \tan\frac{1}{2}\theta$, [1] becomes

$$\tan\theta \equiv \frac{2t}{1 - t^2}$$

From the diagram, and using Pythagoras' theorem,

$$r = 1 + t^2$$

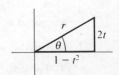

Hence we can express $\sin\theta$, $\cos\theta$ and $\tan\theta$ in terms of t, i.e.

$$\sin\theta = \frac{2t}{1+t^2}$$

$$\cos\theta = \frac{1-t^2}{1+t^2} \qquad \text{where} \quad t = \tan\tfrac{1}{2}\theta$$

$$\tan\theta = \frac{2t}{1-t^2}$$

These identities are often called the 'little t' formulae and they allow all the trig ratios of any one angle to be expressed in terms of a common variable t. This group can be helpful in problems where it is not possible to apply any of the identities used previously.

It is important to appreciate that t represents the tangent of an angle half as big as the original angle. This is not always $\tfrac{1}{2}\theta$.

There is one other group of identities that can be useful, and these are derived from the different forms of the cosine double angle formulae.

Starting with $\cos 2A \equiv 2\cos^2 A - 1$ we have

$$\cos^2 A \equiv \tfrac{1}{2}(1 + \cos 2A)$$

Similarly starting with $\cos 2A \equiv 1 - 2\sin^2 A$, we get

$$\sin^2 A \equiv \tfrac{1}{2}(1 - \cos 2A)$$

Examples 5g

1. Express $\sin\theta + 2\cos\theta$ in terms of t, where $t = \tan\tfrac{1}{2}\theta$. Hence solve the equation $\sin\theta + 2\cos\theta = 1$ for values of θ in the range $0 \leqslant \theta \leqslant 360°$

$$\sin\theta + 2\cos\theta \equiv \frac{2t}{1+t^2} + 2\left(\frac{1-t^2}{1+t^2}\right) \equiv \frac{2 + 2t - 2t^2}{1+t^2}$$

Hence $\qquad \sin\theta + 2\cos\theta = 1 \qquad \Rightarrow \qquad \dfrac{2 + 2t - 2t^2}{1+t^2} = 1$

$$\Rightarrow \qquad 3t^2 - 2t - 1 = 0$$

$$\Rightarrow \qquad (3t + 1)(t - 1) = 0$$

$\therefore \qquad\qquad\qquad\qquad t = -\tfrac{1}{3} \quad \text{or} \quad t = 1$

i.e. $\tan\tfrac{1}{2}\theta = -\tfrac{1}{3}$ or 1

The general solution is $\frac{1}{2}\theta = -18.43° + 180n°$ or $\frac{1}{2}\theta = 45° + 180n°$

$$\theta = -36.9° + 360n° \quad \text{or} \quad \theta = 90° + 360n°$$

$\therefore$ in the specified range, $\theta = 90°, 323.1°$

The next worked example deals with a very similar equation and it illustrates that extra care is needed when using this method to solve equations.

---

2. Find the general solution of the equation $\sin\theta - \cos\theta = 1$

Using $t = \tan\frac{1}{2}\theta$, $\sin\theta - \cos\theta = 1$

$$\Rightarrow \qquad \frac{2t}{1+t^2} - \frac{1-t^2}{1+t^2} = 1$$

$$\Rightarrow \qquad 2t - 1 + t^2 = 1 + t^2$$

We know that if we cancel the two t^2 terms, the solution $t = \infty$ is lost. In this problem we cannot ignore this solution as $t = \infty$ corresponds to a real value of θ. So we proceed as follows.

$\Rightarrow \qquad\qquad\qquad t = \infty \quad \text{or} \quad t = 1$

$\therefore \qquad\qquad\qquad \tan\frac{1}{2}\theta = \infty \quad \text{or} \quad \tan\frac{1}{2}\theta = 1$

$\Rightarrow \qquad\qquad\qquad \frac{1}{2}\theta = \frac{1}{2}\pi + n\pi \quad \text{or} \quad \frac{1}{2}\theta = \frac{1}{4}\pi + n\pi$

$$\Rightarrow \qquad\qquad\qquad \theta = \begin{cases} \frac{1}{2}\pi + 2n\pi \\ (2n+1)\pi \end{cases}$$

---

3. If $\tan\theta = \frac{3}{4}$, find the possible values of
   (a) $\tan\frac{1}{2}\theta$ (b) $\cos 2\theta$

(a) Using $\tan\frac{1}{2}\theta = t$ gives $\dfrac{3}{4} = \dfrac{2t}{1-t^2}$

$\Rightarrow \qquad\qquad\qquad 3t^2 + 8t - 3 = 0$

$\Rightarrow \qquad\qquad\qquad (3t-1)(t+3) = 0$

$\Rightarrow \qquad\qquad\qquad t = \frac{1}{3} \text{ or } -3$

i.e. $\tan\frac{1}{2}\theta = \frac{1}{3}$ or $\tan\frac{1}{2}\theta = -3$

(b) The half-angle formulae are valid for any angle where t is the tan of half the given angle. In this case the given angle is 2θ so $t = \tan\theta$

Using $\qquad\qquad \cos 2\theta = \dfrac{1 - t^2}{1 + t^2} \qquad$ with $t = \tan\theta$

$\Rightarrow \qquad\qquad\qquad \cos 2\theta = \dfrac{1 - (\frac{3}{4})^2}{1 + (\frac{3}{4})^2} = \dfrac{7}{25}$

4. Express $\sqrt{\left(\dfrac{1 - \sin 2\theta}{1 + \sin 2\theta} \right)}$ in terms of $\tan\theta$

If we double the angle in the formula for $\sin\theta$ in terms of t, we get

$\sin 2\theta = \dfrac{2t}{1 + t^2} \qquad$ where $t = \tan\theta$

$\therefore \qquad\qquad 1 - \sin 2\theta = 1 - \dfrac{2t}{1 + t^2} = \dfrac{1 + t^2 - 2t}{1 + t^2} = \dfrac{(1 - t)^2}{1 + t^2}$

and $\qquad\qquad 1 + \sin 2\theta = 1 + \dfrac{2t}{1 + t^2} = \dfrac{1 + t^2 + 2t}{1 + t^2} = \dfrac{(1 + t)^2}{1 + t^2}$

Hence $\qquad \dfrac{1 - \sin 2\theta}{1 + \sin 2\theta} = \dfrac{(1 - t)^2}{1 + t^2} \Big/ \dfrac{(1 + t)^2}{1 + t^2} = \left(\dfrac{1 - t}{1 + t} \right)^2$

$\therefore \qquad \sqrt{\left(\dfrac{1 - \sin 2\theta}{1 + \sin 2\theta} \right)} = \dfrac{1 - t}{1 + t} = \dfrac{1 - \tan\theta}{1 + \tan\theta}$

EXERCISE 5g

1. If $\tan\theta = \frac{4}{3}$, find the possible values of

(a) $\sin 2\theta$ (b) $\cot 2\theta$ (c) $\tan\frac{1}{2}\theta$ (d) $\cos\frac{1}{2}\theta$

Expressing the following in terms of t where $t = \tan\frac{1}{2}\theta$

2. $\dfrac{1 - \cos\theta}{1 + \cos\theta}$

3. $\dfrac{\sin\theta}{1 - \cos\theta}$

4. $\cot\theta \cot\frac{1}{2}\theta$

5. $\dfrac{\cos^2(\frac{1}{2}\theta)}{3\sin\theta + 4\cos\theta - 1}$

6. Prove that $\csc A + \cot A \equiv \cot \frac{1}{2} A$

7. If $\sec \theta - \tan \theta = x$ prove that $\tan \frac{1}{2}\theta = \dfrac{1-x}{1+x}$

Using $t = \tan \frac{1}{2}\theta$, solve the following equations for values of θ in the interval $[-180°, 180°]$.

8. $3 \cos \theta + 2 \sin \theta = 3$ **9.** $5 \cos \theta - \sin \theta + 4 = 0$

10. $\cos \theta + 7 \sin \theta = 5$ **11.** $2 \cos \theta - \sin \theta = 1$

MIXED EXERCISE 5

1. Eliminate θ from the equations $x = \sin \theta$ and $y = \cos 2\theta$

2. Prove the identity $\dfrac{\sin 2\theta}{1 + \cos 2\theta} \equiv \tan \theta$

3. Prove that $\tan (\theta + \frac{1}{4}\pi) \tan (\frac{1}{4}\pi - \theta) \equiv -1$

4. If $\cos A = \frac{4}{5}$ and $\cos B = \frac{5}{13}$ find the possible values of $\cos (A + B)$

5. Eliminate θ from the equations $x = \cos 2\theta$ and $y = \cos^2 \theta$

6. Solve the equation $8 \sin \theta \cos \theta = 3$ for values of θ from $-180°$ to $180°$

7. Find the general solution of the equation $\cos^2\theta - \sin^2\theta = 1$

8. Prove the identity $\cos^4 \theta - \sin^4 \theta \equiv \cos 2\theta$

9. Simplify the expression $\dfrac{1 + \cos 2x}{1 - \cos 2x}$

10. Find the values of A between 0 and $360°$ for which $\sin (60° - A) + \sin (120° - A) = 0$

11. Find the general solution of the equation,
$$2 \sin (2\theta - \tfrac{1}{4}\pi) = -1$$

12. Find the general solution of the equation $\sin (\frac{1}{4}\theta - \pi) = 1$

Hence find the smallest positive value of θ for which $\sin (\frac{1}{4}\theta - \pi) = 1$

CHAPTER 6

INEQUALITIES

RULES FOR INEQUALITIES

If $a > b$

then $\qquad a + k > b + k \quad$ for all real values of k

$\qquad ak > bk \qquad$ for $k > 0$

$\qquad ak < bk \qquad$ for $k < 0$

Solving Inequalities using Graphical Methods

In Module A we saw how to use sketch graphs to solve quadratic inequalities.

For example, to solve $(x - 2)(x - 3) < 0$, we start with a sketch of the curve $y = (x - 2)(x - 3)$

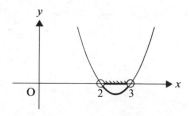

It is now clear that y is negative, i.e. $(x - 2)(x - 3) < 0$, for values of x between 2 and 3

Inequalities involving other functions can be solved just as easily by using sketch graphs.
When each side of an inequality is a function whose graph is recognised, a quick sketch of the graphs leads to an easy solution.
Sometimes a minor rearrangement of the inequality is necessary to produce easily sketched functions.

Examples 6a

1. Find the range of values of x for which $\dfrac{3}{x-2} < x$

We start with sketches of $y = \dfrac{3}{x-2}$ and $y = x$. Using a different colour for each graph makes the diagram clearer.

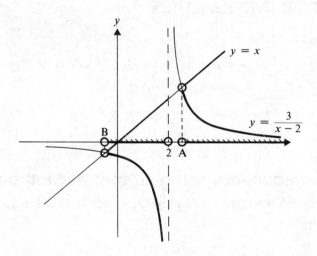

From the diagram we see that, $\dfrac{3}{x-2} < x$ for values of x lying between B and 2 and those beyond A.

At A and B

$$\frac{3}{x-2} = x$$

$\Rightarrow$

$$3 = x(x-2)$$

We can multiply by $x-2$ because we know that $x \neq 2$

$\Rightarrow$

$$x^2 - 2x - 3 = 0$$

$\Rightarrow$

$$(x-3)(x+1) = 0$$

$\Rightarrow$

$$x = 3 \quad \text{or} \quad x = -1$$

$\therefore \quad \dfrac{3}{x-2} < x$ for $-1 < x < 2$ and $x > 3$

2. Find the set of values of x for which $|x - 3| + |x| < 5$

$$|x - 3| + |x| < 5 \quad \Rightarrow \quad |x - 3| < 5 - |x|$$

We can now sketch the graphs $y = |x - 3|$ and $y = 5 - |x|$.

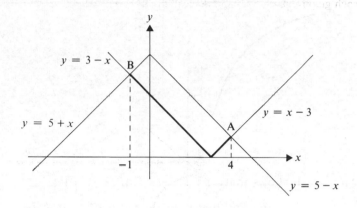

At A
$$x - 3 = 5 - x \quad \Rightarrow \quad x = 4$$

and at B
$$3 - x = 5 + x \quad \Rightarrow \quad x = -1$$

From the graph we see that $|3 - x| < 5 - |x|$ for $-1 < x < 4$

---

3. Solve the inequality $\dfrac{x - 1}{x + 1} > 0$

When an inequality involves a rational function, expressing it in a form that does not involve improper fractions often makes a graphical solution easier.

$$\frac{x - 1}{x + 1} > 0 \quad \Rightarrow \quad 1 - \frac{2}{x + 1} > 0$$

$$\Rightarrow \quad 1 > \frac{2}{x + 1}$$

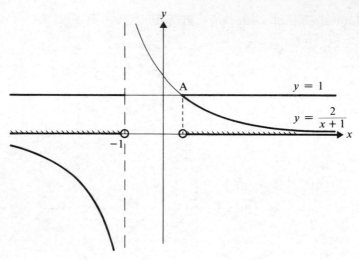

The sketch shows that, $1 > \dfrac{2}{x+1}$ for $x < -1$

and for $x >$ 'the value of x at A'

At A, $1 = \dfrac{2}{x+1}$ $\Rightarrow$ $x = 1$

$\therefore \dfrac{x-1}{x+1} > 0$ for $x < -1$ and $x > 1$

Alternatively we can solve the inequality in the last example, without any rearrangement, as follows.

Since $\dfrac{x-1}{x+1}$ is a fraction, then

$\dfrac{x-1}{x+1} > 0$ when $x-1$ and $x+1$ have the same sign,

i.e. when $x-1$ and $x+1$ are both positive or both negative.
We can find when this is so either by (a) sketching a graph or (b) making a table.

(a)

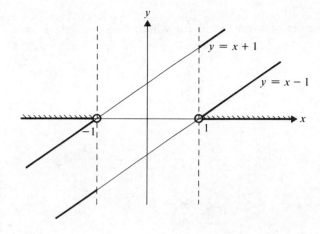

A sketch of the numerator and the denominator shows that they are both negative for $x < -1$ and both positive for $x > 1$

(b) The critical values of x (when the numerator or the denominator is zero) are -1 and 1 so we draw up a table in which the columns are separated by these values, i.e.

	$x < -1$	$-1 < x < 1$	$x > 1$
$x - 1$	$-$	$-$	$+$
$x + 1$	$-$	$+$	$+$
$\dfrac{x - 1}{x + 1}$	$+$	$-$	$+$

Therefore, by either method, $\dfrac{x - 1}{x + 1} > 0$ when $x < -1$ and $x > 1$

These alternative approaches can be used for any rational inequality, *provided that it is first expressed in the form of a single fraction greater than or less than zero.* This is illustrated in the next worked example.

4. Find the set of values of x for which $\dfrac{(x-1)^2}{x+5} < 1$

$$\dfrac{(x-1)^2}{x+5} < 1 \quad \Rightarrow \quad \dfrac{(x-1)^2}{x+5} - 1 < 0$$

$$\Rightarrow \quad \dfrac{(x^2 - 2x + 1) - (x+5)}{x+5} < 0$$

$$\Rightarrow \quad \dfrac{x^2 - 3x - 4}{x+5} < 0$$

$$\Rightarrow \quad \dfrac{(x-4)(x+1)}{x+5} < 0$$

For $\dfrac{(x-4)(x+1)}{x+5} < 0$, $(x-4)(x+1)$ and $(x+5)$ must be such that one is positive and the other is negative. It is a matter of personal choice whether a graph or a table is used to find the range of values of x.

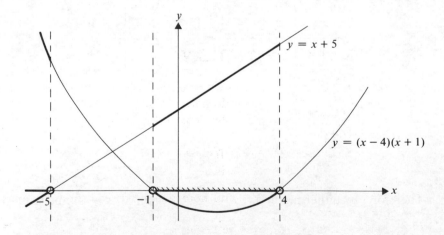

From the sketch, $\dfrac{(x-4)(x+1)}{x+5} < 0$ when the graphs are on

opposite sides of the x-axis, i.e. when $x < -5$ and $-1 < x < 4$

EXERCISE 6a

Solve the following inequalities.

1. $\dfrac{1}{x-1} < x$

2. $\dfrac{1}{x} > x^2$

3. $\dfrac{x}{x-1} < 0$

4. $\dfrac{x+1}{x} < 0$

5. $\dfrac{x+1}{x+2} < 0$

6. $\dfrac{1+x^2}{x} > 0$

7. $|x| < 1 - |x|$

8. $|x-1| < |x+2|$

9. $2|x| < |1-x|$

10. $|x+1| < 2x$

11. $3x - 1 < 1 + |x|$

12. $|3x+2| > 2 - |x+1|$

13. $1 + x^2 > 2x + 1$

14. $1 + x^2 > |2x+1|$

15. $|1 - x^2| < 2x + 1$

16. $1 - |x| > x^2 - 1$

17. $\dfrac{12}{x-3} < x + 1$

18. $\dfrac{x}{x-2} < \dfrac{x}{x-1}$

19. $\dfrac{(x-2)}{(x-1)(x-3)} > 0$

20. $\dfrac{2x}{(x-4)^2} > 1$

21. Find the set of values of x between 0 and 2π for which
$$|\sin x| < |\cos x|$$

THE RANGE OF A FUNCTION

The range of a function f is its set of output values, i.e. all the values that $f(x)$ can have for a given domain. When f is straightforward, the range can usually be written down from knowledge of the basic function,

e.g. if the range of $f(x) = 2\sin x$ for $x \in \mathbb{R}$,
the range is $-2 \leqslant f(x) \leqslant 2$.

Other known facts that are useful here, and when dealing with general inequalities, are
a perfect square can never be negative,
a modulus can never be negative,
the nature of the roots of the quadratic equation $ax^2 + bx + c = 0$
depend upon the value of $b^2 - 4ac$.

The following worked examples give some ideas for tackling problems about the range of a function but they are not comprehensive.

Examples 6b

1. Find the range of f where $f:x \rightarrow \dfrac{1}{2x^2 + 1}$, $x \in \mathbb{R}$.

Using the methods given in Chapter 2 for sketching reciprocal curves, we can start with a sketch of $y = 2x^2 + 1$, then deduce the shape of $y = \dfrac{1}{2x^2 + 1}$ and hence find the required range.

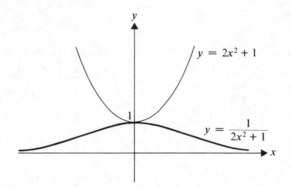

From the sketch, $0 < \dfrac{1}{2x^2 + 1} \leqslant 1$

_____

2. Find the values of p for which $x^2 - 2px + p + 6$ is positive for all real values of x

$x^2 - 2px + p + 6$ is positive for all values of x

i.e. the range is $x^2 - 2px + p + 6 > 0$

Therefore the graph of $f(x)$, where $f(x) = x^2 - 2px + p + 6$, is entirely above the x-axis, i.e.

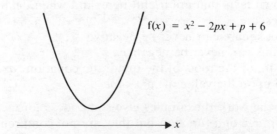

The graph never crosses the x-axis so there are no values of x for which $f(x) = 0$, i.e. $x^2 - 2px + p + 6 = 0$ has no real roots.

$\therefore$ '$b^2 - 4ac$' < 0 $\Rightarrow$ $(-2p)^2 - 4(1)(p + 6) < 0$

$\Rightarrow$ $4p^2 - 4p - 24 < 0$

$\Rightarrow$ $p^2 - p - 6 < 0$

$\Rightarrow$ $(p + 2)(p - 3) < 0$

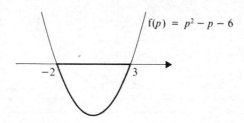

$f(p) = p^2 - p - 6$

From the graph of $f(p) = p^2 - p - 6$ we see that $f(p) < 0$ for values of p between -2 and 3

Therefore $x^2 - 2px + p + 6$ is positive for all real values of x provided that $-2 < p < 3$

---

3. If x is real find the set of possible values of $\dfrac{x^2}{x + 1}$

If we use $y = \dfrac{x^2}{x + 1}$ we are looking for the range of values of y

To make use of the fact that x is real we need a quadratic equation in x

$$y = \frac{x^2}{x + 1} \Rightarrow x^2 - yx - y = 0$$

Since x is real, the roots of this equation are real, so '$b^2 - 4ac$' $\geqslant 0$

i.e. $\qquad\qquad (-y)^2 - 4(1)(-y) \geqslant 0 \quad \Rightarrow \quad y(y+4) \geqslant 0$

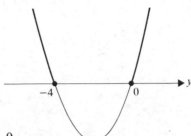

$\therefore \ y \leqslant -4 \ $ or $ \ y \geqslant 0$

Therefore, for real values of x,

$$\frac{x^2}{x+1} \leqslant -4 \quad \text{or} \quad \frac{x^2}{x+1} \geqslant 0$$

EXERCISE 6b

In Questions 1 to 6 find the set of possible values of the given function.

1. $\dfrac{x+1}{2x^2 + x + 1}$

2. $\dfrac{1+x^2}{x}$

3. $\dfrac{x-2}{(x+2)(x-3)}$

4. $\dfrac{x}{1+x^2}$

5. $\dfrac{x-1}{x(x+1)}$

6. $x + 1 + \dfrac{1}{x+1}$

7. Find the set of values of k for which $x^2 + 3kx + k$ is positive for all real values of x.

8. What is the set of values of p for which $p(x^2 + 2) < 2x^2 + 6x + 1$ for all real values of x?

9. Show that, if x is real, the function $\dfrac{2-x}{x^2 - 4x + 1}$ can take any real value.

10. If x is real, find the range of the function $\dfrac{(2x+1)}{(x^2+2)}$.

11. On the same set of axes, sketch the curves whose equations are

$$y = (x - 2)(x - 4) \quad \text{and} \quad y = \frac{1}{(x - 2)(x - 4)}$$

clearly showing any asymptotes and turning points.

Hence find the range of f where $f : x \rightarrow \dfrac{1}{(x - 2)(x - 4)}$

12. Use any method to find the range of the function, f, where

(a) $f(x) = \dfrac{1}{1 - x}$

(b) $f(x) = \dfrac{1}{(x - 2)(x - 6)}$

(c) $f(x) = \dfrac{x}{(1 - x)^2}$

(b) $f(x) = \dfrac{4}{(x - 1)(x - 3)}$

MIXED EXERCISE 6

In Questions 1 to 4 find the set of values of x for which

1. $\dfrac{2x - 4}{x - 1} < 1$

2. $\dfrac{x - 1}{x + 1} \leqslant x$

3. $2 \geqslant \dfrac{x - 1}{x + 1} \geqslant 0$

4. $\dfrac{(x - 1)(x + 1)}{(x + 2)(x - 2)} \leqslant 0$

5. If x is real find the set of possible values of the function

$$\frac{x^2 + 1}{x^2 + x + 1}$$

6. Provided that x is real, prove that the function $\dfrac{2(3x + 1)}{3(x^2 - 9)}$ can take all real values.

7. Sketch the graph of $g(x) = \dfrac{x - 2}{x - 1}$.

On the same set of axes sketch the graph of the function g^{-1} and hence state the relationship between g and g^{-1}.

8. Find the set of values of x for which $|x - 1| > 1 + |1 + x|$

9. Find the range of values of x for which $\dfrac{x^2 - 2}{x} > 1$

10. Find where the curves $y = \dfrac{1}{2x}$ and $y = \dfrac{1}{1+x}$ intersect.

Hence or otherwise find the values of x for which $\dfrac{1}{2x} > \dfrac{1}{1+x}$.

11. Find the values of x for which

   (a) $x < |x + 1|$ (b) $|x| < |x + 1|$

CONSOLIDATION A

SUMMARY

LOGARITHMS

$$\log_a b = c \quad \Longleftrightarrow \quad a = b$$
$$\log_a b + \log_a c = \log_a bc$$
$$\log_a b - \log_a c = \log_a b/c$$
$$\log_a b^n = n \log_a b$$

The base of a logarithm can be changed using the formula

$$\log_a x = \frac{\log_b x}{\log_b a}$$

PARTIAL FRACTIONS

A proper fraction with a denominator which factorises can be expressed in partial fractions as follows.

The numerators which can be found by the cover-up method are screened.

$$\frac{f(x)}{(x-a)(x-b)} = \frac{A}{(x-a)} + \frac{B}{(x-b)}$$

$$\frac{f(x)}{(x-a)(x-b)^2} = \frac{A}{(x-a)} + \frac{B}{(x-b)} + \frac{C}{(x-b)^2}$$

$$\frac{f(x)}{(x-a)(x-b^2)} = \frac{A}{(x-a)} + \frac{Bx+C}{(x^2+b)}$$

FUNCTIONS

The general form of a polynomial function is
$$f(x) = a_n x^n + a_{n-1} x^{n-1} + \ldots + a_0$$
where n is a positive integer and $a_n, a_{n-1}, \ldots$ are constants.

93

The general form of a rational function is $f(x)/g(x)$ where $f(x)$ and $g(x)$ are polynomials.

A *periodic* function has a basic pattern which repeats at regular intervals. The width of the interval is called the period.

The *modulus* function, $f(x) = |x|$ is such that $|x|$ is the positive numerical value of x, e.g. when $x = -3$, $|x| = 3$
The graph of $y = |f(x)|$ is obtained from the graph of $y = f(x)$ by reflecting in the x-axis the parts of the graph for which $f(x)$ is negative. The sections for which $f(x)$ is positive remain unchanged.

TRIGONOMETRY

Compound Angle Identities

$$\sin (A \pm B) \equiv \sin A \cos B \pm \cos A \sin B$$

$$\cos (A \pm B) \equiv \cos A \cos B \mp \sin A \sin B$$

$$\tan (A \pm B) \equiv \frac{\tan A \pm \tan B}{1 \mp \tan A \tan B}$$

Double Angle Identities

$$\sin 2A \equiv 2 \sin A \cos A$$

$$\cos 2A \equiv \begin{cases} \cos^2 A - \sin^2 A \\ 2 \cos^2 A - 1 \\ 1 - 2 \sin^2 A \end{cases} \quad \text{and} \quad \begin{cases} \cos^2 A \equiv \tfrac{1}{2}(1 + \cos 2A) \\ \sin^2 A \equiv \tfrac{1}{2}(1 + \cos 2A) \end{cases}$$

$$\tan 2A \equiv \frac{2 \tan A}{1 - \tan^2 A}$$

'Little t' Identities

$$\left. \begin{array}{l} \tan \theta \equiv \dfrac{2t}{1 - t^2} \\[2mm] \sin \theta \equiv \dfrac{2t}{1 + t^2} \\[2mm] \cos \theta \equiv \dfrac{1 - t^2}{1 + t^2} \end{array} \right\} \quad \text{where} \quad t = \tan \tfrac{1}{2}\theta$$

General Solutions of Trigonometric Equations

If $\sin \theta = c$ has solutions $\theta = \theta_1$ and $\theta = \theta_2$ in the interval $[-\pi, \pi]$

$$\text{the general solution is } \theta = \begin{cases} \theta_1 + 2n\pi \\ \theta_2 + 2n\pi \end{cases}$$

If $\cos \theta = c$ has a solution $\theta = \theta_1$ in the interval $[0, \pi]$

the general solution is $\theta = \pm\theta_1 + 2n\pi$

If $\tan \theta = t$ has a solution $\theta = \theta_1$ in the interval $[-\tfrac{1}{2}\pi, \tfrac{1}{2}\pi]$

the general solution is $\theta = \theta_1 + n\pi$

COORDINATE GEOMETRY

The acute angle α between two
lines with gradients m_1 and m_2
is given by $\tan \alpha = \left| \dfrac{m_1 - m_2}{1 + m_1 m_2} \right|$

The distance from the point (p, q) to the line with equation
$ax + by + c = 0$ is given by $\left| \dfrac{ap + bq + c}{\sqrt{(a^2 + b^2)}} \right|$

Circles

The equation of a circle with centre (a, b) and radius r is
$$(x - a)^2 + (y - b)^2 = r^2$$

The equation $x^2 + y^2 + 2gx + 2fy + c = 0$ represents a circle
with centre $(-g, -f)$ and radius $\sqrt{(g^2 + f^2 - c)}$
provided that $g^2 + f^2 - c > 0$

Two circles with radii r_1 and r_2 touch if the distance between their
centres is equal to $|r_1 \pm r_2|$

MULTIPLE CHOICE EXERCISE A

TYPE I

1. If $\log_x y = 2$ then

   A $\;x = 2y$ C $\;x^2 = y$ E $\;y = \sqrt{x}$
   B $\;x = y^2$ D $\;y = 2x$

2. The fraction $\dfrac{1}{(x + 1)(x - 1)}$ can be expressed as

   A $\;\dfrac{1}{2(x - 1)} - \dfrac{1}{2(x + 1)}$ D $\;\dfrac{1}{x^2} - \dfrac{1}{1}$

   B $\;\dfrac{1}{x + 1} + \dfrac{1}{x - 1}$ E $\;\dfrac{1}{x + 1} - \dfrac{1}{x - 1}$

   C $\;\dfrac{2}{x + 1} - \dfrac{2}{x - 1}$

3. If $\cos \theta = \frac{1}{2}$, the general solution is

   A $\;\theta = 2n\pi \pm \frac{1}{6}\pi$ C $\;\theta = 2n\pi + \frac{1}{3}\pi$ E $\;\theta = n\pi \pm \frac{1}{6}\pi$
   B $\;\theta = n\pi + \frac{1}{3}\pi$ D $\;\theta = 2n\pi \pm \frac{1}{3}\pi$

4. The function $f : \theta \to \cos(2\theta - \frac{1}{2}\pi)$ has a period

   A $\;2\pi$ C $\;\frac{1}{2}\pi$ E $\;$ none of these
   B $\;\pi$ D $\;-\frac{1}{2}\pi$

5. The distance of the point $(2, -1)$ from the line $x - 3y + 4 = 0$ is

   A $\;\dfrac{9}{\sqrt{5}}$ B $\;\dfrac{9}{\sqrt{3}}$ C $\;\dfrac{3}{\sqrt{5}}$ D $\;\dfrac{3}{\sqrt{10}}$ E $\;\dfrac{9}{\sqrt{10}}$

6. The graph of $f(x) = 1 - |x|$ could be

   A    C   E

   B    D

7. The curve $y = \dfrac{x}{x+1}$ has an asymptote with equation

    A $y = 0$ **C** $y = 1$ **E** $x = 0$

    B $y = -1$ **D** $y = \frac{1}{2}$

8. The value of $\log_5 0.04$ is

    A 4   **B** 5   **C** $\frac{1}{2}$ **D** -2 **E** 0.25

9. If $\dfrac{2}{(x+1)(x-1)} = \dfrac{A}{x+1} + \dfrac{B}{x-1}$ then

    A $A = 1,\ B = 1$ **D** $A = 0,\ B = 2$

    B $A = -1,\ B = 1$ **E** $A = x - 1,\ B = x + 1$

    C $A = x,\ B = 1$

10. One of these is not an identity. Which one is it?

    A $\cos^2\theta = 1 - \sin^2\theta$ **D** $1 + \tan^2\theta = \sec^2\theta$

    B $\cos 2\theta = 2\cos^2\theta - 1$ **E** $\tan 2\theta = \dfrac{2\tan\theta}{1 + \tan^2\theta}$

    C $\cos^2\theta = \frac{1}{2}(\cos 2\theta + 1)$

11. If $|x| = |2 - x|$ then x is

    A 0   **B** 2   **C** 1 and -1 **D** -1 **E** 1

TYPE II

12. $f(x) = \dfrac{2}{(x+1)(x-1)}$

    A $f(x) = 0$ has two real roots

    B $f(x) = \dfrac{1}{x-1} - \dfrac{1}{x+1}$

    C $f(x)$ has a maximum value of -2

13. If $\dfrac{x-2}{x-3} < 1$ then

    A $x - 2 < x - 3$

    B $(x-2)(x-3) < (x-3)^2$

    C $-2 < -3$

14. If the equation $ax^2 + by^2 + 2gx + 2fy + c = 0$ represents a circle through the origin, then

A $g = 0$ and $f = 0$
B $c = 0$
C $a = b$

15. If $f(x) = x^2 - 1$ then the curve $y = \dfrac{1}{f(x)}$ has

A a maximum point at $(0, -1)$
B symmetry about the y-axis
C an asymptote $x = -1$

16. In the interval $0 < x < 1$

A $|x + 1| > 0$ **B** $|x - 1| < 0$ **C** $|x + 1| = |x| + 1$

TYPE III

17. $f(x) = \cos x$ is such that $f(x) = f(x + \pi)$.

18. If $f(x) = \dfrac{1}{2x^2 - 6x + 1}$ then $f(x)$ has a maximum value of $-\frac{2}{7}$.

19. $x^2 + y^2 - 2x - 4y + 6 = 0$ is the equation of a circle.

20. The general solution of the equation $\cos\theta = -1$ is $\theta = (2n + 1)\pi$.

21. $\sin 2\theta = \dfrac{2t}{1 + t^2}$ where $t = \tan\frac{1}{2}\theta$.

22. If $\dfrac{1}{x} < 2$ then $\dfrac{1}{2} < x$.

23. $3\log_{10}x + 1 = \log_{10}10x^3$ is an equation.

MISCELLANEOUS EXERCISE A

1. Find the values of A and B for which

$$\frac{x - 2}{(x + 3)(2x - 1)} = \frac{A}{x + 3} + \frac{B}{2x - 1}$$

2. Find the value of   (a) $\log_3 3\sqrt{3}$ (b) $\log_{25} 125\sqrt{5}$

3. Sketch the graph of $y = |x + 2|$ and hence, or otherwise, solve the inequality

$$|x + 2| > 2x + 1 \quad x \in \mathbb{R} \tag{C}$$

4. Find all values of θ, such that $0° \leqslant \theta \leqslant 180°$, which satisfy the equation $2 \sin 2\theta = \tan \theta$ (C)

5. Find the set of values of x for which

$$\frac{2}{x-2} < \frac{1}{x+1}$$

(U of L)

6. Solve the equation

$$2 \log_3 x = 1 + \log_3 (18 - x)$$

(AEB)

7. Solve the equation $3^{2x} = 4^{2-x}$, giving your answer to three significant figures.   (C)

8. Sketch the curve $y = (2x - 1)^2(x + 1)$, showing the coordinates of
   (a) the points where it meets the axes
   (b) the turning points
   (c) the point of inflexion.
   By using your sketch, or otherwise, sketch the graph of

$$y = \frac{1}{(2x - 1)^2(x + 1)}$$

   Show clearly the coordinates of any turning points.   (U of L)

9. (a) Write down the coordinates of the mid-point M of the line joining $A(0, 1)$ and $B(6, 5)$.
   (b) Show that the line $3x + 2y - 15 = 0$ passes through M and is perpendicular to AB.
   (c) Calculate the coordinates of the centre of the circle which passes through A, B and the origin O.   (U of L)

10. Express in partial fractions
    (a) $\dfrac{1}{x^2 - 1}$ (b) $\dfrac{1}{(x - 1)^2(x + 1)}$ (c) $\dfrac{1}{(x - 1)(x^2 + 1)}$

11. Find the complete set of values of x for which

$$\frac{x - 2}{x + 1} < 3$$

(U of L)

12.

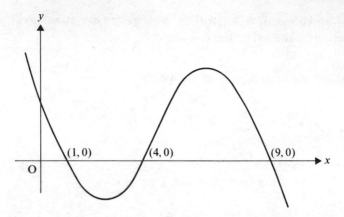

The figure shows the graph of the curve with equation

$$y = f(x)$$

In separate diagrams, sketch the curves with equations

(a) $y = |f(x)|$

(b) $y = f(x + 4)$

(c) $y = f(x^2)$.

Your sketches should show the coordinates of points at which each curve meets the x-axis.   (U of L)

13. Show that $\sin 3x \equiv 3 \sin x - 4 \sin^3 x$.

Find, in radians, the general solution of the equation

$$\sin 3x = 2 \sin x$$ (U of L)

14. Given that $\log(2x - 4) + \log 3 = 3 \log y$ find an expression for x in terms of y

15. Express $\dfrac{3}{x(x + 1)}$ in partial fractions.

16. A circle S is given by the equation

$$x^2 + y^2 - 4x + 6y - 12 = 0$$

Find the radius of S and the coordinates of the centre of S.
Calculate the length of the perpendicular from the centre of S to the line L whose equation is

$$3x + 4y = k$$

where k is a constant. Deduce the values of k for which L is a tangent to S.   (JMB)

17. Find, in terms of π, the general solution of the equation
$$\tan^4 x - 4\tan^2 x + 3 = 0$$
(AEB)

18. Find the set of values of x for which
$$|3x - 2| < 1 - 4x$$
(U of L)

19. Solve the inequality
$$\frac{x + 2}{x - 1} < 3$$
(JMB)

20. Given that $x = 2^p$ and $y = 4^q$, find in terms of p and q,
    (a) $8xy$ as a power of 2
    (b) $\log_2 x^3 y$ in a form not involving logarithms.
(U of L)

21. Given that
$$f(x) \equiv 2 - \frac{3}{x^2 - 2x + 4}$$
show that $f(x)$ is always positive.

Sketch the graph of the curve $y = f(x)$, stating the equations of any asymptotes and the coordinates of any points of intersection with the coordinate axes.
(U of L)

22. Find the general solution of the equation
$$10\cos 2x = 3\cos x - 8$$
giving your answer in degrees to 1 decimal place.
(U of L)

23. Find the set of values of x for which
$$x > |3x - 8|$$
(U of L)

24. The straight lines $3x + 4y = 14$ and $x - 7y = 13$ intersect at the point A, and the line $4x - 3y + 23 = 0$ cuts the other two lines at the points B and C respectively.
    (a) Show that the length of BC is 10 units.
    (b) Calculate the value of the acute angle BAC.
    (c) Find the coordinates of A and calculate the perpendicular distance from A to BC. Hence find the equation of the circle with centre A for which BC is a tangent.
(AEB)

25. (a) Find the values of x which satisfy the equation

$$\log_4 x = 9 \log_x 4$$

(b) By taking $\log_{10} 5 \approx 0.7$, obtain an estimate of the root of the equation

$$10^{y-5} = 5^{y+2}$$

giving your answer to the nearest integer.

26. Given that $x \neq 2$, find the complete set of values of x for which

$$\frac{3x + 1}{x - 2} > 2 \qquad\qquad \text{(U of L)}$$

27. Find all values of θ, such that $0° \leqslant \theta \leqslant 180°$, which satisfy the equation $2 \sin 2\theta = \tan \theta$ (C)

28. The circle with equation $(x - 5)^2 + (y - 7)^2 = 25$ has centre C. The point P(2, 3) lies on the circle. Determine the gradient of PC and hence, or otherwise, obtain the equation of the tangent to the circle at P.

Find also the equation of the straight line which passes through the point C and the point Q(−1, 4). The tangent and the line CQ intersect at R.

Determine the size of angle PRC, to the nearest 0.1°   (AEB)

CHAPTER 7

DIFFERENTIATION OF PRODUCTS AND QUOTIENTS

THE GRADIENT OF A CURVE

Consider two adjacent points, $A(x, y)$ and $B(x + \delta x, y + \delta y)$ on the curve $y = f(x)$.

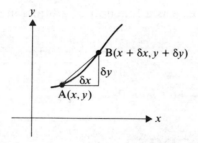

The gradient of the curve at A

$\qquad$ = the gradient of the tangent at A

$\qquad\qquad$ = the limiting value of the gradient of AB

$$= \operatorname*{limit}_{\delta x \to 0} \frac{\delta y}{\delta x}$$

The gradient function can be called the derived function, the derivative, or the differential coefficient, of $f(x)$.

It is denoted by $\dfrac{dy}{dx}$ or $f'(x)$

i.e. $\qquad \dfrac{dy}{dx}$ or $f'(x) = \lim\limits_{\delta x \to 0} \dfrac{\delta y}{\delta x}$ or $\lim\limits_{\delta x \to 0} \dfrac{f(x + \delta x) - f(x)}{\delta x}$

When y is a function of x, the derivative represents the rate at which y is increasing with respect to x.

103

Standard Derivatives

y	$\dfrac{dy}{dx}$
a	0
ax	a
x^n	nx^{n-1}
$(ax + b)^n$	$an(ax + b)^{n-1}$

FUNCTION OF A FUNCTION

If $y = \text{gf}(x)$, i.e. y is a function of a function of x, using $u = \text{f}(x)$ gives $y = \text{g}(u)$

Then
$$\frac{dy}{dx} = \frac{dy}{du} \times \frac{du}{dx}$$

This is known as the *Chain Rule*.

STATIONARY POINTS

At a stationary point on a curve, y is momentarily neither increasing nor decreasing.

So $\dfrac{dy}{dx} = 0$, i.e. the tangent to the curve is horizontal.

At a stationary point, y has a stationary value.

The three types of stationary point are shown in the this diagram.

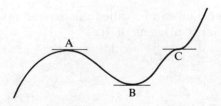

A is a maximum point $\left.\begin{array}{l}\end{array}\right\}$ A and B are turning points
B is a minimum point
C is a point of inflexion.

A stationary point, P, can be located and identified by considering two points close to, and on either side, of P and using either

	Maximum	Minimum	Inflexion
y values on each side of the stationary point	Both smaller	Both larger	One larger and one smaller

or

	Maximum	Minimum	Inflexion
Sign of $\dfrac{dy}{dx}$ at each side of stationary point	$+ \; 0 \; -$	$- \; 0 \; +$	$+ \; 0 \; +$ or $- \; 0 \; -$
Gradient of tangent	$/ \, ^{-} \, \backslash$	$\backslash \, _{-} \, /$	$/ \, ^{-} \, /$ or $\backslash \, _{-} \, \backslash$

There is a third method for distinguishing between the various types of stationary point, which will now be considered.

Note that the points chosen on either side of the stationary point must be such that no *other* stationary points, nor any break in the graph, lies between them.

The Nature of a Stationary Point

The way in which $\dfrac{dy}{dx}$ changes in the region of a stationary point varies with the type of stationary point. This behaviour pattern provides another way to identify the nature of that point.

Now the rate at which $\dfrac{dy}{dx}$ increases with respect to x can be written

$$\frac{d}{dx}\left(\frac{dy}{dx}\right) \text{ which is condensed to } \frac{d^2y}{dx^2}$$

(we say 'd 2 y by d x squared')

If, for example, $\dfrac{dy}{dx}$ is *increasing* as x increases, we can say that

$\dfrac{d^2y}{dx^2}$ is positive.

$\dfrac{d^2y}{dx^2}$ is called the *second derivative* with respect to x of y and,

if $y = f(x)$, the second derivative can also be denoted by $f''(x)$.

Now we can examine the behaviour of $\dfrac{dy}{dx}$ at each type of stationary

point.

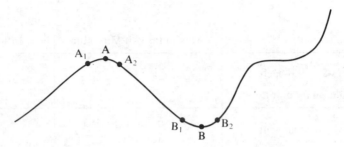

For the maximum point A,

$$\text{at } A_1 \quad \frac{dy}{dx} \text{ is +ve} \qquad \text{and} \qquad \text{at } A_2 \quad \frac{dy}{dx} \text{ is } -\text{ve}$$

so, passing through A, $\dfrac{dy}{dx}$ goes from + to −, i.e. $\dfrac{dy}{dx}$ decreases

$$\Rightarrow \quad \text{at A, } \frac{d^2y}{dx^2} \text{ is negative.}$$

For the minimum point B,

$$\text{at } B_1 \quad \frac{dy}{dx} \text{ is } -\text{ve} \qquad \text{and} \qquad \text{at } B_2 \quad \frac{dy}{dx} \text{ is +ve}$$

so, passing through B, $\dfrac{dy}{dx}$ goes from − to +, i.e. $\dfrac{dy}{dx}$ increases

$$\Rightarrow \quad \text{at B, } \frac{d^2y}{dx^2} \text{ is positive.}$$

At a point of inflexion, although $\dfrac{dy}{dx}$ can become zero, it does not

change sign so $\dfrac{d^2y}{dx^2}$ is zero.

Unfortunately it is also possible for $\dfrac{d^2y}{dx^2}$ to be zero at a turning point

so finding that $\dfrac{d^2y}{dx^2} = 0$ does not provide a definite conclusion.

Summing up we have:

	Maximum	Minimum
Sign of $\dfrac{d^2y}{dx^2}$	Negative (or zero)	Positive (or zero)

Note that, if $\dfrac{d^2y}{dx^2}$ is zero one of the other two methods must be used to determine the nature of a stationary point.

Examples 7a

1. Locate the stationary points on the curve $y = x^3 - 6x^2 + 9x + 5$ and determine the nature of each one. Sketch the curve, marking the coordinates of the stationary points.

$$y = x^3 - 6x^2 + 9x + 5 \quad \Rightarrow \quad \frac{dy}{dx} = 3x^2 - 12x + 9$$

At stationary points $\dfrac{dy}{dx} = 0$

i.e. $3x^2 - 12x + 9 = 0 \quad \Rightarrow \quad 3(x - 1)(x - 3) = 0$

When $x = 1$, $y = 9$ and when $x = 3$, $y = 5$
i.e. the stationary points are $(1, 9)$ and $(3, 5)$.

Differentiating $\dfrac{dy}{dx}$ w.r.t. x gives $\dfrac{d^2y}{dx^2} = 6x - 12$

When $x = 1$, $\dfrac{d^2y}{dx^2}$ is negative, so $(1, 9)$ is a maximum point

When $x = 3$, $\dfrac{d^2y}{dx^2}$ is positive, so $(3, 5)$ is a minimum point

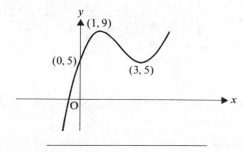

2. Find the stationary values of the function
$$3x^4 - 4x^3 - 6x^2 + 12x + 1$$
and investigate their nature.

$$f(x) = 3x^4 - 4x^3 - 6x^2 + 12x + 1$$

$\Rightarrow$ $\qquad$ $f'(x) = 12x^3 - 12x^2 - 12x + 12$

At stationary values $f'(x) = 0$

i.e. $12x^3 - 12x^2 - 12x + 12 = 0$ $\Rightarrow$ $12(x^3 - x^2 - x + 1) = 0$

Using the Remainder Theorem shows that $(x - 1)$ and $(x + 1)$ are factors.

So $\qquad$ $12(x - 1)(x + 1)(x - 1) = 0$

i.e. there are stationary points where $x = -1$ and $x = 1$

$$x = -1 \quad \Rightarrow \quad f(x) = -10$$

$$x = 1 \quad \Rightarrow \quad f(x) = 6$$

$\Rightarrow$ the stationary values of f(x) are -10 and 6

Differentiating $f'(x)$ w.r.t. x gives $f''(x) = 36x^2 - 24x - 12$

When $x = -1$ $f''(x) = 36 + 24 - 12$ which is positive

$\Rightarrow$ $\qquad$ $f(x) = -10$ is a minimum value

when $x = 1$, $f''(x) = 36 - 24 - 12$ which is zero.

This is inconclusive so we will look at the signs of $f'(x)$ on either side of $x = 1$

x	$\frac{1}{2}$	1	$1\frac{1}{2}$
$f'(x)$	+	0	+
Gradient	/	—	/

From this table we see that the stationary value at $x = 1$,
i.e. 6, is an inflexion.

EXERCISE 7a

Find the stationary point(s) on the following curves and distinguish between them.

1. $y = x^3$ **2.** $y = 12x - 4x^3$ **3.** $y = x(x^2 - 3)$

4. $y = x - x^2$ **5.** $y = x + \dfrac{4}{x}$ **6.** $y = \dfrac{1}{x^2} + x^2$

7. $y = 2x^5 - 5x^3$ **8.** $y = x^4$ **9.** $y = \dfrac{x^2}{2} + \dfrac{1}{x}$

Find the stationary value(s) of each of the following functions and determine their nature.

10. $27x - x^3$ **11.** $\dfrac{1}{x} + x$ **12.** $x^4 - 4x$

13. $2 - 4x + x^2$ **14.** $x^3 - 12x$ **15.** $x^2(3x^2 + 8x + 6)$

16. This is a sketch of the curve $y = x^4$

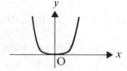

    (a) Describe the behaviour of y as x increases from negative to positive values.

    (b) Describe the behaviour of $\dfrac{dy}{dx}$ as x goes from negative to positive values.

    (c) Find $\dfrac{dy}{dx}$ as a function of x and sketch the graph of $\dfrac{dy}{dx}$ against values of x.

    (d) From your sketch $\left(\text{i.e. do not find } \dfrac{d^2y}{dx^2} \right)$ deduce the value of $\dfrac{d^2y}{dx^2}$ when $x = 0$ and give a reason for your answer.

17. Sketch the curves $y = f(x)$ and $\dfrac{dy}{dx} = f'(x)$ when

    (a) $y = (x - 1)^3$ (b) $y = (x - 1)^6$ (c) $y = 1 - (x - 1)^4$

    In each case use your sketches to deduce the property of y, and the value of $\dfrac{d^2y}{dx^2}$, when $x = 1$

18. Use a graphics calculator, or computer, to draw, on the same set of axes

    (a) $y = x^2$, $y = x^4$, $y = x^6$ (b) $y = x^3$, $y = x^5$, $y = x^7$

    Describe briefly the effect on the shape of the graph of raising the power of x.

DIFFERENTIATING A PRODUCT

In Module A, products were differentiated by multiplying them out and differentiating term by term.
Clearly this is suitable only for fairly simple products and a more general method is needed.

Suppose that $y = uv$ where u and v are both functions of x, e.g. $y = x^2(x^4 - 1)$.

It is dangerously tempting to think that $\dfrac{dy}{dx}$ is given by $\left(\dfrac{du}{dx}\right)\left(\dfrac{dv}{dx}\right)$.

But this is *not so* as is clearly shown by a simple example such as
$y = (x^2)(x^3)$ where, because $y = x^5$, we know that $\dfrac{dy}{dx} = 5x^4$
which is *not* equal to $(2x)(3x^2)$.

i.e.   differentiation is *not* distributive across a product.

Returning to $y = uv$ where $u = f(x)$ and $v = g(x)$, we see that if x increases by a small amount δx then there are corresponding small increases of δu, δv and δy in the values of u, v and y

$$\therefore \qquad\qquad y + \delta y = (u + \delta u)(v + \delta v)$$

$$= uv + u\delta v + v\delta u + \delta u\delta v$$

But $y = uv$ so $\qquad \delta y = u\delta v + v\delta u + \delta u\delta v$

$$\Rightarrow \qquad\qquad \frac{\delta y}{\delta x} = u\frac{\delta v}{\delta x} + v\frac{\delta u}{\delta x} + \delta u\frac{\delta v}{\delta x}$$

Now as $\delta x \to 0$, $\dfrac{\delta v}{\delta x} \to \dfrac{dv}{dx}$, $\dfrac{\delta u}{\delta x} \to \dfrac{du}{dx}$ and $\delta u \to 0$

Therefore $\qquad \dfrac{dy}{dx} = \lim\limits_{\delta x \to 0} \dfrac{\delta y}{\delta x} = u\dfrac{dv}{dx} + v\dfrac{du}{dx} + 0$

i.e.

$$\frac{d}{dx}(uv) = v\frac{du}{dx} + u\frac{dv}{dx}$$

This formula is verified by the simple example we considered above, i.e. $y = (x^2)(x^3)$.

Using $u = x^2$ and $v = x^3$ gives $\dfrac{dy}{dx} = (x^3)(2x) + (x^2)(3x^2) = 5x^4$

which is correct.

Example 7b _____

Differentiate with respect to x

(a) $(x + 1)^3(2x - 5)^2$ (b) $\dfrac{(x - 1)^2}{(x + 2)}$

(a) If $u = (x + 1)^3$, then $\dfrac{du}{dx} = 3(x + 1)^2$

and if $v = (2x - 5)^2$, then $\dfrac{dv}{dx} = \{2(2)(2x - 5)\}$

$\dfrac{d}{dx}(uv) = v\dfrac{du}{dx} + u\dfrac{dv}{dx}$ gives

$\dfrac{d}{dx}(x + 1)^3(2x - 5)^2 = \{(2x - 5)^2\}\{3(x + 1)^2\} + \{(x + 1)^3\}\{2(2)(2x - 5)\}$

$= (2x - 5)(x + 1)^2\{3(2x - 5) + 4(x + 1)\}$

$= (2x - 5)(x + 1)^2(10x - 11)$

(b) If we write $\dfrac{(x - 1)^2}{(x + 2)}$ as $(x - 1)^2(x + 2)^{-1}$

then $\qquad u = (x - 1)^2$ gives $\dfrac{du}{dx} = 2(x - 1)$

and $\qquad v = (x + 2)^{-1}$ gives $\dfrac{dv}{dx} = -(x + 2)^{-2}$

Using $\dfrac{d}{dx}(uv) = v\dfrac{du}{dx} + u\dfrac{dv}{dx}$ we have

$\dfrac{d}{dx}\left[\dfrac{(x - 1)^2}{(x + 2)}\right] = (x + 2)^{-1}\{2(x - 1)\} + (x - 1)^2\{-(x + 2)^{-2}\}$

$= \dfrac{(x - 1)}{(x + 2)^2}\{2(x + 2) - (x - 1)\}$

$= \dfrac{(x - 1)(x + 5)}{(x + 2)^2}$

EXERCISE 7b

Differentiate each function with respect to x.

1. $x(x-3)^2$
2. $(x-6)\sqrt{x}$
3. $(x+2)(x-2)^5$

4. $x(2x+3)^3$
5. $(x+1)(x-1)^4$
6. $\sqrt{x}(x-3)^3$

7. $\dfrac{(x+5)^4}{(x-3)}$
8. $\dfrac{x}{(3x+2)}$
9. $\dfrac{(2x-7)^{3/2}}{x}$

10. $x^3\sqrt{(x-1)}$
11. $x(x+3)^{-1}$
12. $x^2(2x-3)^2$

DIFFERENTIATING A QUOTIENT

To differentiate a function of the form u/v, where u and v are both functions of x, it is sometimes convenient to rewrite the function as uv^{-1} and differentiate it as a product. This method was used in part (b) of the previous worked example but it is not always the neatest way to differentiate a quotient. The alternative is to apply the formula derived below.

When a function is of the form u/v, where u and v are both functions of x, a small increase of δx in the value of x causes corresponding small increases of δu and δv in the values of u and v. Then, as $\delta x \to 0$, δu and δv also tend to zero.

If $y = \dfrac{u}{v}$ then $y + \delta y = \dfrac{(u + \delta u)}{(v + \delta v)}$

$\therefore \qquad \delta y = \dfrac{u + \delta u}{v + \delta v} - \dfrac{u}{v} = \dfrac{v\delta u - u\delta v}{v(v + \delta v)}$

$\therefore \qquad \dfrac{\delta y}{\delta x} = \left(v\dfrac{\delta u}{\delta x} - u\dfrac{\delta v}{\delta x}\right)\Big/v(v + \delta v)$

$\Rightarrow \qquad \dfrac{dy}{dx} = \lim_{\delta x \to 0}\dfrac{\delta y}{\delta x} = \left(v\dfrac{du}{dx} - u\dfrac{dv}{dx}\right)\Big/v^2$

i.e. $\qquad \dfrac{dy}{dx} = \dfrac{v\dfrac{du}{dx} - u\dfrac{dv}{dx}}{v^2}$

Example 7c

If $y = \dfrac{(4x - 3)^6}{(x + 2)}$ find $\dfrac{dy}{dx}$

Using $\qquad u = (4x - 3)^6$ gives $\dfrac{du}{dx} = 24(4x - 3)^5$

and $\qquad v = x + 2$ gives $\dfrac{dv}{dx} = 1$

Then $\qquad \dfrac{dy}{dx} = \left(v\,\dfrac{du}{dx} - u\,\dfrac{dv}{dx}\right)\Big/v^2$

$\qquad\qquad = \dfrac{(x + 2)\{24(4x - 3)^5\} - (4x - 3)^6}{(x + 2)^2}$

$\qquad\qquad = \dfrac{(4x - 3)^5(20x + 51)}{(x + 2)^2}$

EXERCISE 7c

Use the quotient formula to differentiate each of the following functions with respect to x

1. $\dfrac{(x - 3)^2}{x}$

2. $\dfrac{x^2}{(x + 3)}$

3. $\dfrac{(4 - x)}{x^2}$

4. $\dfrac{(x + 1)^2}{x^3}$

5. $\dfrac{4x}{(1 - x)^3}$

6. $\dfrac{2x^2}{(x - 2)}$

7. $\dfrac{x^{5/3}}{(3x - 2)}$

8. $\dfrac{(1 - 2x)^3}{x^3}$

9. $\dfrac{\sqrt{(x + 1)^5}}{x}$

IDENTIFYING THE CATEGORY OF A FUNCTION

Before any of the techniques explained earlier can be used to differentiate a given function, it is important to recognise the category to which the function belongs, i.e. is it a product or a function of a function or, if it is a fraction, is it one which would better be expressed as a product.

A product comprises two parts, each of which is an *independent* function of x, whereas *if one operation is carried out on another function of x* we have a function of a function.

MIXED EXERCISE 7

This exercise contains a mixture of compound functions. In each case first identify the type of function and then use the appropriate method to find its derivative.

1. $x\sqrt{(x + 1)}$ **2.** $(x^2 - 8)^3$ **3.** $x/(x + 1)$

4. $\sqrt[3]{(2 - x^4)}$ **5.** $(x + 1)/(x + 2)$ **6.** $x^2(\sqrt{x} - 2)$

7. $(x^2 - 2)^3$ **8.** $\sqrt{(x - x^2)}$ **9.** $x/(\sqrt{x} + 1)$

10. $x^2\sqrt{(x - 2)}$ **11.** $\sqrt{(x + 1)}/x^2$ **12.** $(x^4 + x^2)^3$

13. $\sqrt{(x^2 - 8)}$ **14.** $x^3(x^2 - 6)$ **15.** $(x^2 - 6)^3$

16. $x/(x^2 - 6)$ **17.** $(x^4 + 3)^{-2}$ **18.** $\sqrt{x}(2 - x)^3$

19. $\sqrt{x}/(2 - x)^3$ **20.** $(x - 1)(x - 2)^2$ **21.** $(2x^3 + 4)^5$

CHAPTER 8

FURTHER TRIGONOMETRIC IDENTITIES

THE FACTOR FORMULAE

The last set of identities considered in this book are called the factor formulae. They convert expressions such as $\sin A + \sin B$ into a product, so *factorising* the expression.

Consider the compound angle identities

$$\sin A \cos B + \cos A \sin B \equiv \sin (A + B)$$
$$\sin A \cos B - \cos A \sin B \equiv \sin (A - B)$$

Adding these gives $\qquad 2 \sin A \cos B \equiv \sin (A + B) + \sin (A - B) \qquad$ [1]

Subtracting gives $\qquad 2 \cos A \sin B \equiv \sin (A + B) - \sin (A - B) \qquad$ [2]

Working similarly with the compound angle identities for $\cos (A + B)$ and $\cos (A - B)$ gives

$$2 \cos A \cos B \equiv \cos (A + B) + \cos (A - B) \qquad [3]$$

and $\qquad -2 \sin A \sin B \equiv \cos (A + B) - \cos (A - B) \qquad$ [4]

115

Identities [1] to [4] can be used when a product has to be expressed as a sum or difference. For example, to express $2 \cos 7\theta \cos 2\theta$ as a sum we would use [3] to give

$$2 \cos 7\theta \cos 2\theta \equiv \cos (7\theta + 2\theta) + \cos (7\theta - 2\theta) \equiv \cos 9\theta + \cos 5\theta$$

However, when a sum or difference has to be expressed as a product, these identities are more easily remembered when they are expressed in an alternative form as follows.

Using $A + B = P$
and $A - B = Q$ gives $A = \frac{1}{2}(P + Q)$ and $B = \frac{1}{2}(P - Q)$

Then equations [1] to [4] become

$$\sin P + \sin Q \equiv 2 \sin \tfrac{1}{2}(P + Q) \cos \tfrac{1}{2}(P - Q) \qquad [5]$$

$$\sin P - \sin Q \equiv 2 \cos \tfrac{1}{2}(P + Q) \sin \tfrac{1}{2}(P - Q) \qquad [6]$$

$$\cos P + \cos Q \equiv 2 \cos \tfrac{1}{2}(P + Q) \cos \tfrac{1}{2}(P - Q) \qquad [7]$$

$$\cos P - \cos Q \equiv -2 \sin \tfrac{1}{2}(P + Q) \sin \tfrac{1}{2}(P - Q) \qquad [8]$$

For example, to express $\sin 6\theta - \sin 4\theta$ as a product, we would use [6], to give

$$\sin 6\theta - \sin 4\theta \equiv 2 \cos \tfrac{1}{2}(6\theta + 4\theta) \sin \tfrac{1}{2}(6\theta - 4\theta) \equiv 2 \cos 5\theta \sin \theta$$

Many people find that these identities are more easily remembered in words than in symbols, for example [5] can be remembered as

sum of two sines $\equiv$ twice sin (half sum) cos (half difference)

However, remembering every one of these identities in detail is not necessary but it is *important* to know that they exist and that they provide a powerful tool for dealing with trig functions. Formulae books provide details when it is known what is being looked for.

This group of identities can now be used to solve equations, simplify expressions, prove further identities and so on.

Examples 8a

1. Find the general solution of the equation $\sin 5x - \sin 3x = 0$

Using identity [6], the LHS of the equation can be expressed as a product.

$$\sin 5x - \sin 3x = 0$$

$$\Rightarrow \qquad 2 \cos \tfrac{1}{2}(5x + 3x) \sin \tfrac{1}{2}(5x - 3x) = 0$$

$$\Rightarrow \qquad \cos 4x \sin x = 0$$

$$\Rightarrow \qquad \cos 4x = 0 \qquad\qquad \text{or} \qquad\qquad \sin x = 0$$

$$\therefore \qquad 4x = \pm\tfrac{1}{2}\pi + 2n\pi$$
$$\Rightarrow \qquad x = \pm\tfrac{1}{8}\pi + \tfrac{1}{2}n\pi \Bigg\} \qquad \text{or} \qquad\qquad x = n\pi$$

The general solution is therefore $x = \tfrac{1}{8}\pi(4n \pm 1),\ n\pi$

2. Factorise $\cos\theta - \cos 3\theta - \cos 5\theta + \cos 7\theta$

As a first step, we take two terms and factorise them, and then take the remaining two terms and factorise them. With a bit of forethought, we can arrange the pairs of terms so that both pairs are in the form $(\cos + \cos)$. In any case, the terms should be rearranged so that this first factorisation results in a common factor.

$(\cos 7\theta + \cos\theta) - (\cos 5\theta + \cos 3\theta)$

$$\equiv 2 \cos 4\theta \cos 3\theta - 2 \cos 4\theta \cos\theta$$

$$\equiv 2 \cos 4\theta (\cos 3\theta - \cos\theta)$$

$$\equiv 2 \cos 4\theta (-2 \sin 2\theta \sin\theta) \equiv -4 \cos 4\theta \sin 2\theta \sin\theta$$

3. Prove that $\dfrac{\sin A + \sin B}{\cos A + \cos B} \equiv \tan \dfrac{A + B}{2}$

$$\text{LHS} \equiv \frac{\sin A + \sin B}{\cos A + \cos B} \equiv \frac{2 \sin \frac{1}{2}(A + B) \cos \frac{1}{2}(A - B)}{2 \cos \frac{1}{2}(A + B) \cos \frac{1}{2}(A - B)}$$

$$\equiv \tan \tfrac{1}{2}(A + B)$$

4. If A, B and C are the angles of a triangle, show that

$$\sin A + \sin B + \sin C = 4 \cos \tfrac{1}{2}A \cos \tfrac{1}{2}B \cos \tfrac{1}{2}C$$

$$\text{LHS} = (\sin A + \sin B) + \sin C$$

$$= 2 \sin \tfrac{1}{2}(A + B) \cos \tfrac{1}{2}(A - B) + 2 \sin \tfrac{1}{2}C \cos \tfrac{1}{2}C$$

Now $\quad A + B + C = 180° \quad \Rightarrow \quad \tfrac{1}{2}(A + B) + \tfrac{1}{2}C = 90°$

i.e. $\tfrac{1}{2}(A + B)$ and $\tfrac{1}{2}C$ are complementary angles

$$\Rightarrow \quad \sin \tfrac{1}{2}(A + B) = \cos \tfrac{1}{2}C \quad \text{and} \quad \sin \tfrac{1}{2}C = \cos \tfrac{1}{2}(A + B)$$

$$\therefore \quad \sin A + \sin B + \sin C$$

$$= 2 \cos \tfrac{1}{2}C \cos \tfrac{1}{2}(A - B) + 2 \cos \tfrac{1}{2}(A + B) \cos \tfrac{1}{2}C$$

$$= 2 \cos \tfrac{1}{2}C \left\{ \cos \tfrac{1}{2}(A - B) + \cos \tfrac{1}{2}(A + B) \right\}$$

$$= 2 \cos \tfrac{1}{2}C \left\{ 2 \cos \tfrac{1}{2}A \cos \left(-\tfrac{1}{2}B\right) \right\}$$

$$= 4 \cos \tfrac{1}{2}A \cos \tfrac{1}{2}B \cos \tfrac{1}{2}C \quad (\cos -\theta = \cos \theta)$$

EXERCISE 8a

1. Express as a product of two trig functions
   (a) $\sin 3\theta + \sin \theta$ (b) $\cos 5\theta + \cos 3\theta$ (c) $\sin 4\theta - \sin 2\theta$
   (d) $\cos 7\theta - \cos \theta$ (e) $\sin 3A - \sin 5A$ (f) $\cos A - \cos 5A$
   (g) $\sin 2A + \sin 20°$ (h) $\sin 2\theta + 1$ (*Hint.* $\sin 90° = 1$)

2. Express as a sum or difference of two trig functions
   (a) $2 \sin 2\theta \cos \theta$ (b) $2 \cos 3\theta \cos 2\theta$ (c) $2 \cos \theta \sin 4\theta$
   (d) $-2 \sin 3\theta \sin \theta$ (e) $2 \sin 4\theta \sin 2\theta$ (f) $\cos \theta \cos 4\theta$

3. Simplify

(a) $\cos(\theta - 60°) + \cos(\theta + 60°)$ (b) $\sin(x - 45°) - \sin(x + 45°)$
(c) $\sqrt{3}\cos\theta - \sin(\theta + 60°) - \sin(\theta + 120°)$

4. Evaluate, leaving your answers in surd form,

(a) $\sin 105° + \sin 15°$ (b) $\cos 15° + \cos 75°$

5. Find the general solutions of the following equations.

(a) $\cos 2x + \cos 4x = 0$ (b) $\sin 3x - \sin x = 0$
(c) $\cos x - \cos 3x = 0$ (d) $\sin 4\theta + \sin 2\theta = 0$

6. Factorise the expression $\cos\theta + \cos 3\theta + \cos 2\theta$. Hence find the general solution of the equation $\cos\theta + \cos 3\theta + \cos 2\theta = 0$

7. Factorise the expression $\sin 3\theta - \sin\theta + \sin 7\theta - \sin 5\theta$. Hence find the general solution of the equation $\sin 3\theta - \sin\theta = \sin 5\theta - \sin 7\theta$.

8. Solve the following equations for angles from 0 to 180°

(a) $\cos x = \cos 2x + \cos 4x$
(b) $\cos x + \cos 3x = \sin x + \sin 3x$
(c) $\sin 3\theta + \sin 6\theta + \sin 9\theta = 0$
(d) $\sin 3\theta - \sin\theta = \cos 2\theta$
(e) $\cos 5\theta - \cos\theta = \sin 3\theta$
(f) $\cos 2x = \cos(30° - x)$

9. Prove the following identities.

(a) $\dfrac{\sin 2A + \sin 2B}{\sin 2A - \sin 2B} \equiv \dfrac{\tan(A+B)}{\tan(A-B)}$

(b) $\dfrac{\cos 2A + \cos 2B}{\cos 2B - \cos 2A} \equiv \cot(A+B)\cot(A-B)$

(c) $\dfrac{\sin A \sin 2A + \sin 3A \sin 6A}{\sin A \cos 2A + \sin 3A \cos 6A} \equiv \tan 5A$

(d) $\sin\theta + \sin 2\theta + \sin 3\theta \equiv (1 + \cos 2\theta)\sin 2\theta$

10. If A, B and C are the angles of a triangle, prove that

(a) $\cos(B+C) = -\cos A$ (b) $\sin C = \sin(A+B)$
(c) $\sin\frac{1}{2}(A+B) = \cos\frac{1}{2}C$ (d) $\sin\frac{1}{2}B = \cos\frac{1}{2}(A+C)$

CHAPTER 9

THE EXPONENTIAL AND LOGARITHMIC FUNCTIONS

THE EXPONENTIAL FUNCTION

The general shape of an exponential curve was seen in Chapter 1. The next diagram shows a few more members of the exponential family.

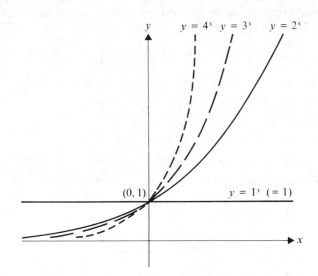

Note that these curves and, in fact, *all* exponential curves, pass through the point $(0, 1)$.

This is because, for any positive base a,

$$\text{when } x \text{ is } 0, \quad y = a^x = a^0 = 1$$

Each exponential curve has a unique property which the reader can discover experimentally by using an accurate plot of the curve $y = 2^x$. Choose three or four points on the curve and, at each one, draw the tangent as accurately as possible and determine its gradient. Then complete the following table.

Point	Gradient of tangent, i.e. $\dfrac{dy}{dx}$	y-coordinate	$\dfrac{dy}{dx} \div y$
1			
2			
3			
4			

An accurate drawing should result in numbers in the last column that are all reasonably close to 0.7

When this experiment is carried out for 3^x and 4^x we find again that $\dfrac{dy}{dx} \div y$ has a constant value;

for 3^x the constant is about 1.1 and for 4^x it is about 1.4

So we have

Base	2	3	4
$\dfrac{dy}{dx} \div y$	0.7	1.1	1.4

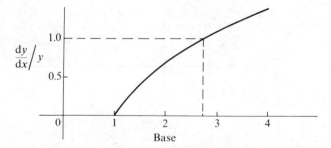

From this graph it can be seen that there is a base, somewhere between 2 and 3, for which $\dfrac{dy}{dx} \div y = 1$, i.e. $\dfrac{dy}{dx} = y$

Calling this base e we have

$$\text{if } y = e^x \text{ then } \frac{dy}{dx} = e^x$$

The function e^x is the only function which is unchanged when differentiated.

In the early eighteenth century, a number of mathematicians, working along different lines of investigation, all discovered the number e at about the same time.

The number e is irrational, i.e. like π, $\sqrt{2}$, etc., it cannot be given an exact decimal value but, to 4 significant figures, e = 2.718 The value of various powers of e, such as e^2, e^3, e^4, can be obtained from a calculator.

Summing up:

> for any value of a $(a > 0)$, a^x is *an* exponential function
>
> for the base e (e $\approx$ 2.718), e^x is *the* exponential function
>
> $$\frac{d}{dx}(e^x) = e^x$$

The following diagrams show sketches of $y = e^x$ and of some simple variations.

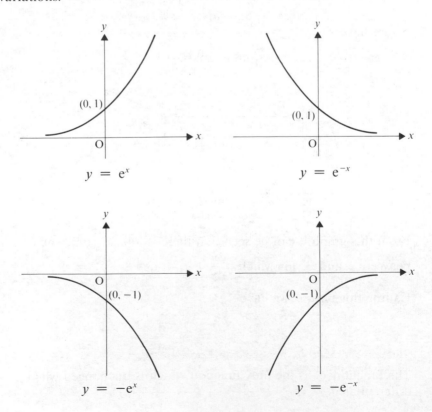

$y = e^x$

$y = e^{-x}$

$y = -e^x$

$y = -e^{-x}$

Example 9a _____

Find the coordinates of the stationary point on the curve $y = e^x - x$, and determine its type. Sketch the curve showing the stationary point clearly.

$$y = e^x - x \quad \Rightarrow \quad \frac{dy}{dx} = e^x - 1$$

At a stationary point, $\dfrac{dy}{dx} = 0$ therefore $e^x - 1 = 0$

i.e. $\qquad\qquad e^x = 1 \quad \Rightarrow \quad x = 0$

When $x = 0$, $y = e^0 - 0 = 1$

Therefore $(0, 1)$ is a stationary point.

$$\frac{d^2y}{dx^2} = e^x \text{ and this is positive when } x = 0$$

Therefore $(0, 1)$ is a minimum point.

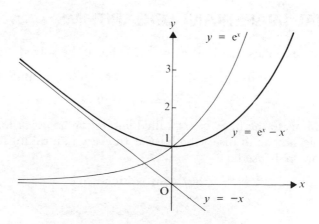

This curve is made up from separate sketches of $y = e^x$ and $y = -x$ by adding their ordinates.

EXERCISE 9a

1. Evaluate, correct to 3 s.f.
   (a) e^2 (b) e^{-1} (c) $e^{1.5}$ (d) $e^{-0.3}$

2. Write down the derivative of
   (a) $2e^x$ (b) $x^2 - e^x$ (c) e^x

In Questions 3 to 5 find the gradient of each curve at the specified value of x

3. $y = e^x - 2x$ where $x = 2$

4. $y = x^2 + 2e^x$ where $x = 1$

5. $y = e^x - 3x^3$ where $x = 0$

6. Find the value of x at which the function $e^x - x$ has a stationary value.

7. Sketch each given curve.
   (a) $y = 1 - e^x$ (b) $y = e^x + 1$ (c) $y = x - e^x$
   (d) $y = 1 - e^{-x}$ (e) $y = 1 + e^{-x}$ (f) $y = x^2 + e^x$

NATURAL (NAPERIAN) LOGARITHMS

Suppose that the equation $e^x = 0.59$ has to be solved.

We know that this equation can be written in logarithmic form

i.e. $x = \log_e 0.59$

Logarithms to the base e are called *natural* or *Naperian* logarithms. To avoid having to insert the base e in every natural logarithm, the notation ln is used, i.e.

$$\log_e a \quad \text{is written} \quad \ln a$$

i.e. $\ln a = b \quad \Longleftrightarrow \quad a = e^b$

Values of natural logs can be found by using a scientific calculator.

Returning to the equation $e^x = 0.59$ we now have

$$e^x = 0.59 \quad \Rightarrow \quad x = \ln 0.59 = -0.528 \quad \text{to 3 s.f.}$$

Logarithms to the base 10 are called *common logarithms* and are denoted by log or lg. They used to be an important tool for calculations but calculators have eliminated their usefulness in that respect.

The laws used for working with logarithms to a general base, given in Chapter 3, apply equally well to natural logarithms, i.e.

$$\ln a + \ln b = \ln ab$$

$$\ln a - \ln b = \ln a/b$$

$$\ln a^n = n \ln a$$

$$\log_a b = \frac{\ln b}{\ln a}$$

The base of an exponential function can also be changed.

Suppose that we wish to express 3^x as a power of e.

Using $\qquad\qquad 3^x = e^p \quad$ gives $\quad x \ln 3 = p$

$\therefore \qquad\qquad\qquad\qquad 3^x = e^{x \ln 3}$

In general $\qquad\qquad a^x = e^{x \ln a}$

Examples 9b

1. Separate $\ln(\tan x)$ into two terms.

$$\ln(\tan x) = \ln\left(\frac{\sin x}{\cos x}\right)$$

$$= \ln \sin x - \ln \cos x$$

2. Express $4\ln(x + 1) - \frac{1}{2}\ln x$ as a single logarithm.

$$4\ln(x + 1) - \frac{1}{2}\ln x = \ln(x + 1)^4 - \ln\sqrt{x}$$
$$= \ln\left(\frac{(x + 1)^4}{\sqrt{x}}\right)$$

EXERCISE 9b

1. Evaluate
    (a) $\ln 3.451$ (b) $\ln 1.201$
    (c) $\ln 17.3$ (d) $\ln 2$

2. Express as a sum or difference of logarithms or as a product
    (a) $\ln \dfrac{x}{x - 1}$ (b) $\ln(5x^2)$
    (c) $\ln(x^2 - 4)$ (d) $\ln \tan x$
    (e) $\ln(\sin^2 x)$ (f) $\ln\sqrt{\left(\dfrac{x + 1}{x - 1}\right)}$

3. Express as a single logarithm
    (a) $\ln x - 2\ln(1 - x)$ (b) $1 - \ln x$
    (c) $\ln \sin x + \ln \cos x$ (d) $2\ln x + \frac{1}{2}\ln(x - 1)$

4. Given that $\ln a = 3$
    (a) express $\log_a x^2$ as a simple natural logarithm
    (b) express as a single logarithm $\ln x^3 + 6\log_a x$

5. Solve the following equations for x
    (a) $e^x = 8.2$ (b) $e^{2x} + e^x - 2 = 0$ (Hint. Use $e^{2x} = (e^x)^2$)
    (c) $e^{2x - 1} = 3$ (d) $e^{4x} - e^x = 0$

6. Given that $\ln a = 2$ solve the following equations for x
    (a) $a^x = e^2$ (b) $a^x = e^6$ (c) $a^x = 1$

7. Use a graphics calculator or a computer with a graph plotting programme. Set the range to $-5 \leqslant x \leqslant 5, \ -5 \leqslant y \leqslant 5$

   (a) Plot the graph of $y = \ln x$.

   (b) State why $x \to \ln x$ is a function and give its domain and range.

   (c) Superimpose the graph of $y = e^x$.

   (d) What do you notice about the two graphs?
      If you see a relationship between the graphs deduce, the relationship between f and g where $f(x) = \ln x$ and $g(x) = e^x$.

THE LOGARITHMIC FUNCTION

Consider the curve with equation $y = f(x)$ where $f(x) = \ln x$

If $y = \ln x$ then $x = e^y$,

i.e. **the logarithmic function is the inverse of the exponential function.**

It follows that the curve $y = \ln x$ is the reflection of the curve $y = e^x$ in the line $y = x$

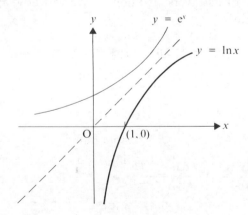

There is no part of the curve $y = \ln x$ in the second and third quadrants. This is because, if $x = e^y$ (i.e. if $y = \ln x$), x is positive for all real values of y. Therefore

$\ln x$ does not exist for negative values of x.

THE DERIVATIVE OF ln x

We know that $y = \ln x \iff x = e^y$ and we also know how to differentiate the exponential function. So a relationship between $\dfrac{d}{dx}(y)$ and $\dfrac{d}{dy}(x)$ would help in finding the derivative of $\ln x$

Consider the equation $y = f(x)$ where $f(x)$ is any function of x.

$$\frac{dy}{dx} = \lim_{\delta x \to 0} \frac{\delta y}{\delta x} = \lim_{\delta x \to 0} \left(1 \Big/ \frac{\delta x}{\delta y}\right)$$

Now $\delta y \to 0$ as $\delta x \to 0$

$\therefore$

$$\frac{dy}{dx} = \lim_{\delta y \to 0} \left(1 \Big/ \frac{\delta x}{\delta y}\right)$$

i.e.

$$\frac{dy}{dx} = 1 \Big/ \frac{dx}{dy}$$

This relationship can be used to find the derivative of *any* function if the derivative of its inverse is known. We will now apply it to differentiate $\ln x$.

$$y = \ln x \iff x = e^y$$

Differentiating e^y w.r.t. y gives

$$\frac{dx}{dy} = e^y = x$$

Therefore $\dfrac{dy}{dx} = 1 \Big/ \dfrac{dx}{dy} = \dfrac{1}{x}$

i.e.

$$\frac{d}{dx} \ln x = \frac{1}{x}$$

This result can be used to differentiate many log functions if they are first simplified by applying the laws given on page 125.

Examples 9c

1. Find the derivative of (a) $\ln(1/x^3)$ (b) $\ln(4\sqrt{x})$

(a) $f(x) = \ln(1/x^3) = \ln(x^{-3}) = -3\ln x$

$$\frac{d}{dx}\{f(x)\} = \frac{d}{dx}\{-3\ln x\} = \frac{-3}{x}$$

(b) $f(x) = \ln(4\sqrt{x}) = \ln 4 + \ln(\sqrt{x}) = \ln 4 + \frac{1}{2}\ln x$

$$\frac{d}{dx}\{f(x)\} = \frac{d}{dx}(\ln 4) + \frac{d}{dx}(\tfrac{1}{2}\ln x)$$

$$= 0 + \frac{\frac{1}{2}}{x} = \frac{1}{2x}$$

---

2. Find $\dfrac{dy}{dx}$ if $y = \log_a x^2$

We only know how to differentiate natural logs, so first we must change the base from a to e.

$$y = \log_a x^2 = \frac{\ln x^2}{\ln a} = \frac{2\ln x}{\ln a}$$

$$\therefore \qquad \frac{dy}{dx} = \frac{2}{x\ln a}$$

EXERCISE 9c

Find the derivative of each of the following functions.

1. $\ln x^3$ 2. $\ln(3x)$ 3. $\ln(x^{-2})$ 4. $\ln(3/\sqrt{x})$

5. $\ln(1/x^5)$ 6. $\ln(2x^{1/2})$ 7. $\ln(x^{-3/2})$ 8. $\ln(x^3/\sqrt{x})$

Locate the stationary points on each curve.

9. $y = \ln x - x$ 10. $y = x^3 - 2\ln x^3$ 11. $y = \ln x - \sqrt{x}$

Sketch each of the following curves.

12. $y = -\ln x$ 13. $y = \ln(-x)$

14. $y = 2 + \ln x$ 15. $y = \ln x^2$

FURTHER DIFFERENTIATION OF EXPONENTIAL AND LOG FUNCTIONS

Methods were introduced in Chapter 7 for differentiating a product, a quotient and a function of a function, i.e.

if $\qquad y = uv$ then $\dfrac{dy}{dx} = v\dfrac{du}{dx} + u\dfrac{dv}{dx}$

if $\qquad y = \dfrac{u}{v}$ then $\dfrac{dy}{dx} = \left(v\dfrac{du}{dx} - u\dfrac{dv}{dx}\right)\Big/v^2$

if $\quad y = g(u)$ and $u = f(x)$ then $\dfrac{dy}{dx} = \dfrac{dy}{du} \times \dfrac{du}{dx}$

(Remember that the substitution can often be done mentally.)

These techniques can now be applied when exponential and log functions are involved.

The function of a function result is particularly useful as many derivatives of this type can be written down directly and easily.

To Differentiate e^u where $u = f(x)$

If $y = e^u$ then $\dfrac{dy}{dx} = \dfrac{dy}{du} \times \dfrac{du}{dx}$ gives

$$\frac{dy}{dx} = e^u \times \frac{du}{dx}$$

This can also be expressed in the form

$$\frac{d}{dx}e^{f(x)} = e^{f(x)}f'(x)$$

i.e. $$\frac{d}{dx}e^{f(x)} = f'(x)e^{f(x)}$$

e.g. $\qquad$ if $y = e^{(x^2 + 1)}$ then $\dfrac{dy}{dx} = 2xe^{(x^2 + 1)}$

The case when u is a linear function of x is particularly useful,

i.e. $$y = e^{(ax + b)} \quad \Rightarrow \quad \frac{dy}{dx} = ae^{(ax + b)}$$

To Differentiate ln u where $u = f(x)$

If $y = \ln u$ then $\dfrac{dy}{dx} = \dfrac{dy}{du} \times \dfrac{du}{dx}$ gives

$$\frac{dy}{dx} = \frac{1}{u} \times \frac{du}{dx}$$

This can also be expressed in the form

$$\frac{d}{dx}\{\ln f(x)\} = \frac{1}{f(x)} \times f'(x) = \frac{f'(x)}{f(x)}$$

e.g.   if $y = \ln(2 + x^3)$ then $\dfrac{dy}{dx} = \dfrac{3x^2}{2 + x^3}$

Again the case when u is $ax + b$ occurs frequently and is worth noting,

i.e. $$y = \ln(ax + b) \quad\Rightarrow\quad \frac{dy}{dx} \stackrel{\smile}{=} \frac{a}{ax + b}$$

Compound Exponential and Logarithmic Functions

For any given function the first step is to identify its category,
e.g. $x^2 e^x$ and $(1 + x)\ln x$ are both products.
Whereas e^{x^2} and $\ln(1 - x^2)$ are both functions of a function.

EXERCISE 9d

In Questions 1 to 9

(a) identify the type of function

(b) express the function in terms of u and/or v, stating clearly the substitutions that have been made.

1. $e^x(x^2 + 1)$ 2. $e^{(x^2 + 1)}$ 3. $x \ln x$

4. $\sqrt{\{e^{(x + 1)}\}}$ 5. $e^x \ln x$ 6. $\ln(3 - x^2)$

7. $(\ln x)^2$ 8. e^{-2x} 9. $1/\ln x$

10. If f and g are the functions defined by $f: x \rightarrow x^2$ and $g: x \rightarrow e^x$ write down the functions $fg(x)$ and $gf(x)$

11. The functions f, g and h are defined as follows

$$f : x \rightarrow x^2 \qquad g : x \rightarrow 1/x \qquad h : x \rightarrow \ln x$$

Write down the functions

(a) fg(x)   (b) hf(x)   (c) hg(x)

(d) fh(x)   (e) hfg(x)   (f) fg^{-1}(x)

Examples 9e

1. Find the derivative of $x^3 e^x$

$y = x^3 e^x$ becomes $y = uv$ if $u = x^3$ and $v = e^x$

$\Rightarrow$ $$\frac{du}{dx} = 3x^2 \quad \text{and} \quad \frac{dv}{dx} = e^x$$

$\therefore$ $$\frac{dy}{dx} = v \frac{du}{dx} + u \frac{dv}{dx} = e^x(3x^2) + x^3(e^x)$$

i.e. $$\frac{dy}{dx} = x^2(3 + x)e^x$$

_____

2. Differentiate $\ln \{x\sqrt{(x^2 - 4)}\}$, w.r.t. x

First we simplify the log expression by changing it into a sum.

$$\ln \{x\sqrt{(x^2 - 4)}\} = \ln x + \ln \sqrt{(x^2 - 4)} = \ln x + \tfrac{1}{2} \ln (x^2 - 4)$$

$\therefore$ $$\frac{d}{dx} \ln \{x\sqrt{(x^2 - 4)}\} = \frac{d}{dx} \ln x + \frac{d}{dx} \{\tfrac{1}{2} \ln (x^2 - 4)\}$$

$$= \frac{1}{x} + \frac{1}{2}\left(\frac{2x}{x^2 - 4}\right)$$

$$= \frac{1}{x} + \frac{x}{x^2 - 4}$$

_____

Simplifying the given function at the start, made the differentiation in this problem much easier. *Before differentiating any function, all possible simplification should be done*, particularly when complicated log expressions are involved.

EXERCISE 9e

Differentiate the following functions with respect to x

1. xe^x

2. $x^2 \ln x$

3. $e^x(x^3 - 2)$

4. $x^2 \ln (x - 2)^6$

5. $(x - 1)e^x$

6. $(x^2 + 4) \ln \sqrt{x}$

7. $x\sqrt{(2 + x)}$

8. $x \ln \sqrt{(x - 5)}$

9. $(x^2 - 2)e^x$

10. $\dfrac{x}{e^x}$

11. $\dfrac{e^x}{x^2}$

12. $\dfrac{(\ln x)}{x^3}$

13. $\dfrac{\sqrt{(x + 1)}}{\ln x}$

14. $\dfrac{e^x}{x^2 - 1}$

15. $\dfrac{e^x}{e^x - e^{-x}}$

16. e^{4x}

17. $\ln (x^2 - 1)$

18. e^{x^2}

19. $6e^{(1 - x)}$

20. $e^{(x^2 + 1)}$

21. $\ln \sqrt{(x + 2)}$

22. $(\ln x)^2$

23. $1/(\ln x)$

24. $\sqrt{(e^x)}$

MIXED EXERCISE 9

In this exercise a variety of functions are to be differentiated. In each case identify the type of function and then use the appropriate method to find the derivative. Some of the given functions can be differentiated by using one of the basic rules so do not assume that special techniques are always needed.

In Questions 1 to 18 differentiate the given function with respect to x

1. $x \ln x$

2. $(4x - 1)^{2/3}$

3. $\dfrac{e^x}{x - 1}$

4. $\dfrac{\sqrt{(1 + x^3)}}{x^2}$

5. $\dfrac{\ln x}{\ln (x - 1)}$

6. 10^{3x}

7. $\dfrac{(1 + 2x^2)}{1 + x^2}$

8. $e^{-2/x}$

9. $\ln (1 - e^x)$

10. $e^{3x}x^3$

11. $\dfrac{2x}{(2x - 1)(x - 3)}$

12. $\dfrac{e^{x/2}}{x^5}$

13. $\ln\left[\dfrac{x^2}{(x+3)(x^2-1)}\right]$ **14.** $\ln 4x^3(x+3)^2$ **15.** $(\ln x)^4$

16. $\dfrac{(x+3)^3}{x^2+2}$ **17.** $\sqrt{(e^x-x)}$ **18.** $4\ln(x^2+1)$

Find and simplify $\dfrac{dy}{dx}$ and hence find $\dfrac{d^2y}{dx^2}$ if

19. $y = \dfrac{(1+2x)}{(1-2x)}$ **20.** $y = \ln\dfrac{x}{x+1}$ **21.** $y = \dfrac{e^x}{e^x-4}$

CHAPTER 10

SERIES

FINDING THE SUM OF A NUMBER SERIES

In Module A we found the sum of two types of number series, i.e. arithmetic progressions and geometric progressions.

It is sometimes possible to find the sum of a series which is neither an AP nor a GP.

Consider the series $\displaystyle\sum_{r=1}^{n} \frac{1}{r(r+1)}$

The general term can be expressed as two terms using partial fractions,

i.e. $\dfrac{1}{r(r+1)} \equiv \dfrac{1}{r} - \dfrac{1}{r+1}$ (using the cover-up method)

Hence $\displaystyle\sum_{r=1}^{n} \frac{1}{r(r+1)} \equiv \sum_{r=1}^{n} \left(\frac{1}{r} - \frac{1}{r+1} \right)$

$$= \left(1 - \tfrac{1}{2}\right) + \left(\tfrac{1}{2} - \tfrac{1}{3}\right) + \left(\tfrac{1}{3} - \tfrac{1}{4}\right) + \ldots + \left(\tfrac{1}{n-1} - \tfrac{1}{n}\right) + \left(\tfrac{1}{n} - \tfrac{1}{n-1}\right)$$

All the terms cancel except for the first and last terms,

therefore $\displaystyle\sum_{r=1}^{n} \frac{1}{r(r+1)} = 1 - \frac{1}{n+1} = \frac{n}{n+1}$

The summation of this series was possible because we were able to express the general term of the given series as the *difference* of two consecutive terms of another series. This is known as the *method of differences* which can be summarised as follows.

135

If the general term, u_r, of a series can be expressed as $f(r) - f(r + 1)$

then $\displaystyle\sum_{r=1}^{n} u_r = \sum_{r=1}^{n} [f(r) - f(r + 1)]$

$$= [f(1) - f(2)] + [f(2) - f(3)] + \ldots$$

$$+ [f(n - 1) - f(n)] + [f(n) - f(n + 1)]$$

$$= f(1) - f(n + 1)$$

Example 10a

Show that $(r + 1)^3 - r^3 \equiv 3r^2 + 3r + 1$. Hence find $\displaystyle\sum_{r=1}^{n} (3r^2 + 3r + 1)$

and deduce that $\displaystyle\sum_{r=1}^{n} r^2 = \tfrac{1}{6}n(n + 1)(2n + 1)$

$\text{LHS} = (r + 1)^3 - r^3 \equiv r^3 + 3r^2 + 3r + 1 - r^3 \equiv 3r^2 + 3r + 1 = \text{RHS}$

i.e. $(r + 1)^3 - r^3 \equiv 3r^2 + 3r + 1$

$\therefore \quad \displaystyle\sum_{r=1}^{n} (3r^2 + 3r + 1) = \sum_{r=1}^{n} \{(r + 1)^3 - r^3\}$

$$= (2^3 - 1^3) + (3^3 - 2^3) + \ldots + \{(n + 1)^3 - n^3\}$$

$$= (n + 1)^3 - 1$$

Now $\displaystyle\sum_{r=1}^{n} (3r^2 + 3r + 1) = \sum_{r=1}^{n} 3r^2 + \sum_{r=1}^{n} 3r + \sum_{r=1}^{n} 1$

$\therefore \quad \displaystyle\sum_{r=1}^{n} 3r^2 + \sum_{r=1}^{n} 3r + \sum_{r=1}^{n} 1 = (n + 1)^3 - 1$ [1]

But $\displaystyle\sum_{r=1}^{n} 3r$ is an AP $(a = 3, \ d = 3)$ so using $S_n = \tfrac{1}{2}n(a + l)$

gives $\displaystyle\sum_{r=1}^{n} 3r = \tfrac{3}{2}n(n + 1)$ and $\displaystyle\sum_{r=1}^{n} 1 = 1 + 1 + \ldots + 1 = n$

Hence [1] becomes $\displaystyle\sum_{r=1}^{n} 3r^2 + \tfrac{3}{2}n(n+1) + n = (n+1)^3 - 1$

$\Rightarrow$ $\displaystyle\sum_{r=1}^{n} 3r^2 = (n+1)^3 - (1+n) - \tfrac{3}{2}n(n+1)$

$$= (n+1)\big[(n+1)^2 - 1 - \tfrac{3}{2}n\big]$$

$$= (n+1)\big[\tfrac{1}{2}n(2n+1)\big]$$

Now $\displaystyle\sum_{r=1}^{n} 3r^2 = 3\sum_{r=1}^{n} r^2$,

$\therefore$ $\displaystyle\sum_{r=1}^{n} r^2 = \tfrac{1}{6}n(n+1)(2n+1)$

EXERCISE 10a

In Questions 1 to 4 express the general term in partial fractions and hence find the sum of the series.

1. $\displaystyle\sum_{r=1}^{n} \frac{1}{r(r+2)}$

2. $\displaystyle\sum_{r=3}^{n} \frac{1}{(r+1)(r+2)}$

3. $\displaystyle\sum_{r=n}^{2n} \frac{1}{r(r+1)}$

4. $\displaystyle\sum_{r=1}^{n} \frac{r}{(2r-1)(2r+1)(2r+3)}$

5. Verify that $4r^3 + r \equiv (r+\tfrac{1}{2})^4 - (r-\tfrac{1}{2})^4$. Hence find $\displaystyle\sum_{r=1}^{n} (4r^3 + r)$

   Deduce that $\displaystyle\sum_{r=1}^{n} r^3 = \tfrac{1}{4}n^2(n+1)^2$

6. If $f(r) \equiv \dfrac{1}{r(r+1)}$, simplify $f(r+1) - f(r)$

   Hence find $\displaystyle\sum_{r=1}^{n} \frac{1}{r(r+1)(r+2)}$

7. If $f(r) \equiv \dfrac{1}{r^2}$, simplify $f(r) - f(r+1)$

   Hence find the sum of the first n terms of the series

   $$\frac{3}{(1^2)(2^2)} + \frac{5}{(2^2)(3^2)} + \frac{7}{(3^2)(4^2)} + \ldots$$

8. Given that $\displaystyle\sum_{r=1}^{n} r^2 = \frac{1}{6}n(n+1)(2n+1)$, use the identity

$$r^4 - (r-1)^4 \equiv 4r^3 - 6r^2 + 4r - 1$$

to find the sum of the cubes of the first n natural numbers,

i.e. $\displaystyle\sum_{r=1}^{n} r^3$

9. If $f(r) \equiv \cos 2r\theta$, simplify $f(r) - f(r+1)$

Use you result to find the sum of the first n terms of the series

$$\sin 3\theta + \sin 5\theta + \sin 7\theta + \ldots$$

NATURAL NUMBER SERIES

The natural numbers are the positive integers, i.e. 1, 2, 3, ...

The series $1 + 2 + 3 + \ldots + n$ is the sum of the first n natural

numbers and can be written $\displaystyle\sum_{r=1}^{n} r$

This series is an AP with $a = 1$ and $d = 1$, so using $S_n = \frac{1}{2}n(a+1)$

gives

$$\sum_{r=1}^{n} r = \frac{1}{2}n(n+1)$$

Now consider the series $1^2 + 2^2 + 3^2 + \ldots + n^2$ which is the sum of the squares of the first n natural numbers.

This series is written $\displaystyle\sum_{r=1}^{n} r^2$ and its sum was found in Example 10a.

i.e.

$$\sum_{r=1}^{n} r^2 = \frac{1}{6}n(n+1)(2n+1)$$

The series $1^3 + 2^3 + 3^3 + \ldots + n^3$ is called the sum of the cubes of the first n natural numbers and we saw in Exercise 10a that

$$\sum_{r=1}^{n} r^3 = \frac{1}{4}n^2(n+1)^2 = \left[\frac{1}{2}n(n+1)\right]^2 = \left[\sum_{r=1}^{n} r\right]^2$$

The results are quotable and can be used to sum other series.

Examples 10b _____

1. Find $\displaystyle\sum_{r=1}^{n} r(r+1)(r+2)$

$$r(r+1)(r+2) \equiv r^3 + 3r^2 + 2r$$

$$\therefore \quad \sum_{r=1}^{n} r(r+1)(r+2) = \sum_{r=1}^{n} r^3 + 3\sum_{r=1}^{n} r^2 + 2\sum_{r=1}^{n} r$$

$$= \tfrac{1}{4}n^2(n+1)^2 + 3\left[\tfrac{1}{6}n(n+1)(2n+1)\right]$$

$$+ 2\left[\tfrac{1}{2}n(n+1)\right]$$

$$= \tfrac{1}{4}n(n+1)(n+2)(n+3)$$

_____

2. Find $\displaystyle\sum_{r=5}^{10} r^2$

$$\sum_{r=5}^{10} r^2 = \sum_{r=1}^{10} r^2 - \sum_{r=1}^{4} r^2 = \tfrac{10}{6}(10+1)(20+1) - \tfrac{4}{6}(4+1)(8+1)$$

$$= 355$$

_____

3. Find the sum of the squares of the first n odd numbers.

The odd numbers can be represented by $2r - 1$ where $r = 1, 2, 3, \ldots$

So we want $\displaystyle\sum_{r=1}^{n} (2r-1)^2$

Now $(2r-1)^2 = 4r^2 - 4r + 1$

$$\therefore \quad \sum_{r=1}^{n} (2r-1)^2 = 4\sum_{r=1}^{n} r^2 - 4\sum_{r=1}^{n} r + \sum_{r=1}^{n} 1$$

$$= 4\left[\tfrac{1}{6}n(n+1)(2n+1)\right] - 4\left[\tfrac{1}{2}n(n+1)\right] + n$$

$$= \tfrac{1}{3}n(4n^2 - 1)$$

_____

EXERCISE 10b

Find the sum of the series.

1. $\displaystyle\sum_{r=1}^{n} r(r+1)$

2. $\displaystyle\sum_{r=1}^{n} r(r+1)(r+2)$

3. $\displaystyle\sum_{r=n}^{2n} r^2(1+r)$

4. $\displaystyle\sum_{r=10}^{20} r^3$

5. $(1)(3) + (2)(4) + (3)(5) + \ldots + (n-1)(n+1)$

6. $1^2 - 2^2 + 3^2 - 4^2 + 5^2 - 6^2 + \ldots - (2n)^2$

(*Hint*. Consider two series, the sum of the squares of even numbers and sum of the squares of odd numbers.)

7. $(1)(3) + (3)(5) + (5)(7) + \ldots + (2n-1)(2n+1)$

GENERAL METHODS FOR SUMMING A NUMBER SERIES

The basic method for summing a number series relies on recognition, and a systematic approach is helpful, i.e.

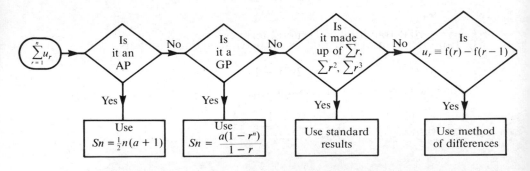

For example, $\displaystyle\sum_{r=n}^{\infty} (\tfrac{1}{3})^r(\tfrac{1}{2})^{r-2}$ can be recognised as a GP with first

term $(\tfrac{1}{3})^n(\tfrac{1}{2})^{n-2}$ and common ratio $(\tfrac{1}{3})(\tfrac{1}{2})$

Similarly, $\displaystyle\sum_{r=1}^{2n} (1+2n)(1-n)$ can be written as $\displaystyle\sum_{r=1}^{2n} (1+n-2n^2)$

when it can be recognised as being made up of natural number series.

THE SUM TO INFINITY OF A NUMBER SERIES

We saw in Module A that if S_n is the sum of the first n terms of a series and if $\lim\limits_{n \to \infty} S_n$ exists, then the series is convergent with a sum to

infinity, S, where
$$S = \lim_{n \to \infty} S_n$$

Note that when evaluating $\lim\limits_{n \to \infty} S_n$ certain assumptions may be made,

i.e. if $n \to \infty$, then $\frac{1}{n} \to 0$,

$a^n \to 0$ if $0 < a < 1$, and $a^n \to \infty$ if $a > 1$

If $S_n \to \dfrac{\infty}{\infty}$ or $\dfrac{0}{0}$, both of which are indeterminate, it may be

possible to evaluate $\lim\limits_{n \to \infty} S_n$ by expressing S_n as a proper fraction,

e.g. if $S_n = \dfrac{n-1}{n+1}$ then $\lim\limits_{n \to \infty} \dfrac{n-1}{n+1}$ is indeterminate,

but $\dfrac{n-1}{n+1} = 1 - \dfrac{2}{n+1}$ so $\lim\limits_{n \to \infty} S_n = \lim\limits_{n \to \infty} \left[1 - \dfrac{1}{n+1} \right] = 1$

Example 10c

Find the sum to infinity of the series $\dfrac{1}{(1)(3)} + \dfrac{1}{(3)(5)} + \dfrac{1}{(5)(7)} + \ldots$

Using partial fractions the general term of this series, i.e.

$\dfrac{1}{(2r-1)(2r+1)}$, becomes $\dfrac{1}{2(2r-1)} - \dfrac{1}{2(2r+1)}$

Therefore the sum of the first n terms of the given series is S_n where

$$S_n = \sum_{r=1}^{\infty} \left[\frac{1}{2(2r-1)} - \frac{1}{2(2r+1)} \right]$$

$$= \tfrac{1}{2}\left[(\tfrac{1}{1} - \tfrac{1}{3}) + (\tfrac{1}{3} - \tfrac{1}{5}) + \ldots + \left(\frac{1}{2n-1} - \frac{1}{2n+1} \right) \right]$$

$$\Rightarrow S_n = \tfrac{1}{2}\left(1 - \frac{1}{2n+1} \right)$$

Now as $n \to \infty$, $\dfrac{1}{2n+1} \to 0$ so $S_n \to \tfrac{1}{2}$

Therefore the sum to infinity of this series is $\tfrac{1}{2}$

EXERCISE 10c

Find the sum of each of the following series.

1. $1 - \frac{1}{3} + \frac{1}{9} - \frac{1}{27} + \ldots$

2. $8 + 27 + 64 + 125 + \ldots + 1000$

3. $\displaystyle\sum_{r=2}^{n} ab^r$

4. $1 + 4 + 9 + 16 + \ldots + 144$

5. $\displaystyle\sum_{r=5}^{2n} 4r$

6. $\displaystyle\sum_{r=1}^{n} r(r^2 + 1)$

7. $\displaystyle\sum_{r=2}^{\infty} \frac{1}{(r-1)(r+1)}$

8. $\ln 3 + \ln 3^2 + \ln 3^3 + \ldots + \ln 3^{20}$

9. $1 + e + e^2 + e^3 + \ldots + e^n$

10. $\displaystyle\sum_{r=1}^{n} r(2r + 1)(r + 2)$

11. $\displaystyle\sum_{r=1}^{\infty} \frac{1}{(r+1)(r+2)}$

12. $\displaystyle\sum_{r=1}^{\infty} \frac{1}{2^{r-1}}$

13. The sum of the squares of the first n even numbers.

14. $\displaystyle\sum_{r=2}^{n} \ln\left(1 - \frac{1}{r}\right)$

THE BINOMIAL THEOREM

In Module A it was shown that, when n is a positive integer,

$$(1 + x)^n = 1 + nx + \binom{n}{2}x^2 + \ldots + x^n$$

Although it cannot be proved at this stage, a very similar expansion of $(1 + x)^n$ exists for *any* real value of n, i.e.

$$(1 + x)^n \equiv 1 + nx + \binom{n}{2}x^2 + \binom{n}{3}x^3 + \ldots \quad \text{for any real value of } n$$

$$\text{provided that } -1 < x < 1$$

Notice that when n is not a positive integer, the binomial expansion of $(1 + x)^n$ does not terminate but carries on to infinity.

Notice also that the expansion is valid only if x is in the range $-1 < x < 1$ and this *range must always be stated*.

Finally it should be noted that the expansion is *not* valid for $(a + x)^n$.

To expand $(a + x)^n$ it must first be written in the form $a^n \left(1 + \dfrac{x}{a} \right)^n$

and this expansion is valid only for $-1 < \dfrac{x}{a} < 1$

Although we cannot prove the binomial expansion of $(1 + x)^n$, it can be verified when $n = -1$, as follows.

Consider the series $1 - x + x^2 - x^3 + x^4 - \ldots$

This is an infinite GP whose common ratio is $-x$
Therefore, provided that $|x| < 1$, the series converges

and the sum to infinity is $\dfrac{1}{1 - (-x)} = (1 + x)^{-1}$.

Now consider the binomial expansion of $(1 + x)^{-1}$

$$(1 + x)^{-1} = 1 + (-1)(x) + \frac{(-1)(-2)}{(2)(1)} x^2 + \frac{(-1)(-2)(-3)}{(3)(2)(1)} x^3 + \frac{(-1)(-2)(-3)(-4)}{(4)(3)(2)(1)} x^4 + \ldots$$

$$= 1 - x + x^2 - x^3 + x^4 - \ldots$$

and we have shown that the sum to infinity of this series is $(1 + x)^{-1}$.

This series occurs frequently and is worth memorising. So also is the series obtained by replacing x by $-x$,
i.e.

$$(1 - x)^{-1} = 1 + x + x^2 + x^3 + \ldots$$

Factorial Notation

To represent $4 \times 3 \times 2 \times 1$ we write 4! and say 'four factorial'.
Similarly 8! means $8 \times 7 \times 6 \times 5 \times 4 \times 3 \times 2 \times 1$

In general, the product of all the natural numbers from any number n down to 1 is called *n factorial* and is written as $n!$

Examples 10d

1. Expand each of the following functions as a series of ascending powers of x up to and including the term in x^3 stating the set of values of x for which each expansion is valid.

   (a) $(1 + x)^{1/2}$ (b) $(1 - 2x)^{-3}$ (c) $(2 - x)^{-2}$

For $|x| < 1$

$$(1 + x)^n = 1 + nx + \frac{n(n - 1)}{2!}x^2 + \frac{n(n - 1)(n - 2)}{3!}x^3 + \dots \qquad [1]$$

(a) Replacing n by $\frac{1}{2}$ in [1] gives

$$(1 + x)^{1/2} = 1 + \frac{1}{2}x + \frac{\frac{1}{2}(\frac{1}{2} - 1)}{2!}x^2 + \frac{\frac{1}{2}(\frac{1}{2} - 1)(\frac{1}{2} - 2)}{3!}x^3 + \dots$$

$$= 1 + \frac{1}{2}x + \frac{\frac{1}{2}(-\frac{1}{2})}{2!}x^2 + \frac{\frac{1}{2}(-\frac{1}{2})(-\frac{3}{2})}{3!}x^3 + \dots$$

$$= 1 + \frac{x}{2} - \frac{x^2}{8} + \frac{x^3}{16} - \dots \quad \text{for } |x| < 1$$

(b) Replacing n by -3 and x by $-2x$ in [1] gives

$$(1 - 2x)^{-3} = 1 + (-3)(-2x) + \frac{(-3)(-4)}{2!}(-2x)^2 + \frac{(-3)(-4)(-5)}{3!}(-2x)^3 + \dots$$

$$= 1 + 6x + 24x^2 + 80x^3 + \dots$$

provided that $-1 < -2x < 1$, i.e. $\frac{1}{2} > x > -\frac{1}{2}$

(c) $(2 - x)^{-2} = 2^{-2}(1 - \frac{1}{2}x)^{-2}$

Replacing n by -2 and x by $-\frac{1}{2}x$ in [1] gives

$$(2 - x)^{-2} = \frac{1}{4}\left[1 + (-2)(-\frac{1}{2}x) + \frac{(-2)(-3)}{2!}(-\frac{1}{2}x)^2 + \frac{(-2)(-3)(-4)}{3!}(-\frac{1}{2}x)^3 + \dots\right]$$

$$= \frac{1}{4}(1 + x + \frac{3}{4}x^2 + \frac{1}{2}x^3 + \dots)$$

$$= \frac{1}{4} + \frac{1}{4}x + \frac{3}{16}x^2 + \frac{1}{8}x^3 + \dots$$

The expansion of $(1 - \frac{1}{2}x)^{-2}$ is valid for $-1 < -\frac{1}{2}x < 1$, i.e. for $2 > x > -2$

Therefore the expansion of $(2 - x)^{-2}$ also is valid for $2 > x > -2$

2. Expand $\dfrac{5}{(1 + 3x)(1 - 2x)}$ as a series of ascending powers of x giving the first four terms and the range of values of x for which the expansion is valid.

Expressing $\dfrac{5}{(1 + 3x)(1 - 2x)}$ in partial fractions gives

$$\frac{5}{(1 + 3x)(1 - 2x)} = \frac{3}{(1 + 3x)} + \frac{2}{(1 - 2x)} = 3(1 + 3x)^{-1} + 2(1 - 2x)^{-1}$$

Now $(1 + x)^{-1} = 1 - x + x^2 - x^3 + \ldots$ for $-1 < x < 1$

Replacing x by $3x$ gives

$$(1 + 3x)^{-1} = 1 - 3x + (3x)^2 - (3x)^3 + \ldots$$

$$= 1 - 3x + 9x^2 - 27x^3 + \ldots \quad \text{for } -1 < 3x < 1$$

Also $(1 - x)^{-1} = 1 + x + x^2 + \ldots$ and replacing x by $2x$ gives

$$(1 - 2x)^{-1} = 1 + (2x) + (2x)^2 + (2x)^3 + \ldots$$

$$= 1 + 2x + 4x^2 + 8x^3 + \ldots \quad \text{for } -1 < -2x < 1$$

Hence $\dfrac{5}{(1 + 3x)(1 - 2x)} = 3(1 + 3x)^{-1} + 2(1 - 2x)^{-1}$

$$= (3 + 2) + (-9 + 4)x + (27 + 8)x^2 + (-81 + 16)x^3 + \ldots$$

provided that $-\frac{1}{3} < x < \frac{1}{3}$ and $-\frac{1}{2} < x < \frac{1}{2}$

Therefore the first four terms of the series are $5 - 5x + 35x^2 - 65x^3$

The expansion is valid for the range of values of x satisfying both $-\frac{1}{3} < x < \frac{1}{3}$ and $-\frac{1}{2} < x < \frac{1}{2}$

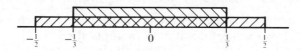

i.e. for $-\frac{1}{3} < x < \frac{1}{3}$

3. Expand $\sqrt{\left(\dfrac{1+x}{1-2x}\right)}$ as a series of ascending powers of x up to and including the term containing x^2

$$\sqrt{\left(\frac{1+x}{1-2x}\right)} \equiv (1+x)^{1/2}(1-2x)^{-1/2}$$

Now $\quad (1+x)^{1/2} = \left[1 + \tfrac{1}{2}x + \dfrac{(\tfrac{1}{2})(-\tfrac{1}{2})}{2!}x^2 + \ldots\right] \quad$ for $\; -1 < x < 1$

and $\quad (1-2x)^{-1/2} = \left[1 + (-\tfrac{1}{2})(-2x) + \dfrac{(-\tfrac{1}{2})(-\tfrac{3}{2})}{2!}(-2x)^2 + \ldots\right]$

$$\text{for} \; -1 < 2x < 1$$

Hence $\sqrt{\left(\dfrac{1+x}{1-2x}\right)} \equiv (1+x)^{1/2}(1-2x)^{-1/2}$

$$= (1 + \tfrac{1}{2}x - \tfrac{1}{8}x^2 + \ldots)(1 + x + \tfrac{3}{2}x^2 + \ldots)$$

$$= 1 + (\tfrac{1}{2}x + x) + (\tfrac{1}{2}x^2 - \tfrac{1}{8}x^2 + \tfrac{3}{2}x^2) + \ldots$$

$$= 1 + \tfrac{3}{2}x + \tfrac{15}{8}x^2 + \ldots$$

provided that $\; -1 < x < 1 \; \textit{and} \; -\tfrac{1}{2} < x < \tfrac{1}{2},$

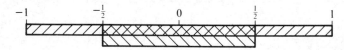

i.e. $\; -\tfrac{1}{2} < x < \tfrac{1}{2}$

It is interesting to compare the methods used in the last two examples.

In Example 2, the function is expressed as the sum of two binomials and the series is obtained by adding two binomial expansions.

In Example 3 the function is expressed as a product of two binomials and the series is obtained by multiplying two binomial expansions.

The first method has the advantage that it is very much easier to add the terms of two series than it is to multiply them.

Therefore, *whenever possible, a compound function should be expressed as a sum of simpler functions before it is expanded as a series* and, when this is not possible, a compound function should be expressed as a product of simpler functions.

Further Approximations

We saw in Module A how a series can be used to find an approximate value of a rational number without having to calculate its exact value. The next example illustrates how a series can be used to find the decimal value, to any required degree of accuracy, of an irrational quantity.

---

4. Use the expansion of $(1 - x)^{1/2}$ with $x = 0.02$ to find the decimal value of $\sqrt{2}$ correct to nine decimal places.

$$(1 - x)^{1/2} = 1 - \tfrac{1}{2}x + \frac{(\tfrac{1}{2})(-\tfrac{1}{2})}{2!}(-x)^2 + \frac{(\tfrac{1}{2})(-\tfrac{1}{2})(-\tfrac{3}{2})}{3!}(-x)^3$$

$$+ \frac{(\tfrac{1}{2})(-\tfrac{1}{2})(-\tfrac{3}{2})(-\tfrac{5}{2})}{4!}(-x)^4 + \ldots$$

$$= 1 - \tfrac{1}{2}x - \tfrac{1}{8}x^2 - \tfrac{1}{16}x^3 - \tfrac{5}{128}x^4 - \tfrac{7}{256}x^5 - \ldots$$

This is valid for $-1 < 1$ and so is valid when $x = 0.02$
Replacing x by 0.02 gives

$(0.98)^{1/2} = 1 - 0.01 - 0.000\ 05 - 0.000\ 000\ 5 - 0.000\ 000\ 006\ 25 - 0.000\ 000\ 000\ 087\ 5 - \ldots$

The next term in the series is 1.3125×10^{-12} and as this does not contribute to the first ten decimal places we do not need it, or any further terms.

i.e. $\qquad\qquad \sqrt{\tfrac{98}{100}} = 0.989\ 949\ 493\ 7$ to 10 d.p.

$\Rightarrow \qquad\qquad \tfrac{7}{10}\sqrt{2} = 0.989\ 949\ 493\ 7$ to 10 d.p.

$\therefore \qquad\qquad \sqrt{2} = 1.414\ 213\ 562$ correct to 9 d.p.

EXERCISE 10d

Expand the following functions as series of ascending powers of x up to and including the term in x^3. In each case give the range of values of x for which the expansion is valid.

1. $(1 - 2x)^{1/2}$

2. $(3 + x)^{-1}$

3. $\left(1 + \dfrac{x}{2}\right)^{-1/2}$

4. $\dfrac{1}{(1 - x)^2}$

5. $\sqrt{\left(\dfrac{1}{1 + x}\right)}$

6. $(1 + x)\sqrt{(1 - x)}$

7. $\dfrac{x+2}{x-1}$
8. $\dfrac{2-x}{\sqrt{(1-3x)}}$
9. $\dfrac{1}{(2-x)(1+2x)}$

10. $\sqrt{\left(\dfrac{1+x}{1-x}\right)}$
11. $\left(1+\dfrac{x^2}{9}\right)^{-1}$

12. $\left(1+\dfrac{1}{x}\right)^{-1}$ $\left[Hint.\ \left(1+\dfrac{1}{x}\right)^{-1} \equiv \left(\dfrac{x+1}{x}\right)^{-1} \equiv \dfrac{x}{1+x}\right]$

13. Expand $\left(1+\dfrac{1}{p}\right)^{-3}$ as a series of descending powers of p, as far as and including the term containing p^{-4}. State the range of values of p for which the expansion is valid.
    (*Hint*. Replace x by $\frac{1}{p}$ in $(1+x)^{-3}$)

14. By substituting 0.08 for x in $(1+x)^{1/2}$ and its expansion find $\sqrt{3}$ correct to four significant figures.

15. By substituting $\frac{1}{10}$ for x in $(1-x)^{-1/2}$ and its expansion find $\sqrt{10}$ correct to six significant figures.

16. Expand $\sqrt{\left(\dfrac{1+2x}{1-2x}\right)}$ as a series of ascending powers of x up to and including the term in x^2

17. If x is so small that x^2 and higher powers of x may be neglected show that $\dfrac{1}{(x-1)(x+2)} \approx -\frac{1}{2} - \frac{1}{4}x$

18. By neglecting x^3 and higher powers of x, find a quadratic function that approximates to the function $\dfrac{1-2x}{\sqrt{(1+2x)}}$ in the region close to $x=0$

19. Find a quadratic function that approximates to
$$f(x) = \dfrac{1}{\sqrt[3]{(1-3x)^2}}$$
for values of x close to zero.

20. Use partial fractions and the binomial series to find a linear approximation for
$$\dfrac{3}{(1-2x)(2-x)}$$

21. If terms containing x^4 and higher powers of x can be neglected, show that

$$\frac{2}{(x+1)(x^2+1)} \approx 2(1-x)$$

22. Show that

$$\frac{12}{(3+x)(1-x)^2} \approx 4 + \tfrac{20}{3}x + \tfrac{88}{9}x^2$$

provided that x is small enough to neglect powers higher than 2

23. If x is very small, find a cubic approximation for

$$\frac{1}{(3-x)^3}$$

MIXED EXERCISE 10

1. Find the first three terms and the last term in the expansion of $(1+2x)^9$ as a series of ascending powers of x. For what values of x is this expansion valid?

2. Expand $\dfrac{1}{1+2x}$ as a series of ascending powers of x, giving the first three terms and the range of values of x for which the expansion is valid.

3. Find the first three terms in the expansion of $\dfrac{x}{1-x}$ as a series of ascending powers of x giving the range of values of x for which the expansion is valid.

4. Express $f(x) = \dfrac{1+x}{1-2x}$ as a series of ascending powers of x, as far as the term in x^2, giving the values of x for which the series converges to $f(x)$.

5. Express $f(x) = \dfrac{1}{(1+x)(1-2x)}$ in partial fractions. Hence find a quadratic function which is approximately equal to $f(x)$ when x is small enough for powers of x greater than x^2 to be ignored.

6. Find the coefficient of x^2 when $\left(\dfrac{1+x}{1-x}\right)^2$ is expanded as a series of ascending powers of x

7. The sum of the first n terms of a series is n^3. Write down the first four terms and the nth term of the series.

8. Show that $\displaystyle\sum_{r=1}^{2n} (r+1)^2 = \tfrac{1}{3}n(8n^2 + 18n + 13)$.

9. Find

   (a) the sum of the cubes of the first n even numbers

   (b) $\displaystyle\sum_{r=2}^{n} \ln\left(1 - \frac{1}{r}\right)$.

10. (a) Expand $\dfrac{1}{\sqrt{(1-3x)}}$ as a series of ascending powers of x, giving the first three terms and the range of values of x for which the series converges.

    (b) Use the expansion in (a) to find an approximate value for $\sqrt{2}$

11. When $\dfrac{1}{(1-ax)^2} - \dfrac{1}{\sqrt{(1-4x)}}$ is expanded as a series of ascending powers of x, the first term is $-3x^2$

    (a) Find the value of a

    (b) Find the second term of the series.

12. Show that $\displaystyle\sum_{r=1}^{n} \frac{1}{r(r+1)} = \frac{n}{n+1}$.

    Hence find the sum to infinity of the series

    (a) $1 + \dfrac{1}{6} + \dfrac{1}{12} + \dfrac{1}{20} + \dfrac{1}{30} + \ldots$

    (b) $\dfrac{1}{3} + \dfrac{1}{6} + \dfrac{1}{10} + \dfrac{1}{15} + \ldots$

CHAPTER 11

REDUCTION OF A RELATIONSHIP TO A LINEAR LAW

THE LINEAR LAW

Any two variables, X and Y say, are said to obey a linear law if they are related by an equation of the form

$$Y = mX + c$$

Now $Y = mX + c$ is the equation of a straight line, so if values of the variable Y are plotted against the corresponding values of the variable X, the points will lie in a straight line.

In this chapter we look at practical applications of these facts.

Linear Relationships

If it is thought that a certain relationship exists between two variable quantities, this hypothesis can be tested by experiment, i.e. by giving one variable certain values and measuring the corresponding values of the other variable.

The experimental data collected can then be displayed graphically. If the graph shows points that lie approximately on a straight line (allowing for experimental error) then a linear relationship between the variables (i.e. a relationship of the form $Y = mX + c$) is indicated.

Further, the gradient m of the line and the vertical axis intercept c can be found from the graph and hence give the values of the constants.

Examples 11a

1. An elastic string is fixed at one end and a variable weight is hung on the other end. It is believed that the length of the string is related to the weight by a linear law. Use the following experimental data to confirm this belief and find the particular relationship between the length of the string and the weight.

Weight (W) in newtons	1	2	3	4	5	6	7	8
Length (l) in metres	0.33	0.37	0.4	0.45	0.5	0.53	0.56	0.6

If l and W are related by a linear law then, allowing for experimental error, we expect that the points will lie on a straight line. Plotting l against W gives the following graph.

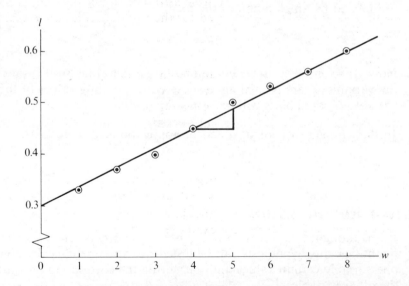

These points do lie fairly close to a straight line.

From the graph, l and W are connected by a linear relationship, i.e. a relationship of the form $l = aW + b$

Now we draw the line of 'closest fit'. This is the line that has the points distributed above and below it as evenly as possible; it is not necessarily the line which goes through the most points.

By measurement from the graph

$$\text{the gradient} = 0.04$$

$$\text{the intercept on the vertical axis} = 0.3$$

So comparing $\left. \begin{matrix} l = aW + b \\ Y = mX + c \end{matrix} \right\}$ with we have $a = 0.04, \quad b = 0.3$

i.e. within the limits of experimental accuracy

$$l = 0.04W + 0.3$$

When the gradient of a line is found from a graph, the increase in a quantity is measured from the *scale used for that quantity* and it is worth noting that the scales used for the two quantities are *not* usually the same.

The values of the constants found from calculating the gradient and intercept from a drawn graph are approximate. Apart from experimental error in the data, selecting the line of best fit is a personal judgement and so is subject to slight variations which affect the values obtained.

There are methods for calculating the equation of the line of best fit; these are called regression lines and computer programmes exist which will give these equations from the data. Using such a programme, the values of a and b in the last example are given as $a = 0.039$ and $b = 0.291$

Now if the relationship is not of a linear form, the points on the graph will lie on a section of a curve. It is very difficult to identify the equation of a curve from a section of it, so the form of a non-linear relationship can rarely be verified in this way.

Non-linear relationships, however, can often be reduced to a linear form. The following examples illustrate some of the relationships which can be verified by plotting experimental data in a form which gives a straight line.

Relationships of the Form $y = ax^n$

A relationship of the form $y = ax^n$ where a is a constant can be reduced to a linear relationship by taking logarithms, since

$$y = ax^n \iff \ln y = n \ln x + \ln a$$

(Although any base can be used, it is sensible to use either e or 10 as these are built into most calculators.)

Comparing $\qquad\qquad \ln y = n \ln x + \ln a$

with $\qquad\qquad\qquad Y = mX + c$

we see that plotting values of $\ln y$ against values of $\ln x$ gives a straight line whose gradient is m and whose intercept on the vertical axis is $\ln a$

Examples 11a (continued)

2. The following data, collected from an experiment is believed to obey a law of the form $p = aq^n$. Verify this graphically and find the values of a and n.

q	1	2	3	4	5	6
p	0.5	0.63	0.72	0.8	0.85	0.9

If the relationship $p = aq^n$ is correct, then $\qquad \ln p = n \ln q + \ln a$

Comparing with $\qquad\qquad\qquad\qquad\qquad y = mx + c$

we see that $\ln p$ and $\ln q$ are related by a linear law.

First a table of values of $\ln p$ and $\ln q$ is needed.

$\ln q$	0	0.69	1.10	1.39	1.61	1.79
$\ln p$	−0.69	−0.46	−0.33	−0.22	−1.16	−0.11

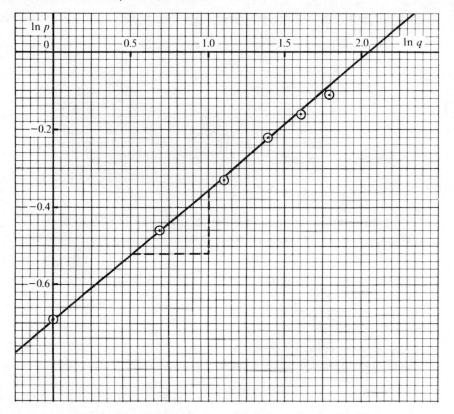

The points lie on a straight line confirming that there is a linear relationship between $\ln q$ and $\ln p$.

From the graph, the gradient of the line is 0.33, $\Rightarrow$ $n = 0.33$ and the intercept on the vertical axis is -0.69, so

$$\ln a = -0.69 \quad \Rightarrow \quad a = 0.5$$

Therefore the data does obey a law of the form $p = aq^n$, where $a \approx 0.5$ and $n \approx 0.33$

(Using the tabulated values of $\ln q$ and $\ln p$ and a computer programme, gives $n = 0.327$ and $\ln a = -0.687$)

An alternative method for investigating relationships of this form uses log–log graph paper, i.e. graph paper where the grids and the scales marked on them are adjusted to represent the logarithms of the numbers being plotted. So to plot $\log S$ against $\log T$ say, values of S and T can be plotted directly.

The diagram illustrates the use of log–log graph paper using the data given in the previous example.

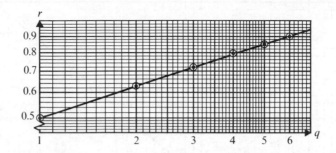

This straight line verifies that the relationship is of the form $p = aq^n$
When $q = 1$, $p = a$. So *from the graph* we see that $a = 0.5$

Reading another pair of values from the graph, (*not* from the table) and substituting these into the relationships gives $q = 2.5$ and $p = 0.68$,

then $0.68 = (0.5)(2.5)^n$ $\Rightarrow$ $n = 0.335$

Relationships of the Form $y = ab^x$

A relationship of the form $y = ab^x$ where a and b are constant can be reduced to a linear relationship by taking logs, since

$$y = ab^x   \iff   \log y = x \log b + \log a$$

Comparing $\log y = x \log b + \log a$

with $Y = mX + c$

we see that plotting values of $\log y$ against corresponding values of x gives a straight line whose gradient is $\log b$ and whose intercept on the vertical axis is $\log a$.

Relationships of the Form $\dfrac{1}{y} + \dfrac{1}{x} = \dfrac{1}{a}$

If a is a constant, $\dfrac{1}{y} + \dfrac{1}{x} = \dfrac{1}{a}$ is a linear relationship between $(1/y)$ and $(1/x)$

i.e. if values of $(1/y)$ are plotted against corresponding values of $(1/x)$, a straight line will result.

By comparing $\qquad (1/y) = -(1/x) + (1/a)$

with $\qquad\qquad\qquad Y = mX + c$

it can be seen that the gradient of the graph should be -1 and the intercept on the $(1/y)$ axis gives the value of $1/a$

Note that in all graphical work the scales should be chosen to give the greatest possible accuracy, i.e. the range of values given in the table should have as much spread as possible. This sometimes means that the horizontal scale does not include zero and the value of c cannot then be read from the graph. In these circumstances, which arise in the next example, we find c by using the equation $Y = mX + c$ together with the measured value of m and the coordinates of any point P on the graph (*not* a pair of values from the table).

Examples 11a (continued)

3. In an experiment, values of a variable y were measured for selected values of a variable x

    The results are shown in the table below. It is believed that x and y are related by a law of the form $2y + 10 = ab^{(x-3)}$. Confirm this graphically and find approximate values for a and b

x	10	12	15	20	21
y	37.5	90	320	2440	3700

If $2y + 10 = ab^{(x-3)}$, taking logs of both sides gives

$$\log(2y + 10) = (x - 3)\log b + \log a$$

which is of the form $\qquad\qquad Y = mX + c$

where $Y = \log(2y + 10)$, $X = x - 3$ and $m = \log b$, $c = \log a$ i.e. $[\log(2y + 10)]$ and $[x - 3]$ obey a linear law.

So we need to tabulate corresponding values of $(x - 3)$ and $\log(2y + 10)$ from the given values of x and y

$x - 3$	7	9	12	17	18
$\log(2y + 10)$	1.9	2.3	2.8	3.7	3.9

Then plotting $\log(2y + 10)$ against $x - 3$ gives the graph below.

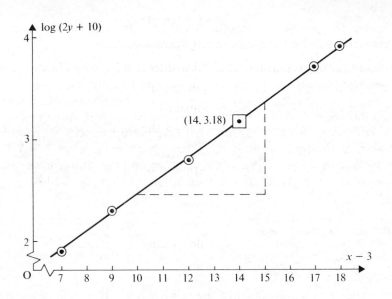

The straight line shows that there is a linear relationship between $\log(2y + 10)$ and $x - 3$, confirming that $2y + 10 = ab^{x-3}$

From the graph, the gradient is 0.175

$$\therefore \qquad \qquad \log b \approx 0.175 \qquad \Rightarrow \qquad b \approx 1.49$$

Using the point $P(14, 3.18)$ and $m = 0.175$ then $Y = mX + c$ gives

$$3.18 = (0.175)(14) + c \qquad \Rightarrow \qquad c = 0.73$$

i.e. $\qquad \qquad \log a \approx 0.73 \qquad \Rightarrow \qquad a \approx 5.37$

When attempting to reduce a relationship between two variables to a form from which a straight line graph can be drawn, the given equation must be expressed in the form

$$Y = mX + c$$

where X and Y are variable terms, values for which must be calculable from the given data, i.e. X and Y must not contain unknown constants. On the other hand m and c must be constants, but may be unknown.

Now X and Y may be functions of one or both variables, as for example

$$f(xy) = m\, g(xy) + c$$

is a linear relationship between $f(xy)$ and $g(xy)$

So to reduce a non-linear relationship to a linear form we:

1) try to express it in a form containing three terms,

2) make one of those terms constant,

3) remove unknown constants from the coefficient of one of the variable terms.

These objectives will now be applied in the next worked example.

Examples 11a (continued) _____

4. It is known that two variables x and y are related by the law

(a) $ae^y = x^2 - bx$ (b) $y = \dfrac{1}{(x-a)(x-b)}$

In each case state how you would reduce the law to a linear form so that a straight line graph could be drawn from experimental data.

(a) $ae^y = x^2 - bx$

This equation has three terms, one of which becomes constant when we divide by x, giving $a\dfrac{e^y}{x} = x - b$

We also have a variable term (x) whose coefficient is a known constant. This equation may now be written as

$$x = a\frac{e^y}{x} + b$$

Comparing with $Y = mX + c$

we see that if values of x are plotted against corresponding values of e^y/x, a straight line will result whose gradient is a and whose intercept on the vertical axis is b

Note that the original equation can be arranged in linear form in a variety of ways

e.g. $\left(\dfrac{e^y}{x^2}\right) = -\dfrac{b}{a}\left(\dfrac{1}{x}\right) + \dfrac{1}{a}$ or $\left(\dfrac{x^2}{e^y}\right) = b\left(\dfrac{x}{e^y}\right) + a$

(b) $y = \dfrac{1}{(x-a)(x-b)}$

Although this form suggests the use of partial fractions, this approach increases the number of times the unknown constants appear. It is better to invert the equation giving

$$\frac{1}{y} = (x-a)(x-b)$$

$\Rightarrow$
$$\frac{1}{y} = x^2 - x(a+b) + ab$$

$\Rightarrow$
$$x^2 - \frac{1}{y} = (a+b)x - ab$$

We can now compare this form with $Y = mX + c$

Thus plotting values of $(x^2 - 1/y)$ against corresponding values of x will give a straight line whose gradient is $a + b$ and whose intercept on the vertical axis is $-ab$

EXERCISE 11a

1. Reduce each of the given relationships to the form $Y = mX + c$. In each case give the functions equivalent to X and Y and the constants equivalent to m and c.

(a) $\dfrac{1}{y} = ax + b$ (b) $y(y-b) = x - a$

(c) $ae^x = y(y-b)$ (d) $x = y^2 + y - k$

(e) $y = ax^{n+2}$ (f) $y^a = e^{x+k}$

In Questions 2 to 7, the table gives sets of values for the related variables and the law which relates the variables. By drawing a straight line graph find approximate values for a and b

2. $y = ax + ab$

x	3	5	7	10
y	−2	2	6	12

3. $s = ab^{-t}$

t	1	2	3	4
s	1.5	0.4	0.1	0.02

4. $r^2 = a\theta - b$

θ	1	4	10	25	40
r	1.6	2	2.6	3.8	4.7

5. $ay = b^x$

x	5	6	7	8
y	1.07	2.13	4.27	8.53

6. $\dfrac{a}{V} + \dfrac{b}{L} = 1$

V	2.5	3	5.5	7	12
L	2.5	1.5	0.79	0.7	0.6

7. $y = (x - a)(x - b)$

x	1	2	3	4	5	6
y	-6	-4	0	6	14	24

8. The variables x and y are believed to satisfy a relationship of the form $y = k(x + 1)^n$. Show that the experimental values shown in the table do satisfy the relationship. Find approximate values for k and n

x	4	8	15	19	24
y	4.45	4.60	4.80	4.89	5.00

9. Two variables s and t are related by a law of the form $s = ke^{-nt}$ The values in the table were obtained from an experiment. Show graphically that these values do verify the relationship and use the graph to find approximate values of k and n

t	1	1.5	2	2.5	3
s	1230	590	260	140	60

10. Two variables x and t are related by a law of the form
$x = \cos(at + \varepsilon)$. The table shows the values of x obtained
experimentally for some different values of t. Show graphically
that these values do NOT satisfy this relationship unless two of
the values of x are assumed to be incorrect.
Estimate the correct values of x

t	0.5	1	1.5	1.75	2	2.25
x	−0.26	−0.90	−0.40	−0.90	−0.71	−0.49

CONSOLIDATION B

SUMMARY

FUNCTIONS

Exponential functions:

$f: x \rightarrow a^x$ is an exponential function

$f: x \rightarrow e^x$ is *the* exponential function: $e = 2.718\,28\ldots$

and $a^x = e^{x \ln a}$

Logarithmic functions:

$f: x \rightarrow \log_n x$,   is a logarithmic function

$f: x \rightarrow \log_e x = \ln x$ is the natural logarithmic function

DIFFERENTIATION

Standard results

$f(x)$	$\dfrac{d}{dx} f(x)$	$f(x)$	$\dfrac{d}{dx} f(x)$
x^n	nx^{n-1}	$(ax+b)^n$	$na(ax+b)^{n-1}$
e^x	e^x	e^{ax}	ae^{ax}
$\ln x$	$1/x$	$\ln(ax)$	$1/x$ (*not* a/x)
a^x	$a^x \ln a$		

$$\frac{dy}{dx} = 1 \bigg/ \frac{dx}{dy}$$

COMPOUND FUNCTIONS

If u and v are both functions of x then

$$y = uv \quad \Rightarrow \quad \frac{dy}{dx} = v\frac{du}{dx} + u\frac{dv}{dx}$$

$$y = \frac{u}{v} \quad \Rightarrow \quad \frac{dy}{dx} = \left(v\frac{du}{dx} - u\frac{dv}{dx}\right) \Big/ v^2$$

THE CHAIN RULE

If $y = f(u)$ and $u = g(x)$ then $\dfrac{dy}{dx} = \dfrac{dy}{du} \times \dfrac{du}{dx}$

CATEGORISING STATIONARY POINTS

There are three methods for categorising stationary points:

		Max	Min	Inflexion
1.	Find value of y on each side of stationary value	Both smaller	Both larger	One smaller One larger
2.	Find sign of $\dfrac{dy}{dx}$ on each side of stationary value	$+\ 0\ -$	$-\ 0\ +$	$+\ 0\ +$ or $-\ 0\ -$
	Gradient	/‾\	_/	/‾/ or \‾\
3.	Find sign of $\dfrac{d^2y}{dx^2}$ at stationary value	$-$ve (or 0)	$+$ve (or 0)	

Method 3 is often the easiest to apply but it fails if $\dfrac{d^2y}{dx^2}$ is zero. In this case use one of the other methods.

TRIGONOMETRY

Factor Formulae

Before using these formulae it is wise to ensure that $A > B$, rearranging the given expression if necessary.

In the form for converting a sum or difference into a product,

$$\sin A + \sin B \;=\; 2 \sin \tfrac{1}{2}(A + B) \cos \tfrac{1}{2}(A - B)$$

$$\sin A - \sin B \;=\; 2 \cos \tfrac{1}{2}(A + B) \sin \tfrac{1}{2}(A - B)$$

$$\cos A + \cos B \;=\; 2 \cos \tfrac{1}{2}(A + B) \cos \tfrac{1}{2}(A - B)$$

$$\cos A - \cos B \;=\; -2 \sin \tfrac{1}{2}(A + B) \sin \tfrac{1}{2}(A - B)$$

In the form for converting a product into a sum or difference,

$$2 \sin A \cos B \;=\; \sin (A + B) + \sin (A - B)$$

$$2 \cos A \sin B \;=\; \sin (A + B) - \sin (A - B)$$

$$2 \cos A \cos B \;=\; \cos (A + B) + \cos (A - B)$$

$$-2 \sin A \sin B \;=\; \cos (A + B) - \cos (A - B)$$

ALGEBRA

Reduction of Relationships to Linear Form

When a non-linear law, containing two unknown constants, connects two variables, the relationship can often be reduced to linear form. The aim is to produce an equation in which one term is constant and another term does not contain a constant. The law can then be expressed in the form

$$Y = mX + C$$

Some common conversions are

$p = a\sqrt{q} + b$; use $Y = p$ and $X = \sqrt{q}$

$p = aq^b$; take logs: $\ln p = \ln a + b \ln q$; use $Y = \ln p$ and $X = \ln q$

$\dfrac{a}{p} + \dfrac{b}{q} = c$; $\Rightarrow$ $\dfrac{1}{p} = \dfrac{c}{a} - \dfrac{b}{a}\dfrac{1}{q}$; use $Y = \dfrac{1}{p}$ and $X = \dfrac{1}{q}$

THE NATURAL NUMBER SERIES

The natural numbers are the positive integers, 1, 2, 3, ...

The sum of the first n natural numbers is $\sum_{r=1}^{n} r = \frac{1}{2}n(n+1)$

The sum of their squares is $\sum_{r=1}^{n} r^2 = \frac{1}{6}n(n+1)(2n+1)$

The sum of their cubes is $\sum_{r=1}^{n} r^3 = \frac{1}{4}n^2(n+1)^2 = \left\{ \sum_{r=1}^{n} r \right\}^2$

THE BINOMIAL THEOREM

If n is a positive integer then $(1+x)^n$ can be expanded as a *finite* series,

where $(1+x)^n = 1 + nx + \binom{n}{2}x^2 + \binom{n}{3}x^3 + \ldots + x^n$

and where $\binom{n}{r} = \dfrac{n(n-1)(n-2)\ldots(n-r+1)}{r!}$

For any real value of n other than the positive integers, $(1+x)^n$ can be expanded as the *infinite* series

$$1 + nx + \binom{n}{2}x^2 + \binom{n}{3}x^3 + \binom{n}{4}x^4 + \ldots \quad \text{provided that } |x| < 1$$

This result can be used to expand $(a+x)^n$ provided that it is expressed in the form $a^n\left(1 + \dfrac{x}{a}\right)^n$

MULTIPLE CHOICE EXERCISE B

TYPE I

1. $\cos(A+B) + \cos(A-B) \equiv$

    A $2\cos A \sin B$ C $2\cos A \cos B$
    B $-2\sin A \cos B$ D $-2\sin A \sin B$

2. $\dfrac{d}{dx} \ln\left(\dfrac{x+1}{2x}\right)$ is

    A $\dfrac{1}{2}$ **C** $\dfrac{2x}{x+1}$

    B $\dfrac{1}{x+1} - \dfrac{1}{2x}$ **D** $\dfrac{1}{x+1} - \dfrac{1}{x}$

3. $\dfrac{d}{dx} a^x$ is

    A xa^{x-1} **B** a^x **C** $x \ln a$ **D** $a^x \ln a$

4. The third term in the binomial expansion of $(1-3x)^{-2}$ is

    A $-108x^3$ **B** $54x^2$ **C** $27x^2$ **D** $6x$

5. If $y = e^x \ln x$ then $\dfrac{dy}{dx}$ is

    A $\dfrac{e^x}{x}$ **B** $e^x \ln x + \dfrac{e^x}{x}$ **C** $\ln x + \dfrac{e^x}{x}$ **D** $e^x + \dfrac{1}{x}$

6. $\cos 3\theta + \cos 5\theta =$

    A $2 \cos 4\theta \cos \theta$ **C** $2 \cos 4\theta \sin \theta$

    B $-2 \sin 4\theta \sin \theta$ **D** $2 \cos 8\theta \cos 2\theta$

7. If $y = \dfrac{e^x}{x}$ then $\dfrac{dy}{dx}$ is

    A e^x **B** $\dfrac{e^x(x+1)}{x^2}$ **C** $\dfrac{e^x}{x^2}$ **D** $\dfrac{e^x(x-1)}{x^2}$

8. The sum of the first $n+1$ natural numbers is

    A $1 + \tfrac{1}{2}n(n+1)$ **C** $n+1$

    B $n(n+1)$ **D** $\tfrac{1}{2}n(n+1)(n+2)$

9. When x is very small, an approximation for $\sqrt{(1+2x)}$ is

    A $1 + 2x$ **B** $1 + x$ **C** $1 - x$ **D** $1 + x\sqrt{2}$

TYPE II

10. If $s \propto t^3$ then a straight line is obtained by plotting

    A s against t **B** s against t^3 **C** $\log s$ against $\log t$

11. If $y = \ln(\ln x)$ and $x > 1$ then

    A $\dfrac{dy}{dx} = \dfrac{1}{\ln x}$ **B** $e^y = \ln x$ **C** $\dfrac{dy}{dx} = \dfrac{1}{x \ln x}$

12. Which of the following relationships gives a straight line when $\ln y$ is plotted against $\ln x$, given that a and b are constants?

    A $ay^3 = bx^2$ **B** $y = a + b^x$ **C** $y^a = b^x$

13. In the expansion of $\dfrac{1}{(1-x)(1+x)}$ as a series of ascending powers of x

    A there are only even powers of x
    B the first three terms are $1 - x^2 + x^4$
    C the expansion is valid for $-1 \leqslant x \leqslant 1$

14. The sum of the first n terms of a series is $\dfrac{2n^2}{n^2 + 1}$

    A The first two terms of the series are 1, $8/5$
    B The sum to infinity of the series is 1
    C The series converges.

15. $\dfrac{(1 + 2x)^3}{(1 - x)}$ is expanded as a series of ascending powers of x.

    A The expansion is valid only for $|x| < \frac{1}{2}$
    B The first two terms are 1, $7x$
    C The expansion has four terms.

TYPE III

16. When x is small, $(1 + x)^{-1} \approx 1 - x + x^2$

17. $2 \sin \theta \cos 3\theta \equiv \sin 2\theta + \sin 4\theta$

18. $\dfrac{d}{dx}(x^2 e^x) = 2xe^x$

19. $\dfrac{d}{dx}\left(\ln \dfrac{x}{x^2 + 1}\right) = \dfrac{1}{x} - \dfrac{2x}{x^2 + 1}$

20. $\dfrac{d}{dx}\left(\dfrac{x^2}{1 + x}\right) = \dfrac{2x(1 + x) + x^2}{(1 + x)^2}$

21. $\dfrac{d}{dx}(uv) = \dfrac{du}{dx} \times \dfrac{dv}{dx}$

MISCELLANEOUS EXERCISE B

1. Find the first four terms in the expansion in ascending powers of x of
$$\frac{1}{(1 + 3x)^{1/3}}$$
State the values of x for which the expansion converges.   (JMB)

2. Find the gradient of the curve $y = \dfrac{x^2 + 1}{x - 1}$ when $x = 2$ (C)

3. Write down $\sin x + \sin 3x$ as a product of factors.
Find, in terms of π, the solutions of the equation
$$\sin x + \sin 3x = \sin 2x$$
which lie in the interval $-\pi < x \leqslant \pi$.
Find also the general solution of the equation.   (JMB)

4. Write down and simplify the expansion of $(1 - 8x)^{1/2}$, in ascending powers of x, up to and including the term in x^3.
State the set of values of x for which the expansion is valid.
          (U of L)

5. Show that
$$\sum_{r=1}^{n} (2r - 1)^2 = \tfrac{1}{3}n(4n^2 - 1)$$ (U of L)

6. Express y, where $y \equiv \dfrac{1}{(1 + x)(1 + 2x)}$, in partial fractions.

Hence, or otherwise, find the value of $\dfrac{d^2y}{dx^2}$ when $x = -2$ (U of L)

7. Find the general solution of the equation
$$\cos \theta + \cos 3\theta = 0$$

8. The first three terms in the expansion of $(1 + 6x)^{1/4}$, in ascending powers of x, are $1, Ax, Bx^2$.
Find the values of the constants A and B.   (U of L)

9. Differentiate the following functions of x:
   (a) $(1 + x^2)^6$
   (b) $\dfrac{x^2}{1 + x^2}$
   (c) $x^2 \ln (1 + x^2)$ (AEB)

10. Given that
$$y = \frac{3x - 14}{(x - 2)(x + 6)}$$

express y as a sum of partial fractions. Hence find $\dfrac{dy}{dx}$ and $\dfrac{d^2y}{dx^2}$.

Show that $\dfrac{dy}{dx} = 0$ for $x = 10$ and for one other value of x.

Find the maximum and minimum values of y, distinguishing between them.   (JMB)

11. In the expansion of
$$\frac{1}{\sqrt{(1 + ax)}} - \frac{1}{1 + 2x}$$

in ascending powers of x, the first non-zero term is the term in x^2. Find the value of the constant a and hence find the terms in x^2 and x^3.   (JMB)

12. Sketch in separate diagrams the curves whose equations are
$$y = \ln x, \quad x > 0$$
$$y = |\ln x|, \quad x > 0$$

Calculate, correct to two decimal places, the values of x for which
$$|\ln x| = 2$$

Giving a reason, state the number of real roots of the equation
$$|\ln x| + x = 3$$
                                                                          (AEB)

13. Pairs of values, x and y, are obtained in an experiment as shown in the table.

x	10	20	40	75	100
y	136	334	824	1850	2710

It is believed that x and y are related by a law of the form
$$\lg y = m \lg x + c$$

where m and c are constants.

Write down a table of values for $\lg x$ and $\lg y$.

By drawing a graph, show that the law is approximately valid.

Estimate values of m and c giving your answers to 2 significant figures.

Using your values of m and c, express y in terms of x.   (U of L)

14. Find the general solution of the equation

$$\sin 3x + \sin x = \cos x$$

15. Show that

$$\sum_{r=1}^{n} \frac{1}{(2r-1)(2r+1)} = \frac{n}{2n+1}$$

Find $\displaystyle\lim_{n \to \infty} \frac{n}{2n+1}$ (U of L)

16. Given that $y = xe^{-2x}$, find $\dfrac{dy}{dx}$ and $\dfrac{d^2y}{dx^2}$.

Find the value of x for which $\dfrac{dy}{dx} = 0,$ and determine whether it gives a maximum or minimum value of y. (C)

17. In an experiment to determine the relationship between the resistance to motion, R newtons, of a plank towed through water and its speed, $V\,\mathrm{ms}^{-1}$, the following data were recorded:

V	1.8	3.7	5.5	9.2	11
R	2.01	7.32	15.1	38.9	56.2

Assuming that $R = kV^n$, where n and k are constants,
(a) obtain a relation between $\ln R$ and $\ln V$.
(b) By drawing a graph of $\ln R$ against $\ln V$ estimate, to 2 significant figures, the values of n and k. (U of L)

18. Write down $2\cos\theta\cos 2\theta$ as a sum of terms. Hence find the general solution of the equation

$$2\cos\theta\cos 2\theta - \cos\theta + 1 = 0$$

19. Find the value of the function $\log_e(1 + 2x) - \dfrac{2x}{1+x}$ when $x = 0$

Differentiate the function and show that the derivative is always positive when $x > 0$

Deduce that $\log_e(1 + 2x) > \dfrac{2x}{1+x}$ when $x > 0$ (WJEC)

20. The following measurements of the volume, $V\,\text{cm}^3$, and the pressure, $p\,\text{cm}$ of mercury, of a given mass of gas were taken.

V	10	50	110	170	230
p	1412.5	151.4	50.3	27.4	18.6

By plotting values of $\log_{10} p$ against $\log_{10} V$, verify graphically the relationship $p = kV^n$ where k and n are constants.

Use your graph to find approximate values for k and n, giving your answers to two significant figures.   (AEB)p

21.

x	2.1	2.8	4.7	6.2	7.3
y	13	32	316	2000	7080

The above table shows corresponding values of variables x and y obtained experimentally. By drawing a suitable graph, show that these values support the hypothesis that x and y are connected by a relationship of the form $y = a^x$, where a is a constant. Use your graph to estimate the value of a to 2 significant figures.   (U of L)

22. (a) Write down and simplify binomial series in ascending powers of x up to and including the terms in x^3 for

$$(1 + x)^{-1}, \ (1 - 2x)^{-1} \quad \text{and} \quad (1 - 2x)^{-2}$$

(b) Express $\dfrac{6 - 11x + 10x^2}{(1 + x)(1 - 2x)^2}$ in partial fractions.

(c) Using your series from (a) expand $\dfrac{6 - 11x + 10x^2}{(1 + x)(1 - 2x)^2}$ in ascending powers of x up to and including the term in x^3.

(d) State the set of values of x for which your series in (c) is valid.
                                                                (AEB)p

23. The function f is defined by

$$f : x \rightarrow \ln (2x - 3), \quad x \in \mathbb{R}, \quad x > \tfrac{3}{2}$$

Write, in a similar form

(a) the inverse function f^{-1}

(b) the derived function $g = \dfrac{df}{dx}$.

(In each case state the domain of the function.)

(c) In separate diagrams, sketch the curves with equations

$$y = f(x)$$
$$y = f^{-1}(x)$$
$$y = g(x)$$

On each diagram write the equations of any asymptotes and the coordinates of any points at which the curve meets the coordinate axes.

(d) Evaluate $f^{-1}gf(8)$, giving your answer to 2 decimal places.

(U of L)

CHAPTER 12

TRIGONOMETRIC FUNCTIONS

$f(\theta) = a\cos\theta + b\sin\theta$

The diagrams below show the graphs of $f(\theta) = a\cos\theta + b\sin\theta$ for a variety of values of a and b.

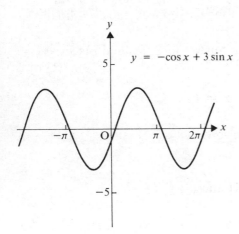

$y = -\cos x + 3\sin x$

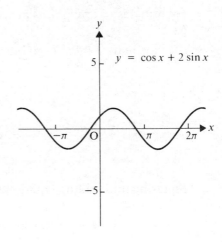

$y = \cos x + 2\sin x$

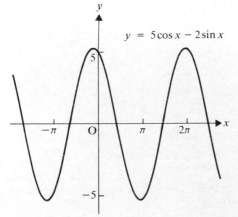

$y = 5\cos x - 2\sin x$

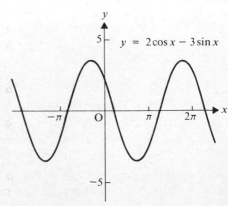

$y = 2\cos x - 3\sin x$

If the reader has access to the appropriate technology, we suggest that the graph of $f(\theta)$ is drawn for some other values of a and b. Each of these graphs is a sine wave, although with differing amplitude and phase shift.

These diagrams suggest that it is possible to express $a \cos \theta + b \sin \theta$ as $r \sin(\theta + \alpha)$ where the values of r and α depend on the values of a and b. This is possible provided that we can find values of r and α such that

$$r \sin(\theta + \alpha) \equiv a \cos \theta + b \sin \theta$$

i.e. $$r \sin \theta \cos \alpha + r \cos \theta \sin \alpha \equiv a \cos \theta + b \sin \theta$$

Since this is an identity we can compare coefficients of $\sin \theta$ and of $\cos \theta$

$$\Rightarrow \qquad\qquad r \sin \alpha = a \qquad\qquad [1]$$

and $$r \cos \alpha = b \qquad\qquad [2]$$

Equations [1] and [2] can now be solved to give r and α in terms of a and b.

Squaring and adding equations [1] and [2] gives

$$r^2(\sin^2\alpha + \cos^2\alpha) = a^2 + b^2 \qquad \Rightarrow \qquad r = \sqrt{(a^2 + b^2)}$$

Dividing equation [1] by equation [2] gives

$$\frac{r \sin \alpha}{r \cos \alpha} = \frac{a}{b} \qquad \Rightarrow \qquad \tan \alpha = \frac{a}{b}$$

Therefore $$r \sin(\theta + \alpha) \equiv a \cos \theta + b \sin \theta$$

$$\text{where} \quad r = \sqrt{(a^2 + b^2)} \quad \text{and} \quad \tan \alpha = \frac{a}{b}$$

It is also possible to express $a \cos \theta + b \sin \theta$ as $r \sin(\theta - \alpha)$ or as $r \cos(\theta \pm \alpha)$, using a similar method.

Examples 12a

1. Express $3 \sin \theta - 2 \cos \theta$ as $r \sin (\theta - \alpha)$

$$3 \sin \theta - 2 \cos \theta \equiv r \sin (\theta - \alpha)$$

$\Rightarrow$ $\qquad 3 \underline{\sin \theta} - 2 \underline{\cos \theta} \equiv r \underline{\sin \theta} \cos \alpha - r \underline{\cos \theta} \sin \alpha$

Comparing coefficients of $\sin \theta$ and of $\cos \theta$ gives

$$\left. \begin{array}{l} 3 = r \cos \alpha \\ 2 = r \sin \alpha \end{array} \right\} \quad \Rightarrow \quad \left\{ \begin{array}{l} 13 = r^2 \quad \Rightarrow \quad r = \sqrt{13} \\ \tan \alpha = \frac{2}{3} \quad \Rightarrow \quad \alpha = 33.7° \end{array} \right.$$

$\therefore \qquad\qquad 3 \sin \theta - 2 \cos \theta = \sqrt{13} \sin (\theta - 33.7°)$

---

2. Find the maximum value of $f(x) = 3 \cos x + 4 \sin x$ and the smallest positive value of x at which it occurs.

Expressing $f(x)$ in the form $r \sin (x + \alpha)$ enables us to 'read' its maximum value, and the values of x at which they occur, from the resulting sine wave. Note also that in this question there is a choice of form in which to express $f(x)$: in this case it is sensible to choose $r \cos (x - \alpha)$ as this fits $f(x)$ better than $r \sin (x + \alpha)$

$$3 \underline{\cos x} + 4 \underline{\sin x} \equiv r \cos (x - \alpha) \equiv r \underline{\cos x} \cos \alpha + r \underline{\sin x} \sin \alpha$$

Hence $\left. \begin{array}{l} r \cos \alpha = 3 \\ r \sin \alpha = 4 \end{array} \right\} \quad \Rightarrow \quad \left\{ \begin{array}{l} r^2 = 25 \quad \Rightarrow \quad r = 5 \\ \tan \alpha = \frac{4}{3} \quad \Rightarrow \quad \alpha = 53.1° \end{array} \right.$

$\therefore \qquad\qquad f(x) \equiv 5 \cos (x - 53.1°)$

The graph of $f(x)$ is a cosine wave with amplitude 5 and phase shift 53.1°

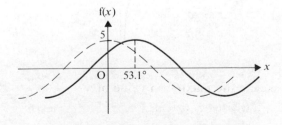

$\therefore$ $f(x)$ has a maximum value of 5 and, from the sketch, the smallest positive value of x at which it occurs is 53.1°

---

3. Find the maximum and minimum values of $\dfrac{2}{\sin x - \cos x}$.

We first express $\sin x - \cos x$ in the form $r \sin (x - \alpha)$ then the given function can be expressed as a cosec function and we can sketch its graph. Note that values of x are not required so we do not need the value of α. Note also that the maximum and minimum values of a function are not necessarily the same as the greatest and least values of that function.

If $f(x) \equiv \underline{\sin x} - \underline{\cos x} \equiv r \sin (x - \alpha) \equiv r \underline{\sin x} \cos \alpha - r \underline{\cos x} \sin \alpha$

then $\left. \begin{array}{l} r \cos \alpha = 1 \\ r \sin \alpha = 1 \end{array} \right\} \quad \Rightarrow \quad r^2 = 2, \text{ i.e. } r = \sqrt{2}$

$\therefore$ $\sin x - \cos x \equiv \sqrt{2} \sin (x - \alpha)$

Hence $\dfrac{2}{f(x)} \equiv \dfrac{2}{\sqrt{2} \sin (x - \alpha)} \equiv \sqrt{2} \operatorname{cosec} (x - \alpha)$

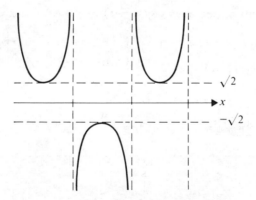

From the sketch, the maximum value of $\dfrac{2}{\sin x - \cos x}$ is $-\sqrt{2}$
and the minimum value is $\sqrt{2}$

Note that $\dfrac{2}{f(x)}$ has a *greatest* value of ∞ and a *least* value of $-\infty$

THE EQUATION $a \cos x + b \sin x = c$

One way of solving an equation of this type was covered in
Chapter 5; it uses the half angle formulae involving t where
$t = \tan \frac{1}{2}x$. An alternative method is to express the LHS of the
equation in the form $r \cos (x + a)$. This method, which has the
advantage that solutions are not easily lost, is illustrated in the next
example.

Examples 12a (continued) ─────────────────────────────

4. Find, in radians, the general solution of the equation
$$\sqrt{3} \cos x + \sin x = 1$$

If $\sqrt{3} \underline{\cos x} + \underline{\underline{\sin x}} \equiv r \cos (x - a) \equiv r \underline{\cos x} \cos a + r \underline{\underline{\sin x}} \sin a$

then $\quad \begin{cases} r \cos a = \sqrt{3} \\ r \sin a = 1 \end{cases} \Rightarrow \quad \begin{cases} r^2 = 4 \\ \tan a = \frac{1}{\sqrt{3}} \end{cases} \Rightarrow \quad \begin{matrix} r = 2 \\ a = \frac{1}{6}\pi \end{matrix}$

i.e. $\sqrt{3} \cos x + \sin x \equiv 2 \cos (x - \frac{1}{6}\pi)$

$\therefore$ the equation becomes

$$2 \cos (x - \frac{1}{6}\pi) = 1$$

$\Rightarrow \quad \cos (x - \frac{1}{6}\pi) = \frac{1}{2}$

$\Rightarrow \quad x - \frac{1}{6}\pi = \pm \frac{1}{3}\pi + 2n\pi$

$\therefore \quad x = \frac{1}{2}\pi + 2n\pi, \; -\frac{1}{6}\pi + 2n\pi$

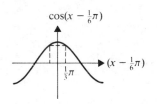

───

EXERCISE 12a

1. Find the values of r and a for which
   (a) $\sqrt{3} \cos \theta - \sin \theta \equiv r \cos (\theta + a)$
   (b) $\cos \theta + 3 \sin \theta \equiv r \cos (\theta - a)$
   (c) $4 \sin \theta - 3 \cos \theta \equiv r \sin (\theta - a)$

2. Express $\cos 2\theta - \sin 2\theta$ in the form $r \cos (2\theta + a)$

3. Express $2 \cos 3\theta + 5 \sin 3\theta$ in the form $r \sin(3\theta + \alpha)$

4. Express $\cos\theta - \sqrt{3}\sin\theta$ in the form $r \sin(\theta - \alpha)$. Hence sketch the graph of $f(\theta) = \cos\theta - \sqrt{3}\sin\theta$. Give the maximum and minimum values of $f(\theta)$ and the values of θ between 0 and 360° at which they occur.

5. Express $7 \cos\theta - 24 \sin\theta$ in the form $r \cos(\theta + \alpha)$. Hence sketch the graph of $f(\theta) = 7 \cos\theta - 24 \sin\theta + 3$ and give the maximum and minimum values of $f(\theta)$ and the values of θ between 0 and 360° at which they occur.

6. Find the greatest and least values of $\cos x + \sin x$. Hence find the maximum and minimum values of $\dfrac{1}{\cos x + \sin x}$

7. Find the maximum and minimum values of $\dfrac{\sqrt{2}}{\cos\theta - \sqrt{2}\sin\theta}$

   State the greatest and least values.

8. Find the general solution of the following equations.
   (a) $\cos x + \sin x = \sqrt{2}$ (b) $7 \cos x + 6 \sin x = 2$
   (c) $\cos x - 3 \sin x = 1$ (d) $2 \cos x - \sin x = 2$

THE EQUATION $\cos A = \cos B$

An equation such as $\cos 2x = \cos x$ can be solved by using a factor formula, but there is another very simple method of solution, based on consideration of the graph of $\cos x$

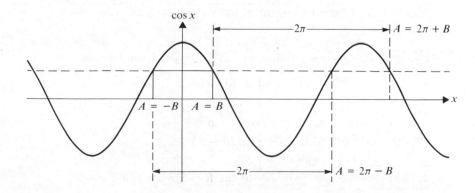

If $$\cos A = \cos B$$

then, from the graph, we see that $$A = 2n\pi \pm B$$

i.e. $$\cos A = \cos B \quad \Rightarrow \quad A = 2n\pi \pm B$$

A similar conclusion is reached from the graphs of $\tan x$ and of $\sin x$

$$\tan A = \tan B \quad \Rightarrow \quad A = B + n\pi$$

$$\sin A = \sin B \quad \Rightarrow \quad A = \begin{cases} 2n\pi + B \\ (2n+1)\pi - B \end{cases}$$

This method can also be used for equations of the form $\cos A = \sin B$, and this is illustrated in Examples 12b, number 2.

Examples 12b

1. Find the general solution of the equation $\cos 4\theta = \cos \theta$

Using only the conclusion to the argument above, we can write

$$4\theta = 2n\pi \pm \theta$$

hence
$$\left. \begin{array}{l} 5\theta = 2n\pi \\ 3\theta = 2n\pi \end{array} \right\} \quad \Rightarrow \quad \theta = \tfrac{2}{5}n\pi, \ \tfrac{2}{3}n\pi$$
or

2. Find the general solution of the equation $\cos 3x = \sin x$

We know that $\sin x = \cos\left(\tfrac{1}{2}\pi - x\right)$, so the equation can be written

$$\cos 3x = \cos\left(\tfrac{1}{2}\pi - x\right)$$

$\therefore$
$$3x = 2n\pi \pm \left(\tfrac{1}{2}\pi - x\right)$$

Hence
$$\left. \begin{array}{l} 4x = 2n\pi + \tfrac{1}{2}\pi \\ 2x = 2n\pi - \tfrac{1}{2}\pi \end{array} \right\} \quad \Rightarrow \quad x = \tfrac{1}{2}n\pi + \tfrac{1}{8}\pi, \ n\pi - \tfrac{1}{4}\pi$$
or

3. Find the values of θ between 0 and $360°$ for which
$\tan(3\theta - 40°) = \tan\theta$

The general solution is $\qquad 3\theta - 40° = 180n° + \theta$

$\Rightarrow \qquad\qquad\qquad\qquad\qquad \theta = 90n° + 20°$

For $0 \leqslant \theta \leqslant 360°$, $n = 0, 1, 2$ and 3

$\Rightarrow \qquad\qquad\qquad \theta = 20°, 110°, 200°, 290°$

EXERCISE 12b

Find the general solution of the following equations.

1. $\cos 4\theta = \cos 3\theta$ **2.** $\tan 7\theta = \tan 2\theta$ **3.** $\cos 3\theta = \sin 2\theta$

4. $\cot 4\theta = \tan 5\theta$ **5.** $\sin 4\theta = \sin 3\theta$ **6.** $\cos 5\theta \sec \theta = 1$

7. $\cos(\theta - \frac{1}{4}\pi) = \cos(4\theta + \frac{1}{4}\pi)$ **8.** $\tan 2\theta = \cot\theta$

Find the solutions, from 0 to π inclusive, of the following equations.

9. $\cos 3\theta = \cos 7\theta$ **10.** $\tan 3\theta = \cot 2\theta$

11. $\sin 7\theta = \sin 2\theta$ **12.** $\sec 6\theta = \sec 5\theta$

Solve the following equations giving values from $-180°$ to $+180°$.

13. $\cos(2\theta + 60°) = \cos\theta$ **14.** $\tan 3\theta = \tan(\theta - 50°)$

15. $\sin\theta = \cos(2\theta + 60°)$ **16.** $\cos(30° - \theta) = \cos 2\theta$

17. $\tan(2\theta + 60°) = \tan 4\theta$ **18.** $\sin 3\theta = \sin(\theta + 80°)$

THE INVERSE TRIGONOMETRIC FUNCTIONS

Consider the function given by $f: x \rightarrow \sin x$ for $x \in \mathbb{R}$

The inverse mapping is given by $\sin x \rightarrow x$ but this is not a function because one value of $\sin x$ maps to many values of x, i.e. $f(x) = \sin x$ does not have an inverse function for the domain $x \in \mathbb{R}$

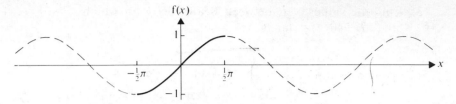

However, if we now consider the function $f: x \to \sin x$ for $-\frac{1}{2}\pi \leqslant x \leqslant \frac{1}{2}\pi$ then the reverse mapping, $\sin x \to x$, is such that one value of $\sin x$ maps to only one value of x
Therefore $f: x \to \sin x$ for $-\frac{1}{2}\pi \leqslant x \leqslant \frac{1}{2}\pi$ does have an inverse, i.e. f^{-1} exists.

Now the equation of the graph of f is $y = \sin x$ for $-\frac{1}{2}\pi \leqslant x \leqslant \frac{1}{2}\pi$ and the curve $y = f^{-1}(x)$ is obtained by reflecting $y = \sin x$ in the line $y = x$

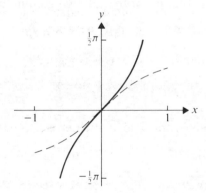

Therefore interchanging x and y gives the equation of this curve, i.e. $\sin y = x$, so $y =$ the angle between $-\frac{1}{2}\pi$ and $\frac{1}{2}\pi$ whose sine is x

Using *arcsin* to mean 'the angle between $-\frac{1}{2}\pi$ and $\frac{1}{2}\pi$ whose sine is', we have $y = \arcsin x$

Thus

if	$f: x \to \sin x$	$-\frac{1}{2}\pi \leqslant x \leqslant \frac{1}{2}\pi$
then	$f^{-1}: x \to \arcsin x$	$-1 \leqslant x \leqslant 1$

It is important to realise that $\arcsin x$ is an angle, and further that this angle is in the interval $[-\frac{1}{2}\pi, \frac{1}{2}\pi]$

Thus, for example, $\arcsin 0.5$ is the angle between $-\frac{1}{2}\pi$ and $\frac{1}{2}\pi$ whose sine is 0.5, i.e. $\arcsin 0.5 = \frac{1}{6}\pi$

Note that the alternative notation $\sin^{-1}$ is sometimes used to mean 'the angle whose sine is', and if this notation is adopted, care is needed not to confuse it with $1/\sin x$

Now consider the function given by $f: x \to \cos x, \ 0 \leqslant x \leqslant \pi$

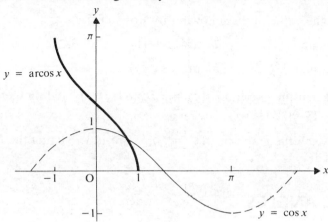

From the diagram, we see that f^{-1} exists and it is denoted by arccos where $\arccos x$ means 'the angle between 0 and π whose cosine is x'

Thus

if	$f: x \to \cos x$	$0 \leqslant x \leqslant \pi$
then	$f^{-1}: x \to \arccos x$	$-1 \leqslant x \leqslant 1$

Note that $\arccos x$ is an angle in the range $0 \leqslant x \leqslant \pi$

Thus, for example, $\quad \arccos(-0.5) = x \quad \Rightarrow \quad x = \frac{2}{3}\pi$

Similarly, if $f: x \to \tan x$ for $-\frac{1}{2}\pi \leqslant x \leqslant \frac{1}{2}\pi$, then f^{-1} exists and is written *arctan* where $\arctan x$ means 'the angle between $-\frac{1}{2}\pi$ and $\frac{1}{2}\pi$ whose tangent is x'

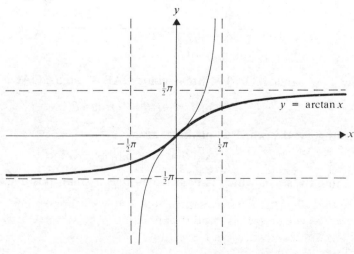

EXERCISE 12c

Find the value of the following in terms of π.

1. $\arctan(\sqrt{3})$ **2.** $\arcsin(-1)$ **3.** $\arccos 0$

4. $\arcsin\left(-\tfrac{1}{2}\sqrt{3}\right)$ **5.** $\arccos\left(-\tfrac{1}{2}\right)$ **6.** $\arctan(-1)$

7. Sketch the graph of $f(x) = \arcsin(x + 1)$, stating the domain and range of $f(x)$.

8. Sketch the graph of $f(x) = \tfrac{1}{4}\pi + \arcsin x$, stating the range of $f(x)$.

SMALL ANGLES

Using a calculator to find the sine and tangent of small angles measured in radians, we find that $\sin\theta$ and $\tan\theta$ are approximately equal to θ. For example, correct to three significant figures
$\sin 0.1 = 0.100$ and $\tan 0.1 = 0.100$

This can be proved as follows.
In the diagram a small angle, θ radians, is subtended by the arc AB at the centre O of a circle of radius r. AB is a chord of the circle and AC is the tangent to the circle at A, cutting OB produced at C.

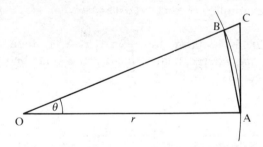

Now   area $\triangle$OAB $<$ area sector OAB $<$ area $\triangle$OAC

i.e. $\tfrac{1}{2}r^2\sin\theta < \tfrac{1}{2}r^2\theta < \tfrac{1}{2}r^2\tan\theta$

Dividing by $\tfrac{1}{2}r^2$, which is positive, gives

$$\sin\theta < \theta < \tan\theta \qquad\qquad [1]$$

Dividing by $\sin\theta$, which is positive as θ is a small positive angle, gives

$$1 < \frac{\theta}{\sin\theta} < \sec\theta$$

Now as $\theta \to 0$, $\sec \theta \to 1$ so $\dfrac{\theta}{\sin \theta}$ lies between 1 and a number that approaches 1 as θ gets smaller,

i.e.

$$\lim_{\theta \to 0} \left(\frac{\theta}{\sin \theta} \right) = 1$$

Similarly, dividing [1] by $\tan \theta$, we can show that

$$\lim_{\theta \to 0} \left(\frac{\theta}{\tan \theta} \right) = 1$$

These limiting values verify that, for small positive values of θ,

$$\sin \theta \approx \theta \quad \text{and} \quad \tan \theta \approx \theta$$

To find an approximate value for $\cos \theta$ when θ is small, we use the identity $\qquad \cos \theta \equiv 1 - 2 \sin^2 \tfrac{1}{2}\theta$

But if $\tfrac{1}{2}\theta$ is small, $\sin \tfrac{1}{2}\theta \approx \tfrac{1}{2}\theta$, so $\cos \theta \approx 1 - 2(\tfrac{1}{2}\theta)^2$

i.e. $\qquad\qquad\qquad\qquad\qquad \cos \theta \approx 1 - \tfrac{1}{2}\theta^2$

Collecting these results, we have

when θ is small and measured in radians,

$\sin \theta \approx \theta$, $\tan \theta \approx \theta$ and $\cos \theta \approx 1 - \tfrac{1}{2}\theta^2$

These approximations are correct to 3 s.f. for angles in the range $-0.105 \text{ rad} < \theta < 0.105 \text{ rad}$, i.e. $-6° < \theta < 6°$

Example 12d

Find, as a rational function of θ, an approximation for the expression $\dfrac{\sin 3\theta}{1 + \cos 2\theta}$ when 3θ is small.

When 3θ is small, $\sin 3\theta \approx 3\theta$

and 2θ is also small, so $\cos 2\theta \approx 1 - \dfrac{(2\theta)^2}{2}$

$\therefore \qquad\qquad\qquad \dfrac{\sin 3\theta}{1 + \cos 2\theta} \approx \dfrac{3\theta}{2(1 - \theta^2)}$

EXERCISE 12d

1. If θ is small enough to regard 4θ as small, find approximations for the following expressions.

   (a) $\dfrac{2\theta}{\sin 4\theta}$ (b) $\dfrac{\theta \sin \theta}{\cos 2\theta}$ (c) $\sin \frac{1}{2}\theta \sec \theta$

   (d) $\dfrac{\theta \tan \theta}{1 - \cos \theta}$ (e) $\dfrac{2 \sin \frac{1}{2}\theta}{\theta}$ (f) $\dfrac{\sin \theta \tan \theta}{\theta^2}$

2. If θ is small enough for θ^2 to be neglected (i.e. $\theta^2 \approx 0$), show that

   (a) $2 \cos \left(\frac{1}{3}\pi + \theta\right) \approx 1 - \theta\sqrt{3}$ (b) $4 \sin \left(\frac{1}{4}\pi - \theta\right) \approx 2\sqrt{2}(1 - \theta)$

3. Find the limiting value as $\theta \to 0$ of the following expressions.

   (a) $\dfrac{\sin \theta}{2\theta}$ (b) $\dfrac{\tan 2\theta}{\sin 3\theta}$

   (c) $\sin \left(\frac{1}{3}\pi + \theta\right)$ (d) $\tan \left(\frac{1}{4}\pi + \theta\right)$

SOLVING TRIGONOMETRIC EQUATIONS

Equations of various forms have appeared at intervals throughout this book, usually when some new mathematics has provided the means to solve a different form of equation. In these situations, the method of solution was usually obvious. However, equations often arise from mathematical models of practical situations and may not be connected obviously to the mathematical knowledge needed to solve them.

Successful solutions of trig equations depend on three factors: recognising and knowing the trig identities, correctly classifying the equation so that the first attempt at solution is likely to be successful and, finally, experience.

There are two general approaches to trig equations. First see whether the equation can be factorised, either in its given form or by applying an appropriate identity. If this is not possible (and often it is not) try to reduce the equation to a form which involves only one variable, i.e. one trig ratio of one angle.

In this section we give most of the common categories of trig equation followed by an appropriate method of solution. This list is neither exhaustive nor infallible, but it covers most forms of equation met at this level.

A. Equations Containing One Angle Only

Form of Equation	Method
1. $a \cos \theta + b \sin \theta = 0$	Divide by $\cos \theta$, provided that $\cos \theta \neq 0$
2. $a \cos \theta + b \sin \theta = c$	Write LHS as $R \cos (\theta + \alpha)$ or use the 'little t' identities.
3. $a \cos^2 \theta + b \sin \theta = c$ $a \sin^2 \theta + b \cos \theta = c$ $a \tan^2 \theta + b \sec \theta = c$	Use the Pythagorean identities to express in terms of one ratio only.
4. $a \cos \theta + b \tan \theta = 0$ $a \sin \theta + b \tan \theta = 0$	Multiply by $\cos \theta$

Note that any of the equations in Section A can be solved by using the 'little t' identities. However this is often not the simplest method and sometimes leads to a polynomial in t whose roots are far from obvious.

B. Equations Containing Multiples of One Angle

Form of Equation	Method
1. $a \cos \theta + b \cos 2\theta = c$ $a \sin \theta + b \cos 2\theta = c$	Use the double angle formulae to reduce to Section A type.
2. $\cos \theta = \sin a\theta$	Express $\sin a\theta$ as $\cos (\frac{1}{2}\pi - a\theta)$ and use $\cos A = \cos B$
3. $\cos a\theta + \cos b\theta + \cos c\theta = 0$ $\sin a\theta + \sin b\theta$ $\qquad + \sin c\theta + \sin d\theta = 0$	Use the factor formulae on a pair(s) of terms to give a common factor.

C. Equations Containing Different Angles

Form of Equation	Method
1. $\cos \theta + \sin (\theta - \alpha) = 0$ etc.	Reduce to $\cos A = \cos B$ or to $\tan A = \tan B$ and quote solution, or use factor formulae to factorise.
2. $\cos \theta \sin (\theta - \alpha) = c$	Use compound angle formulae to reduce to a section A type equation.

Note that equations of the form given in B and C can often be reduced to equations involving one angle only, by using double and compound angle formulae. However this rarely leads to an easy solution so should only be tried as a last resort.

It is important to realise that a method given in this list does not represent the only way of solving a particular equation, nor does it always lead to the quickest solution. Sometimes an equation can be simplified quickly when part of it is recognised as part of a trig identity. Sometimes it may be necessary to classify each side of an equation independently. This is illustrated in the next worked example.

Example 12e _____

Find the general solution of the equation

$$\cos 2\theta \sin \theta + \sin 2\theta \cos \theta = \cos 4\theta$$

$$\cos 2\theta \sin \theta + \sin 2\theta \cos \theta = \cos 4\theta$$

The LHS is recognised as the expansion of $\sin(2\theta + \theta)$, i.e. $\sin 3\theta$, giving

$$\sin 3\theta = \cos 4\theta$$

$\sin 3\theta$ may be expressed as $\cos(\tfrac{1}{2}\pi - 3\theta)$, giving

$$\cos(\tfrac{1}{2}\pi - 3\theta) = \cos 4\theta$$

Using the solution to $\cos A = \cos B$ gives

$$\tfrac{1}{2}\pi - 3\theta = 2n\pi \pm 4\theta$$

$\therefore \quad \theta = \tfrac{1}{14}\pi(1 - 4n) \quad$ or $\quad \tfrac{1}{2}\pi(4n - 1)$

_____

The equation in the worked example illustrates that one approach does not always lead directly to the solution. The situation should be reappraised at each step.

EXERCISE 12e

Find the general solution in radians of each equation.

1. $\sin 2x \cos x + \cos 2x \sin x = 1$
2. $\cos 3x = \sin x$
3. $2 \sin^2 x + \cos x = 1$
4. $5 \cos x + 12 \sin x = 13$
5. $2 \sin x + \sin 2x = 0$
6. $\cos x + \cos 2x + \cos 3x = 0$
7. $\cos^2 x + 2 \sin^2 x = 2$
8. $4 \sin x - 5 = 3 \cos x$
9. $\sin^2 x = 2 \cos x + 1$
10. $\cos x = 2 \tan x$

Solve each equation for $0 \leqslant x < 360°$ giving answers correct to 1 d.p. where necessary.

11. $\cos 2x + \sin x = 1$
12. $2 \sin x - \cos x = 1$
13. $\sin x + \tan x = 0$
14. $\cos 2x + 2 \sin x = 0$
15. $\tan x + 2 \sec^2 x = 3$
16. $2 \cos x + \cos 2x = 2$

17. $\cos (x + 30°) + \sin x = 0$
18. $\sin x + \sin 2x + \sin 3x + \sin 4x = 0$
19. $\cos x \cos 2x - \sin x \sin 2x = \frac{1}{2}$

The next exercise contains questions that may require any of the identities from Chapters 5 and 8 as well as those from this one.

MIXED EXERCISE 12

1. If A is acute and $\sin A = \frac{1}{2}$ and if B is obtuse and $\sin B = \frac{1}{3}$, find in surd form the value of
   (a) $\tan (A + B)$ (b) $\cos (A - B)$

2. Find the values of θ in the range $-90° \leqslant \theta \leqslant 90°$ for which $\tan^2 \theta + \sec \theta = 11$

3. Find the general solution of the equation $\sin 3x - \sin x = \cos 2x$

4. By writing $\frac{3}{10}\pi$ as $\frac{1}{2}\pi - \frac{1}{5}\pi$, show that $\cos \frac{3}{10}\pi = \sin \frac{1}{5}\pi$

5. Find the general solution of the equation $\tan 2\theta + 2 \sin \theta = 0$

6. Prove that if $\sec \alpha = \cos \beta + \sin \beta$ then $\tan^2 \alpha = \sin 2\beta$

7. Find the values of θ from 0 to 180° for which
   $\sin 5\theta - \sin \theta = \cos 3\theta$

8. Prove the identity $\dfrac{1}{\sin x} + \dfrac{1}{\tan x} \equiv \cot \frac{1}{2}x$ and hence find in
   surd form the value of $\tan \frac{1}{8}\pi$

9. Find the Cartesian equation of the curve whose parametric
   equations are $x = \cos 2\theta$ and $y = \sin \theta$

10. Solve the simultaneous equations $\cos x - \cos y = 1$ and
    $\sin^2 x - \sin^2 y = 0$ for values of x and of y from 0 to 360°

11. Prove the identity $\cos 3\theta \equiv 4\cos^3\theta - 3\cos \theta$

12. Find the general solution of the equation $\tan x \sec x = \sqrt{2}$

13. Express $4\sin \theta - 3\cos \theta$ in the form $r\sin(\theta - \alpha)$. Hence find
    the maximum and minimum values of $\dfrac{7}{4\sin \theta - 3\cos \theta + 2}$
    State the greatest and least values.

14. Express $\sin 2\theta - \cos 2\theta$ in the form $r\sin(2\theta - \alpha)$. Hence find
    the smallest positive value of θ for which $\sin 2\theta - \cos 2\theta$ has a
    maximum value.

15. Find the values of x from 0 to 360° for which
    $4\cos x - 6\sin x = 5$

16. Find the Cartesian equation of the curve whose parametric
    equations are $x = 2\sin(\theta + \frac{1}{3}\pi)$ and $y = \cos(\theta - \frac{1}{6}\pi)$

17. Find the values of θ from 0 to 360° for which $\sin 2\theta = \sin 3\theta$

18. Find the Cartesian equation of the curve given parametrically by
    $x = a\cos 2t$ and $y = a\sin t$

19. Find the values of θ from 0 to 360° for which
    $\cos \theta = 2\sin(\theta - 30°)$.

20. Use the substitution $t = \tan \theta$ to show that
    $$\frac{1}{3 + 5\sin 2\theta} = \frac{1 + t^2}{(3t + 1)(t + 3)}$$

21. Find the general solution of the equation $\sin 2x - \sin 4x = 0$

22. Express $\sin \theta$ in the form $\cos \alpha$ giving α in terms of θ. Hence solve the equation $\cos 3\theta = \sin \theta$ for values of θ between 0 and 360°

23. Find the limiting value as $\theta \to 0$ of $\dfrac{\frac{1}{2}(1 - \sin \theta - \cos \theta)}{2\theta}$

24. Express $\cos x + \sin x$ in the form $r \cos (x - \alpha)$. Hence find the smallest positive value of x for which $\dfrac{1}{(\cos x + \sin x)}$ has a minimum value.

25. Find all the values of x in the range $0 \leqslant x \leqslant 2\pi$ for which $\cos 3x = \cos 2x$

26. Find all the values of x between 0 and 180° for which $\sin (x - 30°) = \cos 4x$

27. Show that $x = \frac{1}{14}\pi$ is a solution of the equation $\sin 3x = \cos 4x$

28. Find the general solution of the equation $\tan 5\theta = \cot 2\theta$

29. Express $3 \cos x - 4 \sin x$ in the form $r \cos (x + \alpha)$. Hence express $4 + \dfrac{10}{3 \cos x - 4 \sin x}$ in the form $4 + k \sec (x + \alpha)$ and sketch the graph of $y = 4 + \dfrac{10}{3 \cos x - 4 \sin x}$

CHAPTER 13

DIFFERENTIATION OF TRIGONOMETRIC FUNCTIONS

THE DERIVATIVE OF sin x

The derivative of any new function, $f(x)$, can be found by differentiating from first principles. This involves finding the limit of the gradient of the chord joining any two neighbouring points on the curve $y = f(x)$

Consider the curve $y = \sin x$

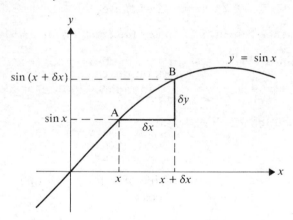

A is any point (x, y) on the curve $y = \sin x$ and at the nearby point B the x-coordinate is $x + \delta x$

At A, $y = \sin x$ and at B, $y = \sin (x + \delta x)$

Now
$$\frac{\delta y}{\delta x} = \frac{\sin (x + \delta x) - \sin x}{\delta x}$$

$$= \frac{2 \cos (x + \tfrac{1}{2}\delta x) \sin (\tfrac{1}{2}\delta x)}{\delta x} \qquad \text{(using a factor formula)}$$

$$= \frac{\cos (x + \tfrac{1}{2}\delta x) \sin (\tfrac{1}{2}\delta x)}{\tfrac{1}{2}\delta x}$$

Then, *provided that x is measured in radians,*

$$\text{as} \quad \delta x \to 0, \quad \cos(x + \tfrac{1}{2}\delta x) \to \cos x \quad \text{and} \quad \frac{\sin(\tfrac{1}{2}\delta x)}{\tfrac{1}{2}\delta x} \to 1$$

$$\therefore \qquad \frac{dy}{dx} = \lim_{\delta x \to 0} \frac{\delta y}{\delta x} = \lim_{\delta x \to 0} \frac{\cos(x + \tfrac{1}{2}\delta x)\sin(\tfrac{1}{2}\delta x)}{\tfrac{1}{2}\delta x} = \cos x$$

i.e.
$$\text{if} \quad y = \sin x \quad \text{then} \quad \frac{dy}{dx} = \cos x$$

In a similar way it can be shown that

$$\text{if} \quad y = \cos x \quad \text{then} \quad \frac{dy}{dx} = -\sin x$$

These two results can be quoted whenever they are needed.

It is important to realise that they are valid only when x is measured in radians.

Throughout all subsequent work on the calculus of trig functions in this book, the angle is measured in radians unless it is stated otherwise.

Examples 13a

1. Find the smallest positive value of x for which there is a stationary value of the function $x + 2\cos x$

$$f(x) = x + 2\cos x \quad \Rightarrow \quad f'(x) = 1 - 2\sin x$$
$$\text{where} \quad f'(x) = \frac{d}{dx}f(x)$$

For stationary values $f'(x) = 0$

i.e. $\qquad 1 - 2\sin x = 0 \quad \Rightarrow \quad \sin x = \tfrac{1}{2}$

The smallest positive angle with a sine of $\tfrac{1}{2}$ is $\tfrac{1}{6}\pi$

NOTE that the answer *must* be given in radians because the rule used to differentiate $\cos x$ is valid only for an angle in radians.

2. Find the smallest positive value of θ for which the curve
   $y = 2\theta - 3\sin\theta$ has a gradient of $\frac{1}{2}$

$$y = 2\theta - 3\sin\theta \quad \text{gives} \quad \frac{dy}{d\theta} = 2 - 3\cos\theta$$

when $\dfrac{dy}{d\theta} = \frac{1}{2}$ $\qquad 2 - 3\cos\theta = \frac{1}{2}$

$$3\cos\theta = \frac{3}{2}$$

$$\cos\theta = \frac{1}{2}$$

The smallest positive value of θ for which $\cos\theta = \frac{1}{2}$, is $\frac{1}{3}\pi$

EXERCISE 13a

1. By differentiating from first principles, show that
   $$\frac{d}{dx}(\cos x) = -\sin x$$

2. Write down the derivative of each of the following expressions.
   (a) $\sin x - \cos x$ (b) $\sin\theta + 4$ (c) $3\cos\theta$
   (d) $5\sin\theta - 6$ (e) $2\cos\theta + 3\sin\theta$ (f) $4\sin x - 5 - 6\cos x$

3. Find the gradient of each curve at the point whose x-coordinate is
   given.
   (a) $y = \cos x;\ \frac{1}{2}\pi$ (b) $y = \sin x;\ 0$
   (c) $y = \cos x + \sin x;\ \pi$ (d) $y = x - \sin x;\ \frac{1}{2}\pi$
   (e) $y = 2\sin x - x^2;\ -\pi$ (f) $y = -4\cos x;\ \frac{1}{2}\pi$

4. For each of the following curves find the smallest positive value of
   θ at which the gradient of the curve has the given value.
   (a) $y = 2\cos\theta;\ -1$ (b) $y = \theta + \cos\theta;\ \frac{1}{2}$
   (c) $y = \sin\theta + \cos\theta;\ 0$ (d) $y = \sin\theta + 2\theta;\ 1$

5. Considering only positive values of x, locate the first two turning
   points on each of the following curves and determine whether they
   are maximum or minimum points.
   (a) $2\sin x - x$ (b) $x + 2\cos x$
   In each case illustrate your solution by a sketch.

6. Find the equation of the tangent to the curve $y = \cos\theta + 3\sin\theta$ at the point where $\theta = \frac{1}{2}\pi$

7. Find the equation of the normal to the curve $y = x^2 + \cos x$ at the point where $x = \pi$

8. Find the coordinates of a point on the curve $y = \sin x + \cos x$ at which the tangent is parallel to the line $y = x$

COMPOUND FUNCTIONS

The variety of functions which can be handled when they occur in products, quotients and functions of a function, now includes the sine and cosine ratios.

Differentiation of $\sin f(x)$

If $y = \sin f(x)$ then using $u = f(x)$ gives $y = \sin u$

Then $\qquad \dfrac{dy}{dx} = \dfrac{dy}{du} \times \dfrac{du}{dx} \qquad \Rightarrow \qquad \dfrac{dy}{dx} = \cos u \, \dfrac{du}{dx}$

i.e. $\qquad \dfrac{d}{dx}\{\sin f(x)\} = f'(x)\cos f(x)$

Similarly $\qquad \dfrac{d}{dx}\{\cos f(x)\} = -f'(x)\sin f(x)$

e.g. $\qquad \dfrac{d}{dx}\sin e^x = e^x \cos e^x \quad$ and $\quad \dfrac{d}{dx}\cos \ln x = -\dfrac{1}{x}\sin \ln x$

In particular $\qquad \dfrac{d}{dx}(\sin ax) = a\cos ax$

and $\qquad \dfrac{d}{dx}(\cos ax) = -a\sin ax$

These results are quotable.

Examples 13b

1. Differentiate $\cos\left(\frac{1}{6}\pi - 3x\right)$ with respect to x

   Using $\cos\left(\frac{1}{6}\pi - 3x\right) = \cos f(x)$ $\Rightarrow$ $f'(x) = -3$

   $$\frac{d}{dx}\left\{\cos\left(\frac{1}{6}\pi - 3x\right)\right\} = -(-3)\sin\left(\frac{1}{6}\pi - 3x\right)$$

   $$= 3\sin\left(\frac{1}{6}\pi - 3x\right)$$

---

2. Find the derivative of $\dfrac{e^x}{\sin x}$

   $$y = \frac{e^x}{\sin x} = \frac{u}{v}$$

   where $u = e^x$ and $v = \sin x$

   $\Rightarrow$ $\dfrac{du}{dx} = e^x$ and $\dfrac{dv}{dx} = \cos x$

   $$\frac{dy}{dx} = \left(v\frac{du}{dx} - u\frac{dv}{dx}\right)\Big/v^2 = \frac{e^x\sin x - e^x\cos x}{\sin^2 x}$$

   $\therefore$ $$\frac{d}{dx}\left(\frac{e^x}{\sin x}\right) = \frac{e^x}{\sin^2 x}(\sin x - \cos x)$$

---

3. Find $\dfrac{dy}{d\theta}$ if $y = \cos^3\theta$

   $\therefore$ $$y = \cos^3\theta = [\cos\theta]^3$$

   $$y = u^3 \quad \text{where} \quad u = \cos\theta$$

   $$\frac{dy}{d\theta} = \frac{dy}{du}\frac{du}{d\theta} = (3u^2)(-\sin\theta) = 3(\cos\theta)^2(-\sin\theta)$$

   $\therefore$ $$y = \cos^3\theta \quad \Rightarrow \quad \frac{dy}{d\theta} = -3\cos^2\theta\sin\theta$$

This is one example of a general rule, i.e.

and

$$\text{if } y = \cos^n x \quad \text{then} \quad \frac{dy}{dx} = -n \cos^{n-1} x \sin x$$

$$\text{if } y = \sin^n x \quad \text{then} \quad \frac{dy}{dx} = n \sin^{n-1} x \cos x$$

EXERCISE 13b

Differentiate each of the following functions with respect to x

1. $\sin 4x$

2. $\cos(\pi - 2x)$

3. $\sin(\frac{1}{2}x + \pi)$

4. $\dfrac{\sin x}{x}$

5. $\dfrac{\cos x}{e^x}$

6. $\sqrt{\sin x}$

7. $\sin^2 x$

8. $\sin x \cos x$

9. $e^{\sin x}$

10. $\ln(\cos x)$

11. $e^x \cos x$

12. $x^2 \sin x$

13. $\sin x^2$

14. $e^{\cos x}$

15. $\ln \sin^3 x$

16. $\sec x$, i.e. $\dfrac{1}{\cos x}$

17. $\tan x$, i.e. $\dfrac{\sin x}{\cos x}$

18. $\operatorname{cosec} x$

19. $\cot x$

Using the answers to Questions 9 to 12, we can now make a complete list of the derivatives of the basic trig functions:

Function	Derivative
$\sin x$	$\cos x$
$\cos x$	$-\sin x$
$\tan x$	$\sec^2 x$
$\cot x$	$-\operatorname{cosec}^2 x$
$\sec x$	$\sec x \tan x$
$\operatorname{cosec} x$	$-\operatorname{cosec} x \cot x$

DIFFERENTIATING INVERSE TRIG FUNCTIONS

In the previous section the derivative of a log function was found, without differentiating from first principles, by treating it as the inverse of an exponential function. A similar approach is used to find the derivative of each of the inverse trig functions.

Differentiation of arcsin x

If $y = \arcsin x$, the inverse relationship is $x = \sin y$ and this can be differentiated.

We can also use the property proved on page 128

that
$$\frac{dy}{dx} = 1 \bigg/ \frac{dx}{dy}$$

Hence if $\quad x = \sin y \quad$ then $\quad \dfrac{dx}{dy} = \cos y \quad \Rightarrow \quad \dfrac{dy}{dx} = \dfrac{1}{\cos y}$

Now $\cos^2 y = 1 - \sin^2 y = 1 - x^2$

Therefore
$$\frac{dy}{dx} = \frac{1}{\sqrt{(1 - x^2)}}$$

i.e.
$$\frac{d}{dx}(\arcsin x) = \frac{1}{\sqrt{(1 - x^2)}}$$

Remember that $y = \arcsin x$ exists only within the range $-\frac{1}{2}\pi \leqslant y \leqslant \frac{1}{2}\pi$. For these values of y, $\cos y \geqslant 0$, therefore $\cos y = \sqrt{(1 - x^2)}$ and not $-\sqrt{(1 - x^2)}$

Similarly
$$\frac{d}{dx}(\arccos x) = -\frac{1}{\sqrt{(1 - x^2)}}$$

and
$$\frac{d}{dx}(\arctan x) = \frac{1}{1 + x^2}$$

It is left to the reader to prove these results in the next exercise.

Example 13c

Differentiate with respect to x,

(a) $\arcsin e^x$ (b) $x \arctan x$

(a) $\arcsin e^x$ is 'an inverse trig function' of 'an exponential function' so the chain rule can be used.

$$\frac{d}{dx}(\arcsin e^x) = \frac{1}{\sqrt{\{1-(e^x)^2\}}} e^x$$

$$= \frac{e^x}{\sqrt{(1-e^{2x})}}$$

Note that some readers may prefer to use the substitution method in full, i.e. $u = e^x$ and $y = \arcsin u$.

(b) $\frac{d}{dx}(x \arctan x) = \arctan x + x\frac{d}{dx}\arctan x$

$$= \arctan x + \frac{x}{1+x^2}$$

EXERCISE 13c

1. Show that $\dfrac{d}{dx}(\arccos x) = -\dfrac{1}{\sqrt{(1-x^2)}}$

2. Show that $\dfrac{d}{dx}(\arctan x) = \dfrac{1}{1+x^2}$

Differentiate each function with respect to x

3. $x^2 \arcsin x$ 4. $(\arctan x)^2$ 5. $\arcsin 3x$

6. $e^{\arctan x}$ 7. $\ln(\arcsin x)$ 8. $\arccos x^3$

MIXED EXERCISE 13

This exercise contains a wide variety of functions. Consider carefully what method to use in each case and do not forget to check first whether a given function has a standard derivative.

Find the derivative of each function in Questions 1 to 22.

1. (a) $-\sin 4\theta$ (b) $\theta - \cos \theta$ (c) $\sin^3\theta + \sin 3\theta$

2. (a) $x^3 + e^x$ (b) $e^{(2x+3)}$ (c) $e^x \sin x$

3. (a) $\ln \frac{1}{3}x^{-3}$ (b) $\ln 2/x^2$ (c) $\ln \sqrt{x}/4$

4. (a) $3 \sin x - e^{-x}$ (b) $\ln x^{1/2} - \frac{1}{2} \cos x$
    (c) $x^4 + 4e^x - \ln 4x$ (d) $\frac{1}{2}e^{-x} + x^{-1/2} - \ln \frac{1}{2}x$

5. $(x+1)\ln x$ **6.** $\sin^2 3x$ **7.** $(4x-1)^{2/3}$

8. $(3\sqrt{x} - 2x)^2$ **9.** $\dfrac{(x^4-1)}{(x+1)^3}$ **10.** $\sin x \cos^3 x$

11. $\ln \cot x$ **12.** $x^2 \sin x$ **13.** $\dfrac{e^x}{x-1}$

14. $\dfrac{1 + \sin x}{1 - \sin x}$ **15.** $x^2\sqrt{(x-1)}$ **16.** $(1-x^2)(1-x)^2$

17. $\ln \sqrt{\dfrac{(x+3)^3}{(x^2+2)}}$ **18.** $e^{\arcsin x}$ **19.** $(\arccos x)^3$

20. $\arctan e^x$ **21.** xa^x **22.** $\ln (x^2 + 1)$

23. Find the value(s) of x for which the following functions have stationary values.
    (a) $3x - e^x$ (b) $x^2 - 2\ln x$ (c) $\ln 1/x + 4x$

In each Question from 24 to 27, find
(a) the gradient of the curve at the given point,
(b) the equation of the tangent to the curve at that point,
(c) the equation of the normal to the curve at that point.

24. $y = \sin x - \cos x; \; x = \frac{1}{2}\pi$ **25.** $y = x + e^x; \; x = 1$
26. $y = 1 + x + \sin x; \; x = 0$ **27.** $y = 3 - x^2 + \ln x; \; x = 1$

28. Considering only positive values of θ, locate the first two turning points, if there are two, on each of the following curves and determine whether they are maximum or minimum points.

(a) $y = 1 - \sin x$ (b) $y = \frac{1}{2}x + \cos x$ (c) $y = e^x - 3x$

29. Find the coordinates of a point on the curve where the tangent is parallel to the given line.

(a) $y = 3x - 2\cos x$; $y = 4x$ (b) $y = 2\ln x - x$; $y = x$

CHAPTER 14

DIFFERENTIATING IMPLICIT AND PARAMETRIC FUNCTIONS

IMPLICIT FUNCTIONS

All the differentiation carried out so far has involved equations that could be expressed in the form $y = f(x)$

However the equations of some curves, for example $x^2 - y^2 + y = 1$, cannot easily be written in this way, as it is too difficult to isolate y. A relationship of this type, where y is not given explicitly as a function of x, is called an *implicit function*, i.e. it is *implied* in the equation that $y = f(x)$.

TO DIFFERENTIATE AN IMPLICIT FUNCTION

The method we use is to differentiate, term by term, with respect to x, but first we need to know how to differentiate terms like y^2 with respect to x.

If $\qquad g(y) = y^2$ and $y = f(x)$

then $\qquad g(y) = \{f(x)\}^2$ which is a function of a function

Using the mental substitution $u = f(x)$ we have

$$\frac{d}{dx}\{f(x)\}^2 = 2\{f(x)\}\left(\frac{d}{dx}f(x)\right) = 2y\left(\frac{dy}{dx}\right) = \left(\frac{d}{dy}g(y)\right)\left(\frac{dy}{dx}\right)$$

In general, $\qquad \boxed{\dfrac{d}{dx}g(y) = \left(\dfrac{d}{dy}g(y)\right)\left(\dfrac{dy}{dx}\right)}$

e.g. $\dfrac{d}{dx}y^3 = 3y^2\dfrac{dy}{dx}$ and $\dfrac{d}{dx}e^y = e^y\dfrac{dy}{dx}$

We can now differentiate, term by term with respect to x, the example considered above, i.e.

if
$$x^2 - y^2 + y = 1$$

then
$$\frac{d}{dx}(x^2) - \frac{d}{dx}(y^2) + \frac{dy}{dx} = \frac{d}{dx}(1)$$

$\Rightarrow$
$$2x - 2y\frac{dy}{dx} + \frac{dy}{dx} = 0$$

Hence
$$2x = \frac{dy}{dx}(2y - 1) \qquad \Rightarrow \qquad \frac{dy}{dx} = \frac{2x}{(2y - 1)}$$

Examples 14a

1. Differentiate each equation with respect to x and hence find $\dfrac{dy}{dx}$ in terms of x and y.

   (a) $x^3 + xy^2 - y^3 = 5$ (b) $y = xe^y$

(a) If $x^3 + xy^2 - y^3 = 5$ then, differentiating term by term,

$$\frac{d}{dx}(x^3) + \frac{d}{dx}(xy^2) - \frac{d}{dx}(y^3) = \frac{d}{dx}(5)$$

The term xy^2 is a product so we differentiate it using the product rule, i.e.

$$\frac{d}{dx}(xy^2) = y^2\frac{d}{dx}(x) + x\frac{d}{dx}(y^2) = y^2 + (x)(2y)\frac{dy}{dx}$$

$\therefore$
$$3x^2 + y^2 + 2xy\frac{dy}{dx} - 3y^2\frac{dy}{dx} = 0$$

Hence
$$\frac{dy}{dx} = \frac{(3x^2 + y^2)}{y(3y - 2x)}$$

(b) If $y = xe^y$ then
$$\frac{dy}{dx} = \frac{d}{dx}(xe^y)$$

$$= e^y\frac{d}{dx}(x) + x\frac{d}{dx}(e^y)$$

$\Rightarrow$
$$\frac{dy}{dx} = e^y + xe^y\frac{dy}{dx}$$

Hence
$$\frac{dy}{dx} = \frac{e^y}{1 - xe^y}$$

2. If $e^x y = \sin x$ show that $\dfrac{d^2y}{dx^2} + 2\dfrac{dy}{dx} + 2y = 0$

In a problem of this type it is tempting to express $e^x y = \sin x$ in the form $y = e^{-x}\sin x$, find $\dfrac{dy}{dx}$ and $\dfrac{d^2y}{dx^2}$ and show that they satisfy the given equation, which is called a *differential equation*. However it is much more direct to differentiate the implicit equation as given.

Differentiating $e^x y = \sin x$ w.r.t. x gives

$$e^x y + e^x \dfrac{dy}{dx} = \cos x$$

Differentiating again w.r.t. x gives

$$\left(e^x y + e^x \dfrac{dy}{dx}\right) + \left(e^x \dfrac{dy}{dx} + e^x \dfrac{d^2y}{dx^2}\right) = -\sin x = -e^x y$$

Hence $e^x \dfrac{d^2y}{dx^2} + 2e^x \dfrac{dy}{dx} + 2e^x y = 0$

There is no finite value of x for which $e^x = 0$ so we can divide the equation by e^x.

i.e. $\dfrac{d^2y}{dx^2} + 2\dfrac{dy}{dx} + 2y = 0$

---

3. Find the equation of the tangent at the point (x_1, y_1) to the curve with equation $x^2 - 2y^2 - 6y = 0$

To find the equation of the tangent we need the gradient of the curve and in this case it must be found by implicit differentiation.

$$x^2 - 2y^2 - 6y = 0 \quad \Rightarrow \quad 2x - 2\left(2y\dfrac{dy}{dx}\right) - 6\dfrac{dy}{dx} = 0$$

$$\Rightarrow \quad \dfrac{dy}{dx} = \dfrac{x}{(3 + 2y)}$$

$\therefore$ the gradient of the tangent at the point (x_1, y_1) is $\dfrac{x_1}{(3 + 2y_1)}$ and the

equation of the tangent is $y - y_1 = \dfrac{x_1}{(3 + 2y_1)}(x - x_1)$ which

simplifies to $xx_1 - 2yy_1 - 3(y + y_1) = x_1^2 - 2y_1^2 - 6y_1$

Now (x_1, y_1) is on the given curve, so $x_1^2 - 2y_1^2 - 6y_1 = 0$, and the equation of the tangent becomes

$$xx_1 - 2yy_1 - 3(y + y_1) = 0$$

Note that in the last example the equation of the curve can be converted into the equation of the tangent by changing x^2 into xx_1, y^2 into yy_1 and y into $\frac{1}{2}(y + y_1)$.

In fact, for any curve whose equation is of degree two, the equation of the tangent at (x_1, y_1) can be written down directly by making the replacements listed above, together with two more, i.e.

$$x \;\rightarrow\; \tfrac{1}{2}(x + x_1) \quad \text{and} \quad xy \;\rightarrow\; \tfrac{1}{2}(xy_1 + x_1 y)$$

This property can be applied to advantage when the numerical values of the coordinates of the point of contact are known,

e.g. the equation of the tangent at the point $(1, -1)$ to the curve

$$3x^2 - 7y^2 + 4xy - 8x = 0$$

can be written down as

$$3x(1) - 7y(-1) + 2\{x(-1) + (1)y\} - 4(x + 1) = 0$$

i.e. $\qquad\qquad\qquad\qquad 9y - 3x = 4$

Question 18 in the following exercise gives the reader the opportunity to justify using these mechanical replacements in the equation of a curve, to give the equation of a tangent.

Note that, although this method allows the equation of a tangent to be *written down*, its use is not suitable when the *derivation* of the equation is required (except as a check on the result).

EXERCISE 14a

Differentiate the following equations with respect to x

1. $x^2 + y^2 = 4$

2. $x^2 + xy + y^2 = 0$

3. $x(x + y) = y^2$

4. $\dfrac{1}{x} + \dfrac{1}{y} = e^y$

5. $\dfrac{1}{x^2} + \dfrac{1}{y^2} = \dfrac{1}{4}$

6. $\dfrac{x^2}{4} - \dfrac{y^2}{9} = 1$

7. $\sin x + \sin y = 1$

8. $\sin x \cos y = 2$

9. $xe^y = x + 1$

10. $\sqrt{\{(1 + y)(1 + x)\}} = x$

11. Find $\dfrac{dy}{dx}$ as a function of x if $y^2 = 2x + 1$

12. Find $\dfrac{d^2y}{dx^2}$ as a function of x if $\sin y + \cos y = x$

13. Find the gradient of $x^2 + y^2 = 9$ at the points where $x = 1$

14. If $y \cos x = e^x$ show that $\dfrac{d^2y}{dx^2} - 2 \tan x \dfrac{dy}{dx} - 2y = 0$

15. Find the equation of the tangent to
    (a) $x^2 - 3y^2 = 4y$ (b) $x^2 + xy + y^2 = 3$
    at the point (x_1, y_1).

16. Show that the equation of the tangent to $x^2 + xy + y = 0$ at the point (x_1, y_1) is
$$x(2x_1 + y_1) + y(x_1 + 1) + y_1 = 0$$

17. Write down the equation of the tangent at $(1, \frac{1}{3})$ to the curve whose equation is
$$2x^2 + 3y^2 - 3x + 2y = 0$$

18. Show that the equation of the tangent at (x_1, y_1) to the curve $ax^2 + by^2 + cxy + dx = 0$ is
$$axx_1 + byy_1 + \tfrac{1}{2}c(xy_1 + yx_1) + \tfrac{1}{2}d(x + x_1) = 0$$

19. Given that $\sin y = 2 \sin x$ show that $\left(\dfrac{dy}{dx}\right)^2 = 1 + 3 \sec^2 y$. By differentiating this equation with respect to x show that

$$\frac{d^2y}{dx^2} = 3 \sec^2 y \tan y$$

and hence that $\cot y \dfrac{d^2y}{dx^2} - \left(\dfrac{dy}{dx}\right)^2 + 1 = 0$

LOGARITHMIC DIFFERENTIATION

The advantage of simplifying a logarithmic expression before attempting to differentiate it has already been noted.

We are now going to examine some equations which are awkward to differentiate as they stand but which are much easier to deal with if we first take logs of both sides of the equation. The types of equation where this method is particularly important are

1) those in which the variable is an index

2) complicated functions involving fractions, roots, products, etc.

The process used is called *logarithmic differentiation*.

Examples 14b

1. Differentiate x^x with respect to x

$$y = x^x$$

$$\ln y = x \ln x$$

Differentiating both sides w.r.t. x gives

$$\frac{1}{y}\frac{dy}{dx} = x\frac{1}{x} + \ln x$$

$$\Rightarrow \qquad \frac{dy}{dx} = y(1 + \ln x)$$

Therefore $\qquad \dfrac{d}{dx}(x^x) = x^x(1 + \ln x)$

2. If $y = \dfrac{x\sqrt{(x^2 - 1)}}{x + 2}$ find $\dfrac{dy}{dx}$

$y = \dfrac{x\sqrt{(x^2 - 1)}}{x + 2} \quad \Rightarrow \quad \ln y = \ln x + \tfrac{1}{2}\ln(x^2 - 1) - \ln(x + 2)$

$\therefore \qquad\qquad \dfrac{1}{y}\dfrac{dy}{dx} = \dfrac{1}{x} + \dfrac{1}{2}\left(\dfrac{2x}{x^2 - 1}\right) - \dfrac{1}{x + 2}$

$\Rightarrow \qquad\qquad \dfrac{dy}{dx} = (y)\left(\dfrac{x^3 + 4x^2 - 2}{x(x^2 - 1)(x + 2)}\right)$

$\qquad\qquad\qquad = \left(\dfrac{x\sqrt{(x^2 - 1)}}{x + 2}\right)\left(\dfrac{x^3 + 4x^2 - 2}{x(x^2 - 1)(x + 2)}\right)$

i.e. $\qquad\qquad \dfrac{dy}{dx} = \dfrac{x^3 + 4x^2 - 2}{(x + 2)^2\sqrt{(x^2 - 1)}}$

---

3. Differentiate the equation $x = y^x$ with respect to x

$\qquad\qquad x = y^x \quad \Rightarrow \quad \ln x = x \ln y$

$\therefore \qquad\qquad\qquad \dfrac{1}{x} = \ln y + (x)\left(\dfrac{1}{y}\dfrac{dy}{dx}\right)$

i.e. $\qquad\qquad\qquad x^2\dfrac{dy}{dx} + xy \ln y = y$

Note that in the third example it is not easy to express $\dfrac{dy}{dx}$ as a function of x. This is because it is difficult in the first place to find y in terms of x. So although the usual practice is to give a derived function in terms of x it is not always possible, or sensible, to do so.

Differentiation of a^x where a is a Constant

The basic rule for differentiating an exponential function applies when the base is e but not for any other base. So for a^x, where we need another approach, we use logarithmic differentiation.

Using $\qquad y = a^x$ gives $\ln y = x \ln a$

Differentiating w.r.t. x gives $\qquad \dfrac{1}{y} \dfrac{dy}{dx} = \ln a$

Hence $\qquad\qquad\qquad \dfrac{dy}{dx} = y \ln a = a^x \ln a$

i.e. $\qquad\qquad\qquad \boxed{\dfrac{d}{dx} a^x = a^x \ln a}$

This result is quotable.

EXERCISE 14b

Differentiate each equation with respect to x

1. $x^y = e^x$ **2.** $x^y = (y + 1)$ **3.** $y = (x + x^2)^x$

Find $\dfrac{dy}{dx}$ if

4. $y = \dfrac{x}{(x + 2)(x - 4)}$ **5.** $y = \dfrac{x^2}{(x - 1)(x - 3)}$

6. $y = (1 - x)^5(x^2 + 2)$ **7.** $y = \sqrt{\{(x + 1)(x - 3)^3\}}$

8. $y = \dfrac{x}{(x + 2)^2(x^2 - 1)}$ **9.** $y = \dfrac{1}{\sqrt{\{(x^2 + 4)(3x - 2)\}}}$

PARAMETRIC EQUATIONS

Sometimes a direct relationship between x and y is awkward to analyse; in such cases it is often easier to express x and y each in terms of a third variable, called a *parameter*.

Consider, for example, the equations

$$x = t^3$$
$$y = t^2 - t$$

The direct relationship between x and y can be found by eliminating t from these two *parametric equations*. It is $y = x^{2/3} - x^{1/3}$

While the gradient and general shape of the curve, as well as the equation of a tangent or a normal, can be obtained from the Cartesian equation, they are often more simply derived from the parametric equations.

Sketching a Curve Given in Parametric Form

To get an idea of the shape of the curve whose parametric equations are

$$x = t^3 \quad \text{and} \quad y = t^2 - t$$

we can

1) *plot* a small number of points by calculating the values of x and y that correspond to certain chosen values of t,

t	-2	-1	0	1	2
x	-8	-1	0	1	8
y	6	2	0	0	2

2) examine the behaviour of the curve as $t \to \pm\infty$,

$$\text{as} \quad t \to \infty, \quad x \to \infty \quad \text{and} \quad y \to \infty$$
$$\text{as} \quad t \to -\infty, \quad x \to -\infty \quad \text{and} \quad y \to \infty$$

There is no finite value of t for which either x or y is undefined so it is reasonable to assume that the curve is continuous.

Based on all this information, a sketch of the curve can now be made.

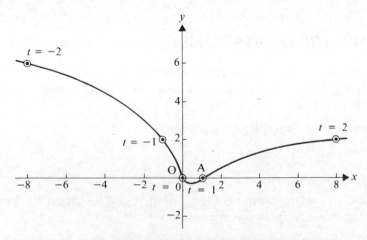

The location of turning points is a further aid to curve sketching.
To use this we need to be able to find the gradient of a curve given parametrically.

FINDING THE GRADIENT FUNCTION USING PARAMETRIC EQUATIONS

If both x and y are given as functions of t then a small increase of δt in the value of t results in corresponding small increases of δx and δy in the values of x and y

As $\delta t \to 0$, δx and δy also approach zero, therefore

$$\frac{dy}{dx} = \frac{dy}{dt} \times \frac{dt}{dx}$$

But $\dfrac{dt}{dx} = 1 \bigg/ \dfrac{dx}{dt}$

Therefore
$$\frac{dy}{dx} = \frac{dy}{dt} \bigg/ \frac{dx}{dt}$$

Hence, for the parametric equations considered opposite, i.e. $x = t^3$ and $y = t^2 - t$, we have

$$\frac{dy}{dx} = \frac{(2t - 1)}{3t^2}$$

Each point on the curve is defined by a value of t which also gives the value of $\dfrac{dy}{dx}$ at that point. Similarly, the value(s) of t where $\dfrac{dy}{dx}$ has a special value lead to the coordinates of the relevant point(s) on the curve.

In this case we can see that, when $t = 0$, both x and y are zero and $\dfrac{dy}{dx}$ is infinitely large, so the curve passes through O and the tangent there is vertical.

At turning point(s) $\quad \dfrac{dy}{dx} = 0 \quad \Rightarrow \quad 2t - 1 = 0 \quad \Rightarrow \quad t = \tfrac{1}{2}$

$t = \tfrac{1}{2} \Rightarrow x = \tfrac{1}{8}$ and $y = -\tfrac{1}{4}$ therefore $(\tfrac{1}{8}, -\tfrac{1}{4})$ is a turning point.

In this case the curve sketched before these facts were known is now only marginally improved, showing that it is not always necessary to carry out this investigation. Some curves, on the other hand, would be difficult to draw without finding their turning points.

Examples 14c _____

1. Find the Cartesian equation of the curve whose parametric
   equations are

   (a) $x = t^2$ (b) $x = \cos\theta$ (c) $x = 2t$

   $\quad\quad y = 2t$ $\quad\quad y = \sin\theta$ $\quad\quad y = 2/t$

(a) $y = 2t \quad\Rightarrow\quad t = \tfrac{1}{2}y$

$\therefore \quad\quad x = t^2 \quad\Rightarrow\quad x = (\tfrac{1}{2}y)^2 = \tfrac{1}{4}y^2 \quad\Rightarrow\quad y^2 = 4x$

(b) Using $\cos^2\theta + \sin^2\theta = 1$ where $\cos\theta = x$ and $\sin\theta = y$
gives

$$x^2 + y^2 = 1$$

(c) $y = 2/t \quad\Rightarrow\quad t = 2/y$

$\therefore \quad\quad x = 2t \quad\Rightarrow\quad x = 4/y \quad\Rightarrow\quad xy = 4$

_____

2. Find the stationary point on the curve whose parametric equations
   are $x = t^3$, $y = (t + 1)^2$ and determine its nature. Sketch the
   curve, showing the stationary point and the behaviour of the curve
   as $x \to \pm\infty$

$$\frac{dy}{dx} = \frac{dy}{dt}\bigg/\frac{dx}{dt} = \frac{2(t + 1)}{3t^2}$$

At stationary points $\dfrac{dy}{dx} = 0$ i.e. $t = -1$

When $t = -1$, $x = -1$ and $y = 0$

Therefore the stationary point is $(-1, 0)$

To determine the nature of the stationary point we examine the sign of $\dfrac{dy}{dx}$ in the
neighbourhood of the point. Remember that, when choosing the points, between
them there must not be any other stationary points or breaks in the curve.

(We use this method because it is often difficult to find the second derivative for
parametric equations.)

The equations $x = t^3$ and $y = (t + 1)^2$ show that there is no finite value of t for which either x or y is undefined, so the curve is continuous, and there is no other stationary point.

Value of t	-2	-1	1
Sign of $\dfrac{dy}{dx}$	$-$ \	0 $-$	$+$ /

Hence $(-1, 0)$ is a minimum point.

Note that the point where $t = 0$ was not chosen as $\dfrac{dy}{dx}$ is infinitely large and its sign is not obvious.

Also

t	-3	-2	1	2	3
x	-27	-8	1	8	27
y	4	1	4	9	16

Further we see that,

$$\text{as } t \to \infty, \quad x \to \infty \quad \text{and} \quad y \to \infty$$

and $$\text{as } t \to -\infty, \quad x \to -\infty \quad \text{and} \quad y \to \infty$$

and we know that, when $t = 0$, $\dfrac{dy}{dx}$ is infinite so the tangent at that point is vertical.

The curve can now be sketched.

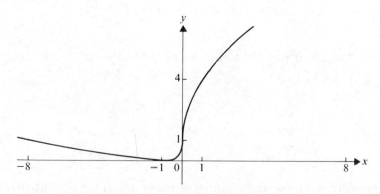

3. The parametric equations of a curve are

$$x = \sin^2\theta, \quad y = 1 + 2\sin\theta$$

Show that the equation of the tangent to the curve at the point $P(\sin^2\theta, 1 + 2\sin\theta)$ is $y = x + \sin\theta + \sin^2\theta$ and find the point(s) where the tangent is parallel to the y-axis.

$$y = 1 + 2\sin\theta \quad \text{and} \quad x = \sin^2\theta$$

$$\Rightarrow \qquad \frac{dy}{d\theta} = 2\cos\theta \qquad \text{and} \qquad \frac{dx}{d\theta} = 2\sin\theta\cos\theta$$

$$\therefore \qquad \frac{dy}{dx} = \frac{dy}{d\theta} \div \frac{dx}{d\theta} = \frac{2\cos\theta}{2\sin\theta\cos\theta}$$

$$= \frac{1}{\sin\theta} \qquad \text{(provided that } \cos\theta \neq 0\text{)}$$

The equation of the tangent is given by $y - y_1 = m(x - x_1)$, i.e.

$$y - (1 + 2\sin\theta) = \frac{1}{\sin\theta}(x - \sin^2\theta)$$

$$\Rightarrow \qquad y\sin\theta = x - \sin^2\theta + \sin\theta + 2\sin^2\theta$$

i.e. $$\qquad y\sin\theta = x + \sin\theta + \sin^2\theta$$

*This is the **general** equation of the tangent because it is the equation of the tangent to the given curve at **any** point on the curve.*

If the tangent is parallel to the y-axis, the value of $\dfrac{dy}{dx}$ is infinitely large,

i.e. $$\qquad \frac{1}{\sin\theta} \to \infty \qquad \Rightarrow \qquad \sin\theta = 0$$

$$\Rightarrow \qquad x = 0 \quad \text{and} \quad y = 1$$

Therefore the tangent is parallel to the y-axis at the point $(0, 1)$

4. Find the equation of the normal to the curve $x = t^2$, $y = t + 2/t$, at the point where $t = 1$. Show, without sketching the curve, that this normal does not cross the curve again.

$x = t^2$ and $y = t + \dfrac{2}{t}$ give $\dfrac{dy}{dt} = 1 - \dfrac{2}{t^2}$ and $\dfrac{dx}{dt} = 2t$

$\therefore$ $\dfrac{dy}{dx} = \dfrac{dy}{dt} \div \dfrac{dx}{dt} = \dfrac{1 - 2/t^2}{2t} = \dfrac{t^2 - 2}{2t^3}$

When $t = 1$; $x = 1$, $y = 3$ and $\dfrac{dy}{dx} = -\dfrac{1}{2}$

Therefore the gradient of the normal at $P(1, 3)$ is $\dfrac{-1}{-\frac{1}{2}} = 2$

The equation of this normal is $y - 3 = 2(x - 1)$

i.e. $y = 2x + 1$

All points for which $x = t^2$ and $y = t + 2/t$ are on the given curve, therefore, for any point that is on both the curve and the normal, these coordinates also satisfy the equation of the normal,
i.e., at points common to the curve and the normal,

$$t + \frac{2}{t} = 2t^2 + 1 \quad \Rightarrow \quad 2t^3 - t^2 + t - 2 = 0 \qquad [1]$$

If a cubic equation can be factorised, each factor equated to zero gives a root of the equation, just as in the case of a quadratic equation.

Now we know that one point where the curve and normal meet is the point where $t = 1$, so $t = 1$ is a root of [1] and $(t - 1)$ is a factor of the LHS,

i.e. $(t - 1)(2t^2 + t + 2) = 0$

Therefore, at any other point where the normal meets the curve, the value of t is a root of the equation $2t^2 + t + 2 = 0$

Checking the value of $b^2 - 4ac$ shows that this equation has no real roots so there are no more points where the normal meets the curve.

EXERCISE 14c

1. Find the gradient function of each of the following curves in terms of the parameter.

   (a) $x = 2t^2$, $y = t$ (b) $x = \sin\theta$, $y = \cos\theta$

   (c) $x = t$, $y = 4/t$

2. If $x = \dfrac{t}{1-t}$ and $y = \dfrac{t^2}{1-t}$, find $\dfrac{dy}{dx}$ in terms of t. What is the value of dy/dx at the point where $x = 1$?

3. (a) If $x = t^2$ and $y = t^3$, find dy/dx in terms of t

   (b) If $y = x^{3/2}$, find dy/dx

   (c) Explain the connection between these two results.

4. Find the Cartesian equation of each of the curves given in Question 1 and hence find dy/dx. Show in each case that dy/dx agrees with the gradient function found in Question 1.

5. Find the turning points of the curve whose parametric equations are $x = t$, $y = t^3 - t$, and distinguish between them.

6. A curve has parametric equations $x = \theta - \cos\theta$, $y = \sin\theta$. Find the coordinates of the points at which the gradient of this curve is zero.

7. Find the equation of the tangent to the curve $x = t^2$, $y = 4t$ at the point where $t = -1$

8. Find the equation of the general normal to the curve $x = t$, $y = 1/t$

9. Find the equation of the general tangent to the curve $x = t^2$, $y = 4t$

10. Find the equation of the normal to the curve $x = \cos\theta$, $y = \sin\theta$ at the point where $\theta = \frac{1}{4}\pi$. Find the coordinates of the point where this normal cuts the curve again.

11. A curve has parametric equations $x = t^2$, $y = 4t$. Find the equation of the normal to this curve at the point $(t^2, 4t)$. Find the coordinates of the point where this normal cuts the coordinate axes. Hence find, in terms of t, the area of the triangle enclosed by the normal and the axes.

12. The parametric equations of a curve are $x = e^t$ and $y = \sin t$.

   (a) Find the gradient function in terms of t

   (b) Find the equation of the curve in the form $y = f(x)$

   (c) Find $\dfrac{dy}{dx}$

   (d) Eliminate x from the result obtained in part (c) and compare with the answer to part (a).

13. The parametric equations of a curve are $x = t$ and $y = 1/t$
Find the general equation of the tangent to this curve (i.e. the equation at the point $(t, 1/t)$).
Find, in terms of t, the coordinates of the points at which the tangent cuts the x and y axes. Hence show that the area enclosed by this tangent and the coordinate axes is constant.

MIXED EXERCISE 14

1. Differentiate with respect to x

   (a) y^4 (b) xy^2 (c) $1/y$ (d) $x \ln y$

   (e) $\sin y$ (f) e^y (g) $y \cos x$ (h) $y \cos y$

In each Question from 2 to 13, find dy/dx

2. $x^2 - 2y^2 = 4$ 3. $1/x + 1/y = 2$

4. $x^2 y^3 = 9$ 5. $y = \dfrac{x^4(2 - x^2)}{(1 + x)^3}$

6. $yx^5 = \dfrac{(x - 1)^4}{(x + 3)}$ 7. $x^2 y^2 = \dfrac{(y + 1)}{(x + 1)}$

8. $x = t^2, \; y = t^3$ 9. $x = (t + 1)^2, \; y = t^2 - 1$

10. $x = \sin^2 \theta, \; y = \cos^3 \theta$ 11. $x = 4t, \; y = 4/t$

12. $y^2 - 2xy + 3y = 7x$ 13. $x = \dfrac{t}{1 - t}, \; y = \dfrac{t^2}{1 - t}$

14. If $x = \sin t$ and $y = \cos 2t$, find $\dfrac{dy}{dx}$ in terms of x and prove that $\dfrac{d^2 y}{dx^2} + 4 = 0$

15. If $x = e^t - t$ and $y = e^{2t} - 2t$, show that $\dfrac{dy}{dx} = 2(e^t + 1)$

CHAPTER 15

COORDINATE GEOMETRY AND CURVES

INTERSECTION OF CURVES

The points of intersection of any two curves (and/or lines) can be found by solving their equations simultaneously. Each real root then gives a point of intersection. Some roots may be repeated and we will now look at the significance of this situation.

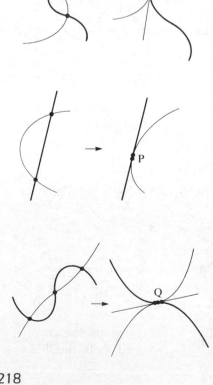

If there are two equal roots the curves meet twice at the same point P, i.e. they touch at P and have a common tangent at P.

In particular, when a line and a curve meet twice at the same point P, the line is a tangent to the curve at P.

If there are three equal roots the curves meet three times at the same point Q. The curves have a common tangent at Q but this time each curve crosses, at Q, to the opposite side of the common tangent.

In particular, when a line and a curve meet at three coincident points Q, the line is a tangent to the curve at Q and the curve has a point of inflexion at Q.

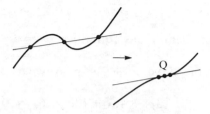

Taking this argument further it becomes clear that,

(a) when the number of coincident points of intersection of a line and a curve is even, the curve touches the line and remains on the same side of the line;

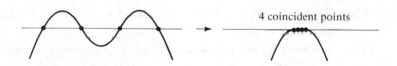

4 coincident points

(b) when the number of coincident points of intersection is odd, the curve touches the line and crosses it. Thus the curve has a point of inflexion.

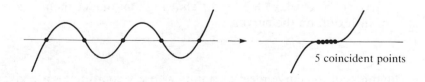

5 coincident points

so if the solution of $\begin{cases} y = mx + c \\ y = f(x) \end{cases}$

has $\begin{cases} \text{distinct roots, the curve crosses the line at distinct points.} \\ \text{repeated roots, the line touches the curve.} \end{cases}$

Examples 15a _____

1. Find the equations of the tangents with gradient 1 to the curve $x^2 + 2y^2 = 6$ and find the coordinates of their points of contact.

Any line with gradient 1 has equation $y = x + c$

This line meets the curve $x^2 + 2y^2 = 6$ where $x^2 + 2(x + c)^2 = 6$

i.e. where $\qquad\qquad 3x^2 + 4cx + (2c^2 - 6) = 0$ $\qquad\qquad$ [1]

For the line to touch the curve, this equation must have equal roots.

So $\qquad\qquad (4c)^2 - 12(2c^2 - 6) = 0 \qquad \Rightarrow \qquad c = \pm 3$

∴ the equations of the tangents are $y = x + 3$ and $y = x - 3$

To find the coordinates of the points of contact we have to go back to equation [1].

When $c = 3$, [1] becomes $x^2 + 4x + 4 = 0 \qquad \Rightarrow \qquad x = -2$

When $c = -3$, [1] becomes $x^2 - 4x + 4 = 0 \qquad \Rightarrow \qquad x = 2$

The corresponding values of y are found from the equations of the tangents.

The points of contact of the two tangents are $(2, -1)$ and $(-2, 1)$

_____

2. Find the points of intersection of the line $y = x + 2$ and the curve $y = x^4 - 2x^3 + 3x + 1$, showing that one of them is a point of inflexion on the curve.

The line and the curve meet at points whose x coordinates are given by

$$x^4 - 2x^3 + 3x + 1 = x + 2 \qquad \Rightarrow \qquad x^4 - 2x^3 + 2x - 1 = 0$$

Using the factor theorem gives $(x - 1)^3(x + 1) = 0$

Thus there are three coincident points where $x = 1$ and one point at $x = -1$

Therefore the line cuts the curve at a point of inflexion (1, 3) and cuts it again at $(-1, 1)$

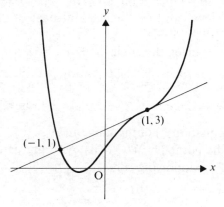

EXERCISE 15a

Investigate the possible intersection of the following lines and curves giving the coordinates of all common points. State clearly those cases where the line touches the curve.

1. $y = x + 1$; $y^2 = 4x$
2. $2y + x = 3$; $x^2 - y^2 - 3y + 3 = 0$
3. $y = x - 5$; $x^2 + 2y^2 = 7$
4. $2y - x = 4$; $x^2 + y^2 - 4x = 4$
5. $y = 0$; $y = x^2 - 3x + 2$
6. $y = 0$; $y = x^3 + 5x^2 + 6x$
7. $y = 0$; $y = (x - 1)^2(x - 2)^2$
8. $y = 0$; $y = (x + 3)^3(x + 2)$
9. $x = 0$; $x = y^4$

Find the value of k such that the given line shall touch the given curve.

10. $y = x + 2$; $y = kx^2$
11. $y = kx + 3$; $xy + 9 = 0$
12. $y = 3x - k$; $x^2 + y^2 = 8$

Find the points of intersection or points of contact (if any) of the following pairs of curves. Illustrate your results by drawing diagrams.

13. $y = 8x^2$; $xy = 1$
14. $x^2 + y^2 + 2y - 7 = 0$; $y = 4x^2$
15. $xy = 2$; $2x^2 + 2y^2 - 6x + 3y - 10 = 0$

16. Find the value(s) or ranges of values of λ for which the line
    $y = 2x + \lambda$
    (a) touches
    (b) cuts in real points
    (c) does not meet, the circle $y^2 + x^2 = 4$

17. Sketch the curves $y = 3x^4$, $y = 4(2-x)^5$, $y = 2(x+3)^7$
    $y = -5x^6$

18. Find the equation(s) of the tangent(s):
    (i) from the point $(1, 0)$ (ii) with gradient $-\frac{1}{2}$
    to each of the following curves,
    (a) $y + 4x^2 = 0$ (b) $xy = 9$ (c) $x^2 = 6y$

THREE MORE CLASSIC CURVES

The geometry of the circle, together with that of three more curves, the parabola, the ellipse and the hyperbola, has been part of mathematical investigation since classical times.

These curves were first defined and studied by Apollonius of Perga (*c*.250–200 BC). He defined them as curves traced out on the surface of a cone when it is cut by a plane; hence these curves are also known as conic sections.

Each of these curves can also be defined as a locus and its properties analysed using coordinate geometry.

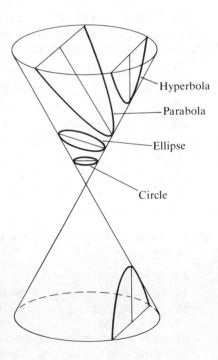

The general form of the Cartesian equation of the circle is now known, and in this section we will look briefly at the equations of the other curves.

The Parabola

If a point, P, is constrained so that its distance from a fixed line is equal to its distance from a fixed point, then the locus of P is a parabola. The curve given by this definition can be seen by plotting some of the possible positions of P.

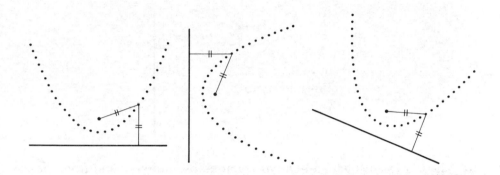

This is a familiar curve; it has an axis of symmetry and, in Module A, we saw that when the axis of symmetry is parallel to Oy, the Cartesian equation of a parabola has the form $y = ax^2 + bx + c$

Similarly, when the axis of symmetry is parallel to Ox, the equation has the form $x = ay^2 + by + c$

The simplest forms of these equations are

$$y = x^2 \qquad\qquad x = y^2, \text{ (i.e. } y^2 = x)$$

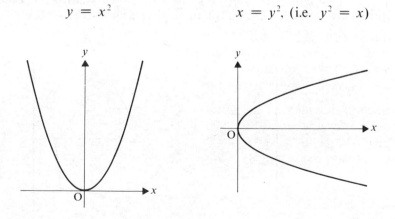

The classical position of the parabola for Cartesian analysis is with its axis parallel to Ox.

The Ellipse

The locus of a point, P, is an ellipse when P moves so that its distance from a fixed point and its distance from a fixed line are in a constant ratio which is less than 1. This can be verified by plotting some positions of P for a value of the constant, say $\frac{1}{2}$

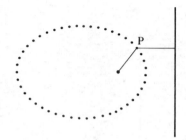

The Cartesian equation of a particular ellipse can be found from the locus definition using the method adopted in Chapter 4.

The standard equation of an ellipse is $\dfrac{x^2}{a^2} + \dfrac{y^2}{b^2} = 1$

The Hyperbola

The locus definition of a hyperbola is very similar to that of the ellipse. A hyperbola is the locus of P when the ratio of the distance of P from a fixed point to the distance of P from a fixed line is constant and greater than 1. The shape of the hyperbola can be seen when some positions of P are plotted.

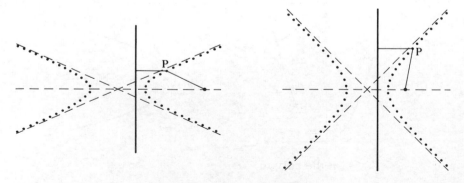

The standard equation of a hyperbola in this position is $\dfrac{x^2}{a^2} - \dfrac{y^2}{b^2} = 1$

All hyperbolas have a line of symmetry and a pair of asymptotes. When these asymptotes are perpendicular, the curve has a familiar shape; it is called a rectangular hyperbola. Further, when the asymptotes are the x and y axes, the equation of the curve has the form $y = c^2/x$ or $xy = c^2$

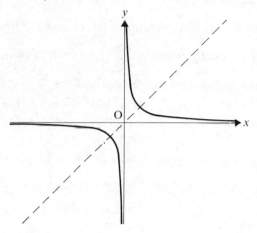

Example 15b

P(x, y) is constrained so that the ratio of its distance from $(1, 0)$ to its distance from the line $x = 4$ is equal to $\frac{1}{2}$. Find the equation of the locus of P.

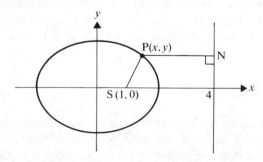

$$\frac{PS}{PN} = \frac{1}{2} \quad \Rightarrow \quad \frac{PS^2}{PN^2} = \frac{1}{4}$$

Now $\qquad PS^2 = (x-1)^2 + y^2 \quad$ and $\quad PN^2 = (x-4)^2$

∴ $\quad$ P(x, y) is on the curve $\qquad \Longleftrightarrow \qquad \dfrac{(x-1)^2 + y^2}{(x-4)^2} = \dfrac{1}{4}$

∴ $\quad$ the equation of the locus of P is $\quad 3x^2 + 4y^2 = 12$

EXERCISE 15b

Find the equation of the locus of P and name the curve in each of the following cases.

1. P is twice as far from the point $(2, 0)$ as it is from the line $x = 8$

2. P is equidistant from the point $(0, 4)$ and the line $y = -4$

3. P is half as far from the point $(2, 0)$ as it is from the line $x = 8$

4. Without plotting them, name each of the following curves.

    (a) $y = x^2 - 3$ (b) $y^2 + x^2 = 9$ (c) $xy = 1$

    (d) $x^2 - 3y^2 = 9$ (e) $y^2 = x + 2$ (f) $4x^2 + y^2 = 16$

5. The equation of a curve is $\dfrac{x^2}{9} + \dfrac{y^2}{4} = 1$. Find the coordinates of the points where the curve cuts the x and y axes. Sketch the curve.

6. Sketch the curve whose equation is $\dfrac{x^2}{a^2} + \dfrac{y^2}{b^2} = 1$

7. On the same set of axes sketch the curves

    $$\frac{x^2}{16} + \frac{y^2}{4} = 1 \text{ and } \frac{x^2}{16} - \frac{y^2}{4} = 1$$

    (a) Give the coordinates of the points where curves cross the coordinate axes.

    (b) At how many points do the curves intersect?

PARAMETERS

The relationship between the x and y coordinates of a point on any standard conic involves at least one term of order 2 (i.e. x^2, y^2, xy). This relationship can often be expressed more simply in the form of two equations, i.e.

$$\left. \begin{array}{l} x = f(t) \\ y = g(t) \end{array} \right\} \text{ where } t \text{ is a parameter}$$

The use of parametric equations to plot curves, find gradients and hence tangents and normals to curves is covered in Chapter 14. In this chapter we look at other ways in which parametric equations can be used, and in particular at the parametric equations for conic sections.

Parametric Equations for a Circle

Consider the circle whose centre is the origin and whose radius is a.

The Cartesian equation of this circle is $x^2 + y^2 = a^2$ [1]

The parametric equations of this circle are $\begin{cases} x = a\cos\theta \\ y = a\sin\theta \end{cases}$ [2]

(Using the identity $\cos^2\theta + \sin^2\theta \equiv 1$ to eliminate θ from the parametric equations verifies that equations [2] are equivalent to equation [1].) The parameter, θ, has graphical significance in this case, as can be seen in the diagram below.

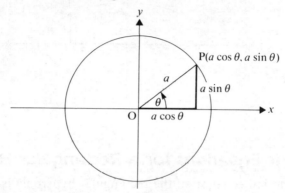

Parametric Equations for a Parabola

Consider the parabola whose vertex is the origin and whose axis is Ox.

If the Cartesian equation of this parabola is $y^2 = 4ax$ then

the parametric equations are $\begin{cases} x = at^2 \\ y = 2at \end{cases}$

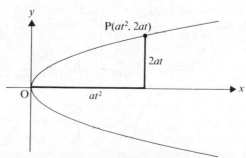

In this case t has no geometrical meaning.

Parametric Equations for an Ellipse

The standard equation of the ellipse is $\dfrac{x^2}{a^2} + \dfrac{y^2}{b^2} = 1$

The parametric equations of this ellipse are $\begin{cases} x = a\cos\theta \\ y = b\sin\theta \end{cases}$

There is no obvious geometrical significance of θ in this case.

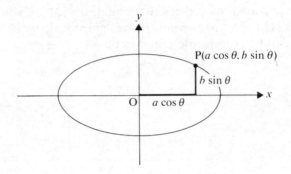

Parametric Equations for a Rectangular Hyperbola

The Cartesian equation of the rectangular hyperbola is $xy = c^2$

The parametric equations of this curve are $\begin{cases} x = ct \\ y = c/t \end{cases}$

Again, t has no obvious significance.

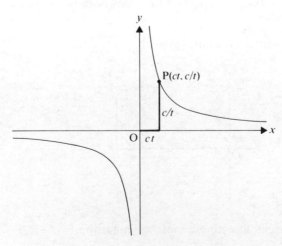

CURVE SKETCHING

Algebraic analysis is usually easier when the parametric equations of a curve are used. However sketching curves, particularly circles, is easier using Cartesian equations which show clearly the important features.

Examples 15c _____

1. On the same set of axes, sketch the curves

$$\begin{cases} x = 2\cos\theta + 1 \\ y = 2\sin\theta \end{cases} \quad \text{and} \quad \begin{cases} x = t^2 \\ y = 4t \end{cases}$$

Hence determine the number of points in which the curves cut.

The first pair of equations can be converted to a Cartesian equation by finding $\cos\theta$ in terms of x and $\sin\theta$ in terms of y and then using the identity $\cos^2\theta + \sin^2\theta \equiv 1$

$$\left. \begin{aligned} x &= 2\cos\theta + 1 \\ y &= 2\sin\theta \end{aligned} \right\} \quad \Rightarrow \quad (x-1)^2 + y^2 = 4$$

This is a circle, centre $(1,0)$ radius 2

$$\left. \begin{aligned} x &= t^2 \\ y &= 4t \end{aligned} \right\} \quad \Rightarrow \quad y^2 = 16x$$

This is a parabola, vertex $(0,0)$ and axis Ox.

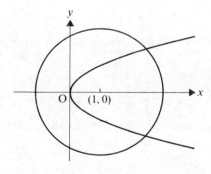

The sketch shows that there are two points of intersection.

_____

2. Find the coordinates of the points where the line $y = 3x - 1$ cuts the curve whose parametric equations are $x = t$, $y = 2t^2$

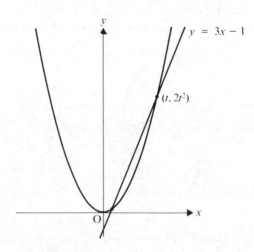

The coordinates of any point on the curve are $(t, 2t^2)$

The line cuts the curve where the coordinates of a point on the curve satisfy the equation of the line, i.e. where

$$2t^2 = 3t - 1$$

$$\Rightarrow \qquad 2t^2 - 3t + 1 = 0 \qquad \Rightarrow \qquad (t - 1)(2t - 1) = 0$$

This is a quadratic equation in t, giving two values of t and therefore giving two distinct points on the curve. This is expected from the sketch which indicates that there are two points of intersection.

So $\qquad\qquad\qquad\qquad\qquad t = 1 \text{ or } \frac{1}{2}$

Therefore the points of intersection are $(1, 2)$ and $(\frac{1}{2}, \frac{1}{2})$

---

3. Form an equation which gives the values of t at points where the line $y = mx + c$ crosses the curve whose parametric equations are $x = 2t$, $y = 2/t$

Find the condition that must be satisfied by m and c if

(a) the line cuts the curve in two places

(b) the line is a tangent to the curve.

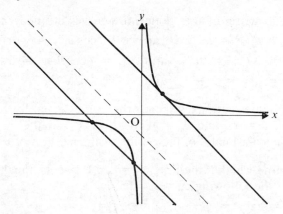

The point $(2t, 2/t)$ is any point on the curve, and the line $y = mx + c$ cuts the curve where $(2t, 2/t)$ is also a point on the line, i.e. where

$$\frac{2}{t} = m(2t) + c \quad \Rightarrow \quad 2mt^2 + ct - 2 = 0 \qquad [1]$$

This equation is quadratic in t, so it can have either two distinct real roots, a repeated root or no real roots.

(a) $y = mx + c$ cuts the curve in two places if equation [1] has real distinct roots,

i.e. if   '$b^2 - 4ac > 0$' $\Rightarrow$ $c^2 - 4(2m)(-2) > 0$

$\Rightarrow$ $c^2 + 16m > 0$

(b) $y = mx + c$ is a tangent to the curve if equation [1] has a repeated root,

i.e. if   '$b^2 - 4ac = 0$' $\Rightarrow$ $c^2 + 16m = 0$

(If equation [1] has no real roots then the line $y = mx + c$ does not meet the curve.)

EXERCISE 15c

1. Sketch the curves given parametrically by

   (a) $x = t^2, \ y = 2t$ (b) $x = 3\cos\theta, \ y = 3\sin\theta$

   (c) $x = t^2 + 1, \ y = t$ (d) $x = 3t, \ y = 3/t$

   (e) $x = 4t, \ y = t^2$ (f) $x = 3\cos\theta, \ y = 4\sin\theta$

   (g) $x = 3t + 2, \ y = t^2 - 1$ (h) $x = \cos\theta + 2, \ y = \sin\theta - 1$

2. On the same set of axes sketch the curves defined parametrically by
   (a) $x = t^2$, $y = t$ and $x = \cos\theta$, $y = \sin\theta$
   (b) $x = 2/t$, $y = 2t$ and $x = 4\cos\theta$, $y = 6\sin\theta$
   (c) $x = t$, $y = 3t - 1$ and $x = 4t^2$, $y = 2t$

3. On the same set of axes sketch the curves given by
   $x = 2t$, $y = t^2$ and $x = 4\cos\theta$, $y = 4\sin\theta$
   Find the coordinates of the points of intersection of these two curves.

4. Determine whether the line $y = 2x + 1$ cuts, touches or misses
   each of the following curves.
   (a) $x = t^2$, $y = 4t$ (b) $x = t^2$, $y = t$
   (c) $x = 2t^2$, $y = 4t$ (d) $x = \cos\theta$, $y = \sin\theta + 1$

5. A curve has parametric equations $x = 2t^2$, $y = 4t$. Find
   (a) the Cartesian equation of the curve,
   (b) the equation of the tangent at the point where $y = 8$
   (c) the equation of the chord joining the points on the curve
       where $t = p$ and $t = q$
   (d) the coordinates of the points where $y = x - 6$ cuts the curve,
   (e) the value of k for which $y = x + k$ is a tangent to the
       curve,
   (f) the coordinates of the point(s) of intersection of the curve and
       the circle $x^2 + y^2 - 2x = 16$
   (g) the coordinates of the point(s) of intersection of the curve and
       the curve given parametrically by $x = 8s$, $y = 8/s$

6. The parametric equations of a curve are $x = t^2$, $y = t^3$.
   (a) Taking values of t from -2 to 2 at intervals of 0.5, find the
       corresponding values of x and y and hence sketch the curve.
   (b) Find $\dfrac{dy}{dx}$ in terms of t and describe the behaviour of $\dfrac{dy}{dx}$ as
       $t \to 0$ from both positive and negative values. Hence describe
       the behaviour of the curve near the origin.

CONSOLIDATION C

SUMMARY

Throughout this summary, a represents a constant quantity.

TRIGONOMETRY

Expressing $a \cos \theta \pm b \sin \theta$ as a Single Term

For various values of a and b, $a \cos \theta \pm b \sin \theta$ can be expressed as

$$r \cos (\theta \pm \alpha) \quad \text{or} \quad r \sin (\theta \pm \alpha)$$

where $r = \sqrt{(a^2 + b^2)}$ and $\tan \alpha$ is either a/b or b/a.

Some General Solutions

If $\cos \theta = \cos \alpha$ then $\theta = 2n\pi \pm \alpha$

If $\tan \theta = \tan \alpha$ then $\theta = n\pi + \alpha$

If $\sin \theta = \sin \alpha$ then $\theta = 2n\pi + \alpha$ or $(2n - 1)\pi - \alpha$

Small Angles

When θ is a small angle measured in radians, then

$$\sin \theta \approx \theta \qquad \tan \theta \approx \theta \qquad \cos \theta \approx 1 - \tfrac{1}{2}\theta^2$$

and $\quad \lim\limits_{\theta \to 0} \dfrac{\sin \theta}{\theta} = 1$

233

Inverse Trigonometric Functions

The inverse trig functions are $\arcsin x$, $\arccos x$, $\arctan x$.

$\arcsin x$ means 'the angle in the range $-\frac{1}{2}\pi \leqslant \theta \leqslant \frac{1}{2}$ whose sine is x'.

$\arccos x$ means 'the angle in the range $0 \leqslant \theta \leqslant \pi$ whose cosine is x'.

$\arctan x$ means 'the angle in the range $0 \leqslant \theta \leqslant \frac{1}{2}\pi$ whose tangent is x'.

DIFFERENTIATION

Standard Results

$f(x)$	$\dfrac{d}{dx}f(x)$	$f(x)$	$\dfrac{d}{dx}f(x)$
$\sin x$	$\cos x$	$\arcsin x$	$\dfrac{1}{\sqrt{(1-x^2)}}$
$\cos x$	$-\sin x$		
$\tan x$	$\sec^2 x$	$\arccos x$	$-\dfrac{1}{\sqrt{(1-x^2)}}$
$\sec x$	$\sec x \tan x$		
$\operatorname{cosec} x$	$-\operatorname{cosec} x \cot x$	$\arctan x$	$\dfrac{1}{(1+x^2)}$
$\cot x$	$-\operatorname{cosec}^2 x$		

Further Quotable Results

$f(x)$	$\dfrac{d}{dx}f(x)$
$\arcsin \dfrac{x}{a}$	$\dfrac{1}{\sqrt{(a^2-x^2)}}$
$\arctan \dfrac{x}{a}$	$\dfrac{a}{(a^2+x^2)}$

Implicit Differentiation

When y cannot be isolated, each term can be differentiated in turn with respect to x,

e.g.

$$\frac{d}{dx}(y^2) = (2y)\left(\frac{dy}{dx}\right) \quad \text{and} \quad \frac{d}{dx}(xy) = y + (x)\left(\frac{dy}{dx}\right) \quad \text{(by product rule)}$$

Logarithmic Differentiation

Sometimes it is easier to differentiate $y = f(x)$ with respect to x if we first take logs of both sides.

When doing this remember that $\quad \dfrac{d}{dx}(\ln y) = \dfrac{1}{y}\dfrac{dy}{dx}$

This process is called logarithmic differentiation. It is *essential* when differentiating functions such as x^x.

Parametric Differentiation

If $y = f(t)$ and $x = g(t)$ then $\quad \dfrac{dy}{dx} = \dfrac{dy}{dt} \div \dfrac{dx}{dt}$

CONIC SECTIONS

The Cartesian equation of this circle is $x^2 + y^2 = a^2$

The corresponding parametric equations are

$$x = a\cos\theta, \quad y = a\sin\theta$$

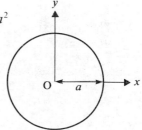

The Cartesian equation of this parabola is $y^2 = 4ax$

The corresponding parametric equations are

$$x = at^2, \quad y = 2at$$

The Cartesian equation of an ellipse with its centre at O is

$$\frac{x^2}{a^2} + \frac{y^2}{b^2} = 1$$

The corresponding parametric equations are

$$x = a \cos \theta, \ y = b \sin \theta$$

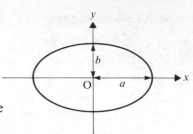

The Cartesian equation of this rectangular hyperbola is $xy = c^2$

The parametric equations are

$$x = ct, \ y = c/t$$

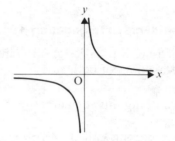

MULTIPLE CHOICE EXERCISE C

TYPE I

1. If $x^2 + y^2 = 4$ then $\dfrac{dy}{dx}$ is

   A $2x + 2y$ B $4 - x^2$ C $-\dfrac{x}{y}$ D $\dfrac{y}{x}$

2. The approximate value, when θ is small, of the expression $\dfrac{2\theta - \sin \theta}{\sin 2\theta - \theta}$ is

   A 1   B 2   C -1 D -2

3. $\dfrac{d}{dx}\left(\dfrac{1}{1+x}\right)$ is

   A $\dfrac{-1}{(1+x)^2}$ C $\ln(1+x)$

   B $\dfrac{1}{1-x}$ D $\dfrac{-1}{1+x^2}$

4. $\dfrac{d}{dx} \ln\left(\dfrac{x+1}{2x}\right)$ is

A $\dfrac{1}{2}$

C $\dfrac{2x}{x+1}$

B $\dfrac{1}{x+1} - \dfrac{1}{2x}$

D $\dfrac{1}{x+1} - \dfrac{1}{x}$

5. $\dfrac{d}{dx} a^x$ is

A xa^{x-1} B a^x C $x \ln a$ D $a^x \ln a$

6. If $x = \cos\theta$ and $y = \cos\theta + \sin\theta$, $\dfrac{dy}{dx}$ is

A $1 - \cot\theta$ C $\cot\theta - 1$
B $1 - \tan\theta$ D $\cot\theta + 1$

7. A point P moves so that is equidistant from A and B. The locus of the set of points P is

A a circle on AB as diameter
B a line parallel to AB
C the perpendicular bisector of AB
D a parabola

8. The curve whose parametric equations are $x = 2\cos\theta$, $y = 3\sin\theta$ is

A a circle
B a parabola
C an ellipse
D a rectangular hyperbola

9. The value of $\sin^{-1}\frac{1}{2}$ is

A 2   B $\frac{1}{3}\pi$ C $\dfrac{1}{2.1}$ D $\frac{1}{6}\pi$

10. The greatest value of $5\cos\theta - 4\sin\theta$ is

A 3   B 1   C $\sqrt{41}$ D ± 5

TYPE II

11. $3 \cos \theta - 4 \sin \theta \equiv$

    A $\quad 5 \cos (\theta + \alpha)$ where $\tan \alpha = \frac{3}{4}$
    B $\quad 5 \sin (\alpha - \theta)$ where $\tan \alpha = \frac{3}{4}$
    C $\quad 5 \cos (\theta + \alpha)$ where $\tan \alpha = \frac{4}{3}$
    D $\quad -5 \cos (\theta - \alpha)$ where $\tan \alpha = \frac{4}{3}$

12. Given that α is a very small angle measured in radians,

    A $\quad \sin (2\pi + \alpha) = \sin \alpha$ $\qquad$ C $\quad \sin (2\pi + \alpha) \approx 2\pi + \alpha$
    B $\quad \sin \alpha \approx \alpha$ $\qquad\qquad\qquad$ D $\quad \cos \alpha \approx \alpha$

13. Given that $y = \arccos x$,

    A $\quad \cos y = x$ $\qquad\qquad$ C $\quad y = \sec x$
    B $\quad 0 \leqslant x \leqslant \pi$ $\qquad\quad$ D $\quad 0 \leqslant y \leqslant \pi$

14. Given the curve with equation $x^2 + y^2 + xy = 0$

    A $\quad$ The curve is a circle $\qquad\qquad$ C $\quad \dfrac{dy}{dx} = 2x + y$

    B $\quad 2x + y + (x + 2y)\dfrac{dy}{dx} = 0 \qquad$ The curve goes through O

15. Given that $x = \cos^2\theta$ and $y = \sin^2\theta$,

    A $\quad x^2 + y^2 = 1$ $\qquad\qquad$ C $\quad 0 \leqslant y \leqslant 1$

    B $\quad \dfrac{dy}{dx} = \tan \theta$ $\qquad\quad$ D $\quad y = x - \frac{1}{2}\pi$

TYPE III

16. $\sin 3\theta = \cos 4\theta \qquad \Rightarrow \qquad \frac{1}{2}\pi - 3\theta = 2n\pi \pm 4\theta$

17. $\dfrac{d}{dx}(x^2 y^2) = 2xy^2 + 2x^2 y$

18. A curve in the xy plane is a circle $\qquad \Rightarrow \qquad x^2 + y^2 = a^2$

19. When $y = \cos 2\theta$ and $x = \sin \theta$, $\dfrac{dy}{dx} = -4x\sqrt{(1 - x^2)}$

20. If $y = f(t)$ and $x = g(t)$ then $\dfrac{dy}{dx} = \dfrac{dy}{dt} \div \dfrac{dx}{dt}$

21. If $y = \arctan ax$ then $\dfrac{dy}{dx} = \left(\dfrac{1}{1 + x^2} \right)\left(\dfrac{1}{1 + a^2} \right)$

MISCELLANEOUS EXERCISE C

1. Differentiate with respect to x

(a) $\dfrac{x}{2x + 1}$ (b) $\arcsin(x^2)$ (C)

2. Differentiate with respect to x

(a) $\sin(2x^2)$ (b) 2^x (U of L)

3. Find $\dfrac{dy}{dx}$ when

(a) $y = \dfrac{1 + \sin x}{1 + \cos x}$ (b) $y = \ln \sqrt{\left(\dfrac{1 + x}{1 - x} \right)}$, $|x| < 1$

and simplify your answers as far as possible.   (U of L)

4. A curve has parametric equations

$$x = 2t + \sin 2t, \quad y = \cos 2t, \quad 0 < t < \tfrac{1}{2}\pi$$

Show that, at the point with parameter t, the gradient of the curve is $-\tan t$.   (C)

5. Show that, when x is small,

$$(1 - \sin^2 3x)\cos 2x \approx 1 - 11x^2$$

6. Evaluate $\dfrac{dy}{dx}$ when $y = 1$, given that

(a) $y(x + y) = 3$ (b) $x = \dfrac{1}{(4 - t)^2}$, $y = \dfrac{t}{4 - t^2}$, $0 < t < 4$

                                                 (U of L)

7. Prove the identity

$$\sqrt{3}\cos \theta + \sin \theta \equiv 2\cos(\theta - \tfrac{1}{6}\pi)$$

Find, in terms of π, the general solution of the equation

$$\sqrt{3}\cos \theta + \sin \theta = 1$$

                                                 (AEB)

8. Given that $y = \sin x + \frac{1}{2}\sin 2x + \frac{1}{3}\sin 3x$, show that

$$\frac{dy}{dx} = (1 + 2\cos x)\cos 2x$$

Find the complete set of values of x for which $\dfrac{dy}{dx} < 0$ in the interval $0 < x < \pi$. (AEB)

9. Differentiate with respect to x
   (a) $(4x - 1)^{20}$ (b) $\arctan(\sqrt{x})$

10. The parametric equations of a curve are $x = 5a\sec\theta$ and $y = 3a\tan\theta$ where $-\frac{1}{2}\pi < \theta < \frac{1}{2}\pi$ and a is a positive constant. Find the coordinates of the point on the curve where the gradient is -1.

11. Given that $y = e^x \ln(1 + \sin 2x)$, $-\frac{1}{2}\pi \leqslant x \leqslant \frac{1}{2}\pi$, find $\dfrac{dy}{dx}$ in terms of x. (U of L)

12. Find all the values of θ for which $0 \leqslant \theta \leqslant \frac{1}{2}\pi$ and $\sin 8\theta = \sin 2\theta$ (U of L)

13. Given that $3\cos\theta + 4\sin\theta \equiv R\cos(\theta - \alpha)$, where $R > 0$ and $0 \leqslant \alpha \leqslant \frac{1}{2}\pi$, state the value of R and the value of $\tan\alpha$.

    For each of the following equations, solve for θ in the interval $0 \leqslant \theta \leqslant 2\pi$ and give your answers in radians correct to one decimal place.
    (a) $3\cos\theta + 4\sin\theta = 2$
    (b) $3\cos 2\theta + 4\sin 2\theta = 5\cos\theta$

    The curve with equation $y = \dfrac{10}{3\cos x + 4\sin x + 7}$, between $x = -\pi$ and $x = \pi$, cuts the y-axis at A, has a maximum point at B and a minimum point at C. Find the coordinates of A, B and C. (AEB)

14. A curve is given by the equation

$$y = \sin x + \frac{1}{2}\sin 2x, \quad 0 \leqslant x \leqslant 2\pi$$

    Find the values of x for which y is zero.

    Find the exact coordinates of the stationary points on the curve and sketch the curve. (JMB)

15. The parametric equations of a curve are

$$x = \cos 2\theta + 2 \cos \theta \quad y = \sin 2\theta - 2 \sin \theta$$

Show that $\dfrac{dy}{dx} = \tan \tfrac{1}{2}\theta$

Find the equation of the normal to the curve at the point where $\theta = \tfrac{1}{2}\pi$ (AEB)

16. A curve has parametric equations

$$x = 2t - \ln(2t), \quad y = t^2 - \ln(t^2)$$

where $t > 0$

Find the value of t at the point on the curve at which the gradient is 2.   (C)

17. The outputs of two signal generators A and B are combined and the resultant waveform displayed on an oscilloscope. Given that the signal from A is given by $y_1 = 2 \sin x$ and that from B by $y_2 = 4 \cos x$, express the resultant, $y = y_1 + y_2$, in the form $R \sin(x + \alpha)$, where $R > 0$ and $0 < \alpha < \tfrac{1}{2}\pi$, giving R and α to 3 significant figures.   (U of L)

18. A curve has parametric equations

$$x = 1 + \sqrt{(32)} \cos \theta, \quad y = 5 + \sqrt{(32)} \sin \theta, \quad 0 \leqslant \theta \leqslant 2\pi$$

Show that the tangent to the curve at the point with parameter θ is given by

$$(y - 5) \sin \theta + (x - 1) \cos \theta = \sqrt{(32)}$$

Find the two values of θ such that this tangent passes through the point $A(1, -3)$. Hence, or otherwise, find the equations of the two tangents to the curve from the point A.   (U of L)

19. The equation of a curve is $\quad 3x^2 + y^2 = 2xy + 8x - 2$

Find an equation connecting x, y and $\dfrac{dy}{dx}$ at all points on the curve. Hence show that the coordinates of all points on the curve at which $\dfrac{dy}{dx} = 2$ satisfy the equation $\quad x + y = 4$

Deduce the coordinates of these points. **(JMB)**

20. A particle P moves in a straight line so that its displacement, s metres, at time t seconds, from a point O on the line is given by

$$s = e^t \sin t, \quad t \geq 0$$

(a) Find $\dfrac{ds}{dt}$ and deduce the speed at time $t = 0$

(b) Find the values of t in the interval $0 \leq t \leq 4\pi$ when P is instantaneously at rest.

(c) Show that the values of s corresponding to the values of t found in (b) are consecutive terms in a geometrical series and give the common ration of this series.

(d) Sketch the graph of $s = e^t \sin t, \quad t \geq 0$ **(U of L)**

21. A curve is defined by the parametric equations $x = \cos^3 t$, $y = \sin^3 t, \ 0 \leq t \leq \frac{1}{4}\pi$.

Show that the equation of the normal to the curve at the point $P(\cos^3 t, \sin^3 t)$ is

$$x \cos t - y \sin t = \cos^4 t - \sin^4 t \qquad \text{(JMB)}$$

22. Find all values of x lying between $0°$ and $360°$ which satisfy the equation

$$3 \sin x + 4 \cos x = 1$$

Give your answers correct to the nearest degree. **(WJEC)**

23. A particle moves along the x-axis such that at time t seconds, its displacement x metres from O is given by

$$x = e^{-t} \cos t, \quad t \geqslant 0$$

(a) Find $\dfrac{dx}{dt}$, and hence determine the initial speed, in ms^{-1}, of the particle.

(b) Find the values of t, in the interval $0 \leqslant t \leqslant 2\pi$, when the particle is instantaneously at rest.

(c) Obtain an expression for the acceleration of the particle at time t seconds.   (U of L)

24.

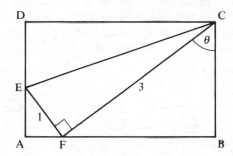

The diagram shows a rectangle ABCD containing a triangle CEF which is right angled at F. Given that $EF = 1$ cm, $FC = 3$ cm and the angle $BCF = \theta$, show that the perimeter of the rectangle is l cm, where

$$l = 8 \cos \theta + 6 \sin \theta$$

Given that $l = 7$, calculate the value of θ correct to the nearest tenth of a degree.   (JMB)

25. (a) Express $3 \cos \theta - 4 \sin \theta$ in the form $r \cos (\theta + \alpha)$, where $r > 0$ and $0 < \alpha < \tfrac{1}{2}\pi$

(b) Deduce the amplitude of the wave motion defined by the equation

$$x = 3 \cos t - 4 \sin t \qquad \text{(U of L)}$$

26. Differentiate the following functions with respect to x:

(a) $x \sin x$ (b) $\tan^2 x$ (WJEC)

27. Given that $xy = 3x^2 + y^2$, find $\dfrac{dy}{dx}$, giving your answer in terms of x and y.   (AEB)p

28. Sketch the curve defined by the parametric equations

$$x = 4\cos\theta, \qquad y = 2\sin\theta, \qquad 0 < \theta < \tfrac{1}{2}\pi$$

Show that the equation of the normal to the curve at the point $P(4\cos\theta, 2\sin\theta)$ is

$$2x\sin\theta - y\cos\theta = 6\sin\theta\cos\theta$$

This normal meets the x-axis at Q and the y-axis at R. The point O is the origin and the point S is such that OQSR is a rectangle. Find the coordinates of S. Show the normal at P and the rectangle OQSR on your sketch.

Show that the perimeter L of the rectangle OQSR may be expressed in the form $r\cos(\theta - \alpha)$, where $r > 0$ and $0 < \alpha < \tfrac{1}{2}\pi$. Give the values, in surd form, of r, $\cos\alpha$ and $\sin\alpha$. State the maximum value of L as θ varies in the interval $0 < \theta < \tfrac{1}{2}\pi$, and find the coordinates of S when L has the maximum value. (JMB)

CHAPTER 16

INTEGRATION

REVERSING DIFFERENTIATION

The process of finding a function from its derivative is called integration. This process reverses the operation of differentiation.

The result of integrating $f'(x)$, which is called the integral of $f'(x)$, is not unique; any constant can be added to the basic integral. So, for any function $f'(x)$, we have

$$\int f'(x)\, dx = f(x) + K$$

where $\int \ldots dx$ means 'the integral of ... w.r.t. x'

and K is called the constant of integration

STANDARD RESULTS

$$\left. \begin{array}{l} \displaystyle\int x^n\, dx = \frac{1}{(n+1)} x^{n+1} + K \\[3mm] \displaystyle\int (ax+b)^n\, dx = \frac{1}{a(n+1)} (ax+b)^{n+1} + K \end{array} \right\} \quad n \neq -1$$

A sum or difference of functions can be integrated term by term.

DEFINITE INTEGRATION

Definite integration requires the value of an integral to be calculated between specified values of the variable,

e.g. $$\int_a^b x^n\, dx = \left[\frac{1}{(n+1)} x^{n+1} \right]_a^b$$

$$= \left\{ \frac{1}{(n+1)} x^{n+1} \right\}_{x=b} - \left\{ \frac{1}{(n+1)} x^{n+1} \right\}_{x=a}$$

The constant of integration is unnecessary in a definite integral as it disappears upon subtraction.

245

INTEGRATING EXPONENTIAL FUNCTIONS

Whenever a function $f(x)$ is *recognised* as the derivative of a function $f(x)$ then

$$\frac{d}{dx} f(x) = f'(x) \quad \Rightarrow \quad \int f'(x)\, dx = f(x) + K$$

Thus any function whose derivative is known can be established as a standard integral.

It is already known that $\dfrac{d}{dx} e^x = e^x$

hence

$$\int e^x\, dx = e^x + K$$

Further, we have $\dfrac{d}{dx}(ce^x) = ce^x$

and $\dfrac{d}{dx} e^{(ax + b)} = ae^{(ax + b)}$

Hence

$$\int ce^x\, dx = ce^x + K$$

and

$$\int e^{(ax + b)} = \frac{1}{a} e^{(ax + b)} + K$$

e.g. $\displaystyle\int 2e^x\, dx = 2e^x + K$ and $\displaystyle\int 4e^{(1 - 3x)}\, dx = (4)(-\tfrac{1}{3})e^{(1 - 3x)} + K$

To integrate an exponential function where the given base is not e but is some other constant, a say, the base must first be changed to e as follows.

Using $a^x = e^z$ and taking logs to the base e we have

$$x \ln a = z$$

Hence $a^x = e^{x \ln a} \quad \Rightarrow \quad \displaystyle\int a^x\, dx = \int e^{x \ln a}\, dx$

$$\Rightarrow \quad \int a^x\, dx = \frac{1}{\ln a} e^{x \ln a} + K$$

i.e.

$$\int a^x\, dx = \frac{1}{\ln a} a^x + K$$

Alternatively, this result can be obtained directly if it is remembered that

$$\frac{d}{dx}(a^x) = (\ln a)a^x$$

Example 16a

Write down the integral of e^{3x} w.r.t. x and hence evaluate $\int_0^1 e^{3x}\,dx$

$$\int e^{3x}\,dx = \tfrac{1}{3}e^{3x} + K$$

The constant of integration disappears when a definite integral is calculated, hence

$$\int_0^1 e^{3x}\,dx = \left[\tfrac{1}{3}e^{3x}\right]_0^1 = \tfrac{1}{3}e^3 - \tfrac{1}{3}e^0$$

i.e.
$$\int_0^1 e^{3x}\,dx = \tfrac{1}{3}(e^3 - 1)$$

EXERCISE 16a

Integrate each function w.r.t. x

1. e^{4x} **2.** $4e^{-x}$ **3.** $e^{(3x-2)}$ **4.** $2e^{(1-5x)}$

5. $6e^{-2x}$ **6.** $5e^{(x-3)}$ **7.** $e^{(2+x/2)}$ **8.** 2^x

9. $4^{(2+x)}$ **10.** $e^{2x} + \dfrac{1}{e^{2x}}$ **11.** $a^{(1-2x)}$ **12.** $2^x + x^2$

Evaluate the following definite integrals.

13. $\displaystyle\int_0^2 e^{2x}\,dx$ **14.** $\displaystyle\int_{-1}^1 2e^{(x+1)}\,dx$

15. $\displaystyle\int_2^3 e^{(2-x)}\,dx$ **16.** $\displaystyle\int_0^2 -e^x\,dx$

FUNCTIONS WHOSE INTEGRALS ARE LOGARITHMIC

To Integrate $\dfrac{1}{x}$

At first sight it looks as though we can write $\dfrac{1}{x} = x^{-1}$ and integrate

by using the rule $\displaystyle\int x^n \, dx = \dfrac{1}{n+1} x^{(n+1)} + K$

However, this method fails when $n = -1$ because the resulting integral is meaningless.

Taking a second look at $\dfrac{1}{x}$ it can be *recognised* as the derivative of $\ln x$.

It must be remembered, however, that $\ln x$ is defined only when $x > 0$. Hence, provided that $x > 0$ we have

$$\dfrac{d}{dx}(\ln x) = \dfrac{1}{x} \quad \Longleftrightarrow \quad \int \dfrac{1}{x}\, dx = \ln x + K$$

Now if $x < 0$ the statement $\displaystyle\int \dfrac{1}{x}\, dx = \ln x$ is not valid because the log of a negative number does not exist.

However, $\dfrac{1}{x}$ exists for negative values of x, as the graph of $y = \dfrac{1}{x}$ shows.

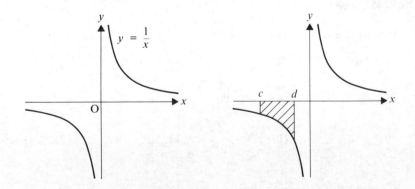

Also, the definite integral $\displaystyle\int_c^d \dfrac{1}{x}\, dx$, which is represented by the shaded area, clearly exists. It must, therefore, be possible to integrate $\dfrac{1}{x}$ when x is negative and we see opposite how to deal with the problem.

If $x < 0$ then $-x > 0$

i.e. $$\int \frac{1}{x} \, dx = \int \frac{-1}{(-x)} \, dx = \ln(-x) + K$$

Thus, when $x > 0$, $\int \frac{1}{x} \, dx = \ln x + K$

and when $x < 0$, $\int \frac{1}{x} \, dx = \ln(-x) + K$

These two results can be combined so that, for both positive and negative values of x, we have

$$\int \frac{1}{x} \, dx = \ln|x| + K$$

where $|x|$ denotes the numerical value of x regardless of sign

e.g. $|-1| = 1$ and $|-4| = 4$

The expression $\ln|x| + K$ can be simplified if K is replaced by $\ln A$, where A is a positive constant, giving

$$\int \frac{1}{x} \, dx = \ln|x| + \ln A = \ln A|x|$$

Further $\frac{d}{dx}(\ln x^c) = \frac{d}{dx}(c \ln x) = \frac{c}{x}$

$\therefore$ $$\int \frac{c}{x} \, dx = c \ln|x| + K \quad \text{or} \quad c \ln A|x|$$

e.g. $$\int \frac{4}{x} \, dx = 4 \ln|x| + K \quad \text{or} \quad 4 \ln A|x|$$

Also $\frac{d}{dx} \ln(ax + b) = \frac{a}{ax + b}$

$\therefore$ $$\int \frac{1}{ax + b} \, dx = \frac{1}{a} \ln|ax + b| + K = \frac{1}{a} \ln A|ax + b|$$

e.g. $$\int \frac{1}{2x + 5} \, dx = \frac{1}{2} \ln|2x + 5| \quad \text{or} \quad \frac{1}{2} \ln A|2x + 5|$$

and $$\int \frac{1}{4 - 3x} \, dx = -\frac{1}{3} \ln|4 - 3x| + K \quad \text{or} \quad -\frac{1}{3} \ln A|4 - 3x|$$

$$= \frac{1}{3} \ln \frac{A}{|4 - 3x|}$$

EXERCISE 16b

Integrate w.r.t. x giving each answer in a form which

(a) uses K (b) uses $\ln A$ and is simplified.

1. $\dfrac{1}{2x}$ **2.** $\dfrac{4}{x}$ **3.** $\dfrac{1}{3x+1}$ **4.** $\dfrac{3}{1-2x}$

5. $\dfrac{6}{2+3x}$ **6.** $\dfrac{3}{4-2x}$ **7.** $\dfrac{4}{1-x}$ **8.** $\dfrac{5}{6-7x}$

Evaluate

9. $\displaystyle\int_{1}^{2}\dfrac{3}{x}\,dx$ **10.** $\displaystyle\int_{1}^{2}\dfrac{1}{2x}\,dx$ **11.** $\displaystyle\int_{4}^{5}\dfrac{2}{x-3}\,dx$ **12.** $\displaystyle\int_{0}^{1}\dfrac{1}{2-x}\,dx$

INTEGRATING TRIGONOMETRIC FUNCTIONS

Knowing the derivatives of the six trig functions, we can recognise the following integrals.

$$\frac{d}{dx}(\sin x) = \cos x \qquad\Longleftrightarrow\qquad \int \cos x\, dx = \sin x + K$$

$$\frac{d}{dx}(\cos x) = -\sin x \qquad\Longleftrightarrow\qquad \int \sin x\, dx = -\cos x + K$$

$$\frac{d}{dx}(\tan x) = \sec^2 x \qquad\Longleftrightarrow\qquad \int \sec^2 x\, dx = \tan x + K$$

$$\frac{d}{dx}(\sec x) = \sec x \tan x \qquad\Longleftrightarrow\qquad \int \sec x \tan x\, dx = \sec x + K$$

$$\frac{d}{dx}(\operatorname{cosec} x) = -\operatorname{cosec} x \cot x \qquad\Longleftrightarrow\qquad \int \operatorname{cosec} x \cot x\, dx = -\operatorname{cosec} x + K$$

$$\frac{d}{dx}(\cot x) = -\operatorname{cosec}^2 x \qquad\Longleftrightarrow\qquad \int \operatorname{cosec}^2 x\, dx = -\cot x + K$$

Remembering the derivatives of some variations of the basic trig functions we also have

$$\int c \cos x \, dx = c \sin x + K$$

and
$$\int \cos(ax + b) \, dx = \frac{1}{a} \sin(ax + b) + K$$

with similar results for the remaining trig integrals,

e.g.
$$\int 3 \sec^2 x \, dx = 3 \tan x + K$$

$$\int \sin 4\theta \, d\theta = -\tfrac{1}{4} \cos 4\theta + K$$

$$\int \operatorname{cosec}^2(2x + \tfrac{3}{4}\pi) \, dx = -\tfrac{1}{2} \cot(2x + \tfrac{3}{4}\pi) + K$$

$$\int \operatorname{cosec} 5\theta \cot 5\theta \, d\theta = -\tfrac{1}{5} \operatorname{cosec} 5\theta + K$$

$$\int \sec(\tfrac{1}{2}\pi - 6x)\tan(\tfrac{1}{2}\pi - 6x) \, dx = -\tfrac{1}{6} \sec(\tfrac{1}{2}\pi - 6x) + K$$

Note that there is no need to *learn* these standard integrals. Knowledge of the standard derivatives is sufficient.

EXERCISE 16c

Integrate each function w.r.t. x

1. $\sin 2x$
2. $\cos 7x$
3. $\sec^2 4x$
4. $\sin(\tfrac{1}{4}\pi + x)$
5. $3\cos(4x - \tfrac{1}{2}\pi)$
6. $\sec^2(\tfrac{1}{3}\pi + 2x)$
7. $\operatorname{cosec}^2 4x$
8. $2\sin(3x - \alpha)$
9. $5\cos(\alpha - \tfrac{1}{2}x)$
10. $5\sec 4x \tan 4x$
11. $\cos 3x - \cos x$
12. $\sec^2 2x - \operatorname{cosec}^2 4x$

Evaluate

13. $\displaystyle\int_0^{\pi/6} \sin 3x \, dx$
14. $\displaystyle\int_{\pi/4}^{\pi/6} \cos(2x - \tfrac{1}{2}\pi) \, dx$
15. $\displaystyle\int_0^{\pi/2} 2\sin(2x - \tfrac{1}{2}\pi) \, dx$
16. $\displaystyle\int_0^{\pi/8} \sec^2 2x \, dx$

FUNCTIONS WHOSE INTEGRALS ARE INVERSE TRIG FUNCTIONS

We know that $\qquad y = \arcsin x \qquad \Rightarrow \qquad \dfrac{dy}{dx} = \dfrac{1}{\sqrt{(1-x^2)}}$

Therefore
$$\int \frac{1}{\sqrt{(1-x^2)}}\,dx = \arcsin x + K$$

Similarly it can be seen that

$$\int \frac{1}{1+x^2}\,dx = \arctan x + K$$

Now consider $\qquad y = \arcsin \dfrac{x}{a} \qquad \Rightarrow \qquad x = a\sin y$

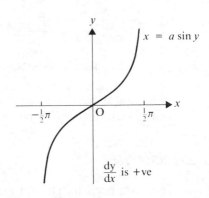

Hence $\qquad \dfrac{dx}{dy} = a\cos y$

$$= a\left(\frac{\sqrt{(a^2-x^2)}}{a}\right)$$

$\Rightarrow \qquad \dfrac{dy}{dx} = \dfrac{1}{\sqrt{(a^2-x^2)}}$

$x = a\sin y$

$\dfrac{dy}{dx}$ is +ve

Therefore
$$\int \frac{1}{\sqrt{(a^2-x^2)}}\,dx = \arcsin\frac{x}{a} + K$$

Similar working shows that if $\quad y = \arctan\dfrac{x}{a} \qquad \Rightarrow \qquad x = a\tan y$

then $\qquad \dfrac{dx}{dy} = a\sec^2 y = a\left(\dfrac{\sqrt{(a^2-x^2)}}{a}\right)^2$

Therefore
$$\int \frac{a}{a^2+x^2}\,dx = \arctan\frac{x}{a} + K$$

EXERCISE 16d

Write down the integral w.r.t. x of each function.

1. $\dfrac{1}{\sqrt{(1 - x^2)}}$ **2.** $\dfrac{2}{x^2 + 4}$ **3.** $\dfrac{1}{\sqrt{(4 - x^2)}}$ **4.** $\dfrac{3}{1 + x^2}$

5. $\dfrac{5}{x^2 + 9}$ **6.** $\dfrac{1}{\sqrt{(9 - x^2)}}$ **7.** $\dfrac{1}{16 + x^2}$ **8.** $\dfrac{5}{\sqrt{(2 - x^2)}}$

EXERCISE 16e

This exercise contains a variety of functions.
The reader is advised always to check that an integral is correct by differentiating it mentally.

Integrate w.r.t. x

1. $\sin(\tfrac{1}{2}\pi - 2x)$ **2.** $e^{(4x - 1)}$

3. $\sec^2 7x$ **4.** $\dfrac{1}{2x - 3}$

5. $\dfrac{1}{\sqrt{(2x - 3)}}$ **6.** $\dfrac{1}{(3x - 2)^2}$

7. 5^x **8.** $\operatorname{cosec}\tfrac{1}{2}x \cot \tfrac{1}{2}x$

9. $(3x - 5)^2$ **10.** $e^{(4x - 5)}$

11. $\sqrt{(4x - 5)}$ **12.** $\operatorname{cosec}^2 3x$

13. $\dfrac{3}{2(1 - x)}$ **14.** $10^{(x + 1)}$

15. $\cos(3x - \tfrac{1}{3}\pi)$ **16.** $\tan^3 2x$

Evaluate

17. $\displaystyle\int_{-1/2}^{1/2} \sqrt{(1 - 2x)}\, dx$ **18.** $\displaystyle\int_{0}^{2} e^{(x/2 + 1)}\, dx$

19. $\displaystyle\int_{\pi/4}^{\pi/2} \sin 4x\, dx$ **20.** $\displaystyle\int_{0}^{1/2} \dfrac{1}{\sqrt{(1 - x^2)}}\, dx$

THE RECOGNITION ASPECT OF INTEGRATION

We have already seen the importance of the recognition aspect of integration in compiling a set of standard integrals.
Recognition is equally important when it is used to avoid serious errors in integration.

Consider, for instance, the derivative of the product $x^2 \sin x$
Using the product formula gives

$$\frac{d}{dx}(x^2 \sin x) = 2x \sin x + x^2 \cos x$$

Clearly the derivative is the sum of two products therefore the integral of a single product is not itself a simple product

i.e.   **integration is not distributive when applied to a product.**

On the other hand, when we differentiate the function of a function $(1 + x^2)^3$ we get

$$\frac{d}{dx}(1 + x^2)^3 = 6x(1 + x^2)^2$$

This time the derivative *is* a product so clearly the integral of a product *may* be a function of a function.

INTEGRATING PRODUCTS

First consider the function e^u where u is a function of x.
Differentiating as a function of a function gives

$$\frac{d}{dx}(e^u) = \left(\frac{du}{dx}\right)(e^u)$$

Thus any product of the form $\left(\frac{du}{dx}\right)e^u$ can be integrated by recognition, since

$$\int \left(\frac{du}{dx}\right)e^u \, dx = e^u + K$$

e.g.
$$\int 2x\, e^{x^2}\, dx = e^{x^2} + K \quad (u = x^2)$$

$$\int \cos x\, e^{\sin x}\, dx = e^{\sin x} + K \quad (u = \sin x)$$

$$\int x^2\, e^{x^3}\, dx = \tfrac{1}{3}\int 3x^2\, e^{x^3}\, dx = \tfrac{1}{3} e^{x^3} + K \quad (u = x^3)$$

In these simple cases the substitution of u for $f(x)$ is done mentally. All the results can be checked by differentiating them mentally.

Similar, but slightly less simple functions, can also be integrated by changing the variable but for these the substitution is written down.

Changing the Variable

Consider a general function $g(u)$ where u is a function of x

$$\frac{d}{dx}g(u) = \frac{du}{dx}g'(u) \quad \text{or} \quad g'(u)\frac{du}{dx}$$

Therefore
$$\int g'(u)\frac{du}{dx}\, dx = g(u) + K \qquad [1]$$

We also know that
$$\int g'(u)\, du = g(u) + K \qquad [2]$$

Comparing [1] and [2] gives

$$\int g'(u)\frac{du}{dx}\, dx = \int g'(u)\, du$$

Replacing $g'(u)$ by $f(u)$ gives

$$\int f(u)\frac{du}{dx}\, dx = \int f(u)\, du$$

i.e.
$$\ldots \frac{du}{dx}\, dx \equiv \ldots\, du \qquad [3]$$

Thus integrating (a function of u) $\dfrac{du}{dx}$ w.r.t. x is *equivalent to* integrating (the same function of u) w.r.t. u

i.e. the relationship in [3] is neither an equation nor an identity but is a pair of equivalent operations.

Suppose, for example, that we want to find $\displaystyle\int 2x(x^2 + 1)^5 \, dx$

Writing the integral in the form $\displaystyle\int (x^2 + 1)^5 \, 2x \, dx$ and making the substitution $u = x^2 + 1$ gives

$$\int (x^2 + 1)^5 \, 2x \, dx = \int u^5(2x) \, dx$$

But $\dfrac{du}{dx} = 2x$ and $\ldots \dfrac{du}{dx} \, dx \equiv \ldots du$

Therefore $\ldots 2x \, dx \equiv \ldots du$

i.e. $\displaystyle\int (x^2 + 1)^5 \, 2x \, dx = \int u^5 \, du$

$$= \tfrac{1}{6}u^6 + K = \tfrac{1}{6}(x^2 + 1)^6 + K$$

In practice we can go direct from $\dfrac{du}{dx} = 2x$ to the equivalent operators $\ldots 2x \, dx \equiv \ldots du$ by 'separating the variables'.

Products which can be integrated by this method are those in which one factor is basically the derivative of the function in the other factor, and we choose to substitute x for the function in the other factor.

Examples 16f

1. Integrate $x^2\sqrt{(x^3 + 5)}$ w.r.t. x

In this product x^2 is basically the derivative of $x^3 + 5$ so we choose the substitution $u = x^3 + 5$

If $u = x^3 + 5$ then $\dfrac{du}{dx} = 3x^2$

$\Rightarrow$ $\ldots du \equiv \ldots 3x^2 \, dx$

Hence $\displaystyle\int x^2\sqrt{(x^3 + 5)} \, dx = \tfrac{1}{3}\int (x^3 + 5)^{1/2}(3x^2 \, dx) = \tfrac{1}{3}\int u^{1/2} \, du$

$$= (\tfrac{1}{3})(\tfrac{2}{3})u^{3/2} + K$$

i.e. $\displaystyle\int x^2\sqrt{(x^3 + 5)} \, dx = \tfrac{2}{9}(x^3 + 5)^{3/2} + K$

2. Find $\int \cos x \sin^3 x \, dx$

Writing the given integral in the form $\cos x (\sin x)^3$ shows that a suitable substitution is $u = \sin x$

If $u = \sin x$ then $\ldots du \equiv \ldots \cos x \, dx$

$$\therefore \qquad \int \cos x \sin^3 x \, dx = \int (\sin x)^3 \cos x \, dx = \int u^3 \, du$$

$$= \tfrac{1}{4} u^4 + K$$

i.e. $\qquad \int \cos x \sin^3 x \, dx = \tfrac{1}{4} \sin^4 x + K$

Applied generally, the method used above shows that

$$\int \cos x \sin^n x \, dx = \frac{1}{n+1} \sin^{(n+1)} x + K$$

and similarly that

$$\int \sin x \cos^n x \, dx = \frac{-1}{n+1} \cos^{(n+1)} x + K$$

3. Find $\int \dfrac{\ln x}{x} \, dx$

Initially this looks like a fraction but once it is recognised as the product of $\dfrac{1}{x}$ and $\ln x$, it is clear that $\dfrac{1}{x} = \dfrac{d}{dx}(\ln x)$ and that we can make the substitution $u = \ln x$

If $u = \ln x$ then $\ldots du \equiv \ldots \dfrac{1}{x} \, dx$

Hence $\qquad \int \dfrac{1}{x} \ln x \, dx = \int u \, du = \tfrac{1}{2} u^2 + K$

i.e. $\qquad \int \dfrac{\ln x}{x} \, dx = \tfrac{1}{2}(\ln x)^2 + K$

Note that $(\ln x)^2$ is *not* the same as $\ln x^2$

EXERCISE 16f

Integrate the following expressions w.r.t. x

1. $4x^3 e^{x^4}$

2. $\sin x \, e^{\cos x}$

3. $\sec^2 x \, e^{\tan x}$

4. $(2x + 1)e^{(x^2 + x)}$

5. $\csc^2 x \, e^{(1 - \cot x)}$

6. $(1 + \cos x) \, e^{(x + \sin x)}$

7. $2x \, e^{(1 + x^2)}$

8. $(3x^2 - 2) \, e^{(x^3 - 2x)}$

Find the following integrals by making the substitution suggested.

9. $\int x(x^2 - 3)^4 \, dx$ $u = x^2 - 3$

10. $\int x\sqrt{(1 - x^2)} \, dx$ $u = 1 - x^2$

11. $\int \cos 2x (\sin 2x + 3)^2 \, dx$ $u = \sin 2x + 3$

12. $\int x^2(1 - x^3) \, dx$ $u = 1 - x^3$

13. $\int e^x \sqrt{(1 + e^x)} \, dx$ $u = 1 + e^x$

14. $\int \cos x \, \sin^4 x \, dx$ $u = \sin x$

15. $\int \sec^2 x \, \tan^3 x \, dx$ $u = \tan x$

16. $\int x^n (1 + x^{n+1})^2 \, dx$ $u = 1 + x^{n+1}$

17. $\int \csc^2 x \, \cot^2 x \, dx$ $u = \cot x$

18. $\int \sqrt{x} \, \sqrt{(1 + x^{3/2})} \, dx$ $u = 1 + x^{3/2}$

By using a suitable substitution, or by integrating at sight, find

19. $\int x^3(x^4 + 4)^2 \, dx$

20. $\int e^x(1 - e^x)^3 \, dx$

21. $\int \sin \theta \sqrt{(1 - \cos \theta)} \, d\theta$

22. $\int (x + 1)\sqrt{(x^2 + 2x + 3)} \, dx$

23. $\int x \, e^{(x^2 + 1)} \, dx$

24. $\int \sec^2 x \, (1 + \tan x) \, dx$

DEFINITE INTEGRATION WITH A CHANGE OF VARIABLE

A definite integral can be evaluated only after the appropriate integration has been performed. Should this require a change of variable, e.g. from x to u, it is usually most convenient to change the limits of integration also from x values to u values.

Example 16g _____

By using the substitution $u = x^3 + 1$, evaluate $\displaystyle\int_0^1 x^2\sqrt{(x^3 + 1)}\,dx$

If $u = x^3 + 1$ then $\quad \ldots du \equiv \ldots 3x^2\,dx$

$$\text{and} \quad \begin{cases} x = 0 \implies u = 1 \\ x = 1 \implies u = 2 \end{cases}$$

Hence

$$\int_0^1 x^2\sqrt{(x^3 + 1)}\,dx = \tfrac{1}{3}\int_1^2 \sqrt{u}\,du$$

$$= \tfrac{1}{3}\left[\tfrac{2}{3}u^{3/2}\right]_1^2$$

$$= \tfrac{2}{9}(2\sqrt{2} - 1)$$

_____

EXERCISE 16g

Evaluate

1. $\displaystyle\int_0^1 x\,e^{x^2}\,dx$

2. $\displaystyle\int_0^{\pi/2} \cos x \sin^4 x\,dx$

3. $\displaystyle\int_1^2 \frac{1}{x}\ln x\,dx$

4. $\displaystyle\int_1^2 x^2(x^3 - 1)^4\,dx$

5. $\displaystyle\int_0^{\pi/4} \sec^2 x\, e^{\tan x}\,dx$

6. $\displaystyle\int_1^2 x(1 + 2x^2)\,dx$

7. $\displaystyle\int_2^3 (x - 1)e^{(x^2 - 2x)}\,dx$

8. $\displaystyle\int_0^{\pi/6} \cos x(1 + \sin^2 x)\,dx$

9. $\displaystyle\int_1^3 \frac{1}{x}(\ln x)^2\,dx$

10. $\displaystyle\int_0^3 x\sqrt{(1 + x^2)}\,dx$

INTEGRATION BY PARTS

It is not always possible to express a product in the form $f(u)\dfrac{du}{dx}$ so an alternative approach is needed.

Looking again at the differentiation of a product uv where u and v are both functions of x we have

$$\frac{d}{dx}(uv) = v\frac{du}{dx} + u\frac{dv}{dx} \quad\Rightarrow\quad v\frac{du}{dx} = \frac{d}{dx}(uv) - u\frac{dv}{dx}$$

Now $v\dfrac{du}{dx}$ can be taken to represent a product which is to be integrated w.r.t. x

Thus $$\int v\frac{du}{dx}\,dx = \int \frac{d}{dx}(uv)\,dx - \int u\frac{dv}{dx}\,dx$$

i.e. $$\int v\frac{du}{dx}\,dx = uv - \int u\frac{dv}{dx}\,dx$$ ✷ change round. us and vs.

At this stage it may appear that the RHS is more complicated than the original product on the LHS.

However, by careful choice of the factor to be replaced by v we can ensure that $u\dfrac{dv}{dx}$ is easier to integrate than $v\dfrac{du}{dx}$

The factor chosen to be replaced by v is usually the one whose derivative is a simpler function. It must also be remembered, however, that the other factor is replaced by $\dfrac{du}{dx}$ and therefore it must be possible to integrate it.

This method for integrating a product is called *integrating by parts*.

Examples 16h

1. Integrate $x\,e^x$ w.r.t. x

Taking $$v = x \quad\text{and}\quad \frac{du}{dx} = e^x$$

gives $$\frac{dv}{dx} = 1 \quad\text{and}\quad u = e^x$$

Then
$$\int v \frac{du}{dx} \, dx = uv - \int u \frac{dv}{dx} \, dx$$

gives
$$\int x \, e^x \, dx = (e^x)(x) - \int (e^x)(1) \, dx$$

$$= x \, e^x - e^x + K$$

---

2. Find $\int x^2 \sin x \, dx$

Taking
$$v = x^2 \quad \text{and} \quad \frac{du}{dx} = \sin x$$

gives
$$\frac{dv}{dx} = 2x \quad \text{and} \quad u = -\cos x$$

Then
$$\int v \frac{du}{dx} \, dx = uv - \int u \frac{dv}{dx} \, dx$$

gives
$$\int x^2 \sin x \, dx = (-\cos x)(x^2) - \int (-\cos x)(2x) \, dx$$

$$= -x^2 \cos x + 2 \int x \cos x \, dx \qquad [1]$$

At this stage the integral on the RHS cannot be found without *repeating* the process of integrating by parts on the term $\int x \cos x \, dx$ as follows.

Taking
$$v = x \quad \text{and} \quad \frac{du}{dx} = \cos x$$

gives
$$\frac{dv}{dx} = 1 \quad \text{and} \quad u = \sin x$$

Then
$$\int x \cos x \, dx = (\sin x)(x) - \int (\sin x)(1) \, dx$$

$$= x \sin x + \cos x + K$$

Hence equation [1] becomes

$$\int x^2 \sin x \, dx = -x^2 \cos x + 2x \sin x + 2 \cos x + K$$

---

3. Find $\int x^4 \ln x \, dx$

Because $\ln x$ can be differentiated but *not integrated*, we must use $\quad v = \ln x$

Taking $\qquad\qquad\qquad\qquad v = \ln x \quad$ and $\quad \dfrac{du}{dx} = x^4$

gives $\qquad\qquad\qquad\qquad \dfrac{dv}{dx} = \dfrac{1}{x} \quad$ and $\quad u = \tfrac{1}{5}x^5$

The formula for integrating by parts then gives

$$\int x^4 \ln x \, dx = (\tfrac{1}{5}x^5)(\ln x) - \int (\tfrac{1}{5}x^5)\left(\dfrac{1}{x}\right) dx$$

$$= \tfrac{1}{5}x^5 \ln x - \tfrac{1}{5}\int x^4 \, dx$$

$\Rightarrow \qquad\qquad \int x^4 \ln x \, dx = \tfrac{1}{5}x^5 \ln x - \tfrac{1}{25}x^5 + K$

Special Cases of Integration by Parts

An interesting situation arises when an attempt is made to integrate $e^x \cos x \quad$ or $\quad e^x \sin x$

---

4. Find $\int e^x \cos x \, dx$

Taking $\qquad\qquad\qquad\qquad v = e^x \quad$ and $\quad \dfrac{du}{dx} = \cos x$

gives $\qquad\qquad\qquad\qquad \dfrac{dv}{dx} = e^x \quad$ and $\quad u = \sin x$

Hence $\qquad\qquad \int e^x \cos x \, dx = e^x \sin x - \int e^x \sin x \, dx \qquad\qquad$ [1]

But since $\int e^x \sin x \, dx \quad$ is very similar to $\quad \int e^x \cos x \, dx \quad$ it seems that we have made no progress. However, if we now apply integration by parts to $\int e^x \sin x \, dx$ an interesting situation emerges.

Taking $$v = e^x \quad \text{and} \quad \frac{du}{dx} = \sin x$$

gives $$\frac{dv}{dx} = e^x \quad \text{and} \quad u = -\cos x$$

so that $$\int e^x \sin x \, dx = -e^x \cos x + \int e^x \cos x \, dx$$

or $$\int e^x \cos x \, dx = e^x \cos x + \int e^x \sin x \, dx \qquad [2]$$

Adding [1] and [2] gives $\quad 2 \int e^x \cos x \, dx = e^x(\sin x + \cos x) + K$

i.e. $$\int e^x \cos x \, dx = \tfrac{1}{2}e^x(\sin x + \cos x) + K$$

Clearly the same two equations can be used to give

$$2 \int e^x \sin x \, dx = e^x(\sin x - \cos x) + K$$

Note that neither of the equations [1] and [2] contains a completed integration process, so the constant of integration is introduced only when these two equations have been combined.

Note also that the same choice of function for v must be made in both applications of integration by parts, i.e. we chose $v = e^x$ each time.

Integration of ln x

So far we have found no way to integrate $\ln x$. Now, however, if $\ln x$ is regarded as the product of 1 and $\ln x$ we can apply integration by parts as follows.

Examples 16h (continued)

5. Find $\int \ln x \, dx$

Taking $$v = \ln x \quad \text{and} \quad \frac{du}{dx} = 1$$

gives $$\frac{dv}{dx} = \frac{1}{x} \quad \text{and} \quad u = x$$

Then
$$\int v \frac{du}{dx} dx = uv - \int u \frac{dv}{dx} dx$$

becomes
$$\int \ln x \, dx = x \ln x - \int x \left(\frac{1}{x}\right) dx$$
$$= x \ln x - x + K$$

i.e.
$$\int \ln x \, dx = x(\ln x - 1) + K$$

This 'trick' of multiplying by 1 to create a product can also be used to integrate arcsin x and other inverse trig functions as will be shown in Examples 16i.

EXERCISE 16h

Integrate the following functions w.r.t. x

1. $x \cos x$
2. $x^2 e^x$
3. $x^3 \ln 3x$
4. $x e^{-x}$
5. $3x \sin x$
6. $e^x \sin 2x$
7. $e^{2x} \cos x$
8. $x^2 e^{4x}$
9. $e^{-x} \sin x$
10. $\ln 2x$
11. $e^x(x + 1)$
12. $x(1 + x)^7$
13. $x \sin (x + \frac{1}{6}\pi)$
14. $x \cos nx$
15. $x^n \ln x$
16. $3x \cos 2x$
17. $2e^x \sin x \cos x$
18. $x^2 \sin x$
19. $e^{ax} \sin bx$

20. By writing $\cos^3 \theta$ as $(\cos^2 \theta)(\cos \theta)$ use integration by parts to find $\int \cos^3 \theta \, d\theta$.

Each of the following products can be integrated either:
    (a) by immediate recognition, or
    (b) by a suitable change of variable, or
    (c) by parts.

Choose the best method in each case and hence integrate each function w.r.t x.

21. $(x - 1)e^{x^2 - 2x + 4}$
22. $(x + 1)^2 e^x$
23. $\sin x(4 + \cos x)^3$

24. $\cos x \, e^{\sin x}$ **25.** $x^4\sqrt{(1 + x^5)}$ **26.** $e^x(e^x + 2)^4$

27. $x \, e^{2x-1}$ **28.** $x(1 - x^2)^9$ **29.** $\cos x \sin^5 x$

DEFINITE INTEGRATION BY PARTS

When using the formula

$$\int v \frac{du}{dx} dx = uv - \int u \frac{dv}{dx} dx$$

it must be appreciated that the term uv on the RHS is fully integrated. Consequently in a definite integration, uv must be *evaluated between the appropriate boundaries*

i.e. $$\int_a^b v \frac{du}{dx} dx = \left[uv \right]_a^b - \int_a^b u \frac{dv}{dx} dx$$

Examples 16i

1. Evaluate $\displaystyle\int_0^1 x \, e^x \, dx$

$$\int x \, e^x \, dx = \int v \frac{du}{dx} dx$$

where $$v = x \quad \text{and} \quad \frac{du}{dx} = e^x$$

Hence $$\int_0^1 x \, e^x \, dx = \left[x e^x \right]_0^1 - \int_0^1 e^x \, dx$$

$$= \left[x e^x \right]_0^1 - \left[e^x \right]_0^1$$

$$= (e^1 - 0) - (e^1 - e^0)$$

$$= e - e + 1$$

i.e. $$\int_0^1 x \, e^x \, dx = 1$$

2. Find the value of $\displaystyle\int_0^{1/2} \arccos x \, dx$

We regard $\arccos x$ as a product, i.e. $(1)(\arccos x)$ and integrate by parts.
As $\arccos x$ cannot yet be integrated, it must be replaced by v

Taking $\qquad\qquad\qquad v = \arccos x \qquad$ and $\qquad \dfrac{du}{dx} = 1$

gives $\qquad\qquad\qquad \dfrac{dv}{dx} = \dfrac{-1}{\sqrt{(1-x^2)}} \qquad$ and $\qquad u = x$

Then $\qquad \displaystyle\int_0^{1/2} (1)(\arccos x)\, dx = \left[x \arccos x \right]_0^{1/2} - \int_0^{1/2} \dfrac{-x}{\sqrt{(1-x^2)}}\, dx$

$$= \{(\tfrac{1}{2})(\tfrac{1}{3}\pi) - 0\} + \int_0^{1/2} \dfrac{x}{\sqrt{(1-x^2)}}\, dx$$

Now $\qquad \displaystyle\int_0^{1/2} \dfrac{x}{\sqrt{(1-x^2)}}\, dx = \int_1^{3/4} \dfrac{1}{\sqrt{t}} (-\tfrac{1}{2}dt)$

where $t = 1 - x^2$ and

x	0	$\tfrac{1}{2}$
t	1	$\tfrac{3}{4}$

$$= \left[-\sqrt{t} \right]_1^{3/4}$$

$$= -\tfrac{1}{2}\sqrt{3} + 1$$

$\therefore \qquad\qquad \displaystyle\int \arccos x \, dx = \tfrac{1}{6}\pi - \tfrac{1}{2}\sqrt{3} + 1$

EXERCISE 16i

Evaluate

1. $\displaystyle\int_0^{\pi/2} x \sin x \, dx$
2. $\displaystyle\int_1^2 x^5 \ln x \, dx$
3. $\displaystyle\int_0^1 (x+1)\, e^x \, dx$

4. $\displaystyle\int_0^\pi e^x \cos x \, dx$
5. $\displaystyle\int_1^2 x\sqrt{(x-1)}\, dx$
6. $\displaystyle\int_0^{\pi/2} x^2 \cos x \, dx$

7. $\displaystyle\int_0^1 \arcsin x \, dx$
8. $\displaystyle\int_0^1 x^2 e^x \, dx$
9. $\displaystyle\int_1^{\sqrt{3}} \arctan x \, dx$

MIXED EXERCISE 16

Integrate the following functions, taking care to choose the best method in each case. (Remember that $\exp x \equiv e^x$.)

1. $x^2 e^{2x}$

2. $2x \exp(x^2)$

3. $\sec^2 x(3 \tan x - 4)$

4. $(x + 1)\ln(x + 1)$

5. $\sec^2 x \tan^3 x$

6. $x^2 \cos x$

7. $\sin x \, e^{\cos x}$

8. $x(2x + 3)^7$

9. $(1 - x)\exp(1 - x)^2$

10. $xe^{(2x - 1)}$

11. $\cos x \sin^5 x$

12. $\sin x(4 + \cos x)^3$

13. $(x - 1)e^{(x^2 - 2x + 3)}$

14. $x^2(1 - x^3)^9$

Evaluate each definite integral.

15. $\displaystyle\int_1^3 e^{3x} \, dx$

16. $\displaystyle\int_0^{\pi/8} \cos 4x \, dx$

17. $\displaystyle\int_0^1 \frac{1}{x - 2} \, dx$

18. $\displaystyle\int_{\pi/4}^{\pi/2} \operatorname{cosec}^2 x \, dx$

19. $\displaystyle\int_0^1 x^2 e^{3x^3} \, dx$

20. $\displaystyle\int_0^{\pi/4} x \cos 2x \, dx$

21. $\displaystyle\int_0^1 \frac{1}{2 - x} \ln (2 - x) \, dx$

22. $\displaystyle\int_1^2 x^2 \ln x \, dx$

CHAPTER 17

FURTHER INTEGRATION

INTEGRATING FRACTIONS

Some expressions have an integral that is a function but there are many others for which an exact integral cannot be found. In this book we are concerned mainly with expressions which *can* be integrated but even with this proviso the reader should be aware that, while the methods suggested usually work, they are not infallible.

There are several different methods for integrating fractions, the appropriate method in a particular case depending upon the form of the fraction. Consequently it is very important that each fraction be categorised carefully to avoid embarking on unnecessary and lengthy working.

Method 1   Using Recognition

Consider the function $\ln u$ where $u = f(x)$

Differentiating with respect to x gives

$$\frac{d}{dx}\ln u = \left(\frac{1}{u}\right)\left(\frac{du}{dx}\right) \quad \text{i.e.} \quad \frac{du/dx}{u}$$

i.e.
$$\frac{d}{dx}\ln f(x) = \frac{f'(x)}{f(x)}$$

Hence
$$\int \frac{f'(x)}{f(x)}\,dx = \ln|f(x)| + K$$

Thus all fractions of the form $f'(x)/f(x)$ can be integrated *immediately* by recognition, e.g.

$$\int \frac{\cos x}{1 + \sin x}\,dx = \ln|1 + \sin x| + K \quad \text{as} \quad \frac{d}{dx}(1 + \sin x) = \cos x$$

$$\int \frac{e^x}{e^x + 4}\,dx = \ln|e^x + 4| + K \quad \text{as} \quad \frac{d}{dx}(e^x + 4) = e^x$$

268

Note, however, that $\displaystyle\int \frac{x}{\sqrt{(1+x)}}\,dx$ is *not* equal to $\ln|\sqrt{(1+x)}| + K$

because $\qquad \dfrac{d}{dx}\sqrt{(1+x)}$ is not equal to x

Method 1 applies only to an integral whose numerator is basically the derivative of *the complete denominator*.

An integral whose numerator is the derivative, not of the complete denominator but of a function *within* the denominator, belongs to the next type.

Method 2 Using Substitution

Consider the integral $\displaystyle\int \frac{2x}{\sqrt{(x^2+1)}}\,dx$

Noting that $2x$ is the derivative of $x^2 + 1$ we make the substitution $u = x^2 + 1$, i.e.

$$\text{if } \; u = x^2 + 1 \quad \text{then} \quad \ldots du \equiv \ldots 2x\,dx$$

By this change of variable the given integral is converted into the simple form $\displaystyle\int \frac{1}{\sqrt{u}}\,du$

Examples 17a _____

1. Find $\displaystyle\int \frac{x^2}{1+x^3}\,dx$

$$\int \frac{x^2}{1+x^3}\,dx = \frac{1}{3}\int \frac{3x^2}{1+x^3}\,dx$$

This integral is of the form $\displaystyle\int \frac{f'(x)}{f(x)}\,dx$ so we use recognition

$$= \tfrac{1}{3}\ln|1+x^3| + K$$

_____

2. By writing $\tan x$ as $\dfrac{\sin x}{\cos x}$, find $\displaystyle\int \tan x \, dx$

$$\int \tan x \, dx = \int \frac{\sin x}{\cos x} \, dx = -\int \frac{f'(x)}{f(x)} \, dx \quad \text{where} \quad f(x) = \cos x$$

so
$$\int \frac{\sin x}{\cos x} \, dx = -\ln|\cos x| + K$$

$\therefore$
$$\int \tan x \, dx = K - \ln|\cos x| \quad \text{or} \quad K + \ln|\sec x|$$

Note that, similarly, $\displaystyle\int \cot x \, dx = \ln|\sin x| + K$

3. Find $\displaystyle\int \frac{e^x}{(1 - e^x)^2} \, dx$

e^x is basically the derivative of $1 - e^x$ but not of $(1 - e^x)^2$ so we make the substitution $u = 1 - e^x$

If $u = 1 - e^x$ then $\ldots du \equiv \ldots -e^x \, dx$

So
$$\int \frac{e^x}{(1 - e^x)^2} \, dx = \int \frac{-1}{u^2} \, du = \frac{1}{u} + K$$

$\therefore$
$$\int \frac{e^x}{(1 - e^x)^2} \, dx = \frac{1}{1 - e^x} + K$$

4. Find $\displaystyle\int \frac{\sec^2 x}{\tan^3 x} \, dx$

$\sec^2 x$ is the derivative of $\tan x$ but not of $\tan^3 x$

Taking $u = \tan x$ gives $\ldots du \equiv \ldots \sec^2 x \, dx$

Then
$$\int \frac{\sec^2 x}{\tan^3 x} \, dx = \int \frac{1}{u^3} \, du = -\tfrac{1}{2} u^{-2} + K$$

i.e.
$$\int \frac{\sec^2 x}{\tan^3 x} \, dx = \frac{-1}{2 \tan^2 x} + K$$

EXERCISE 17a

In Questions 1 to 18 integrate each function w.r.t. x

1. $\dfrac{\cos x}{4 + \sin x}$

2. $\dfrac{e^x}{3e^x - 1}$

3. $\dfrac{x}{(1 - x^2)^3}$

4. $\dfrac{\sin x}{\cos^3 x}$

5. $\dfrac{x^3}{1 + x^4}$

6. $\dfrac{2x + 3}{x^2 + 3x - 4}$

7. $\dfrac{x^2}{\sqrt{(2 + x^3)}}$

8. $\dfrac{\cos x}{(\sin x - 2)^2}$

9. $\dfrac{1}{x \ln x}$ i.e. $\dfrac{1/x}{\ln x}$

10. $\dfrac{\cos x}{\sin^6 x}$

11. $\dfrac{\csc^2 x}{\cot^4 x}$

12. $\dfrac{e^x}{\sqrt{(1 - e^x)}}$

13. $\dfrac{x - 1}{3x^2 - 6x + 1}$

14. $\dfrac{\cos x}{\sin^n x}$

15. $\dfrac{\sin x}{\cos^n x}$

16. $\dfrac{\sec x \tan x}{4 + \sec x}$

17. $\dfrac{\sec^2 x}{(1 - \tan x)^3}$

18. $\dfrac{\sin x}{(3 + \cos x)^2}$

19. By writing $\sec x$ as $\dfrac{\sec x(\sec x + \tan x)}{(\tan x + \sec x)}$ find $\int \sec x \, dx$

20. By writing $\csc x$ as $\dfrac{\csc x(\csc x + \cot x)}{(\cot x + \csc x)}$ find $\int \csc x \, dx$

Evaluate

21. $\displaystyle\int_1^2 \dfrac{2x + 1}{x^2 + x} \, dx$

22. $\displaystyle\int_0^1 \dfrac{x}{x^2 + 1} \, dx$

23. $\displaystyle\int_2^3 \dfrac{2x}{(x^2 - 1)^3} \, dx$

24. $\displaystyle\int_0^1 \dfrac{e^x}{(1 + e^x)^2} \, dx$

25. $\displaystyle\int_{\pi/6}^{\pi/3} \dfrac{\sin 2x}{\cos(2x - \pi)} \, dx$

26. $\displaystyle\int_2^4 \dfrac{1}{x(\ln x)^2} \, dx$

Quotable Results

Some of the integrals found in Examples 17a and Exercise 17a are important enough to be regarded as standard and are listed here

$$\int \tan x \, dx = \ln|\sec x| + K$$

$$\int \cot x \, dx = \ln|\sin x| + K$$

$$\int \sec x \, dx = \ln|\sec x + \tan x| + K$$

$$\int \cosec x \, dx = -\ln|\cosec x + \cot x| + K$$

USING PARTIAL FRACTIONS

If a fraction has not fallen into any of the previous categories, it may be that it is easy to integrate when expressed in partial fractions. Remember however that only proper fractions can be converted directly into partial fractions; an improper fraction must first be divided out until it comprises non-fractional terms and a proper fraction.

It is not very often that actual long division is needed. Usually a simple adjustment in the numerator is all that is required, as the following examples show. When such an adjustment is not obvious, however, long division can always be used.

Examples 17b _____

1. Integrate $\dfrac{2x - 3}{(x - 1)(x - 2)}$ w.r.t. x

Using the cover-up method gives

$$\frac{2x - 3}{(x - 1)(x - 2)} = \frac{1}{x - 1} + \frac{1}{x - 2}$$

$$\therefore \qquad \int \frac{2x - 3}{(x - 1)(x - 2)} \, dx = \int \frac{1}{x - 1} \, dx + \int \frac{1}{x - 2} \, dx$$

$$= \ln|x - 1| + \ln|x - 2| + \ln A$$

$$= \ln A |(x - 1)(x - 2)|$$

2. Find $\displaystyle\int \frac{x^2 + 1}{x^2 - 1}\,dx$

This fraction is improper so, before we can factorise the denominator and use partial fractions we must adjust the given fraction as follows.

$$\frac{x^2 + 1}{x^2 - 1} = \frac{(x^2 - 1) + 2}{x^2 - 1} = 1 + \frac{2}{x^2 - 1} = 1 + \frac{2}{(x - 1)(x + 1)}$$

Then $\displaystyle\int \frac{x^2 + 1}{x^2 - 1}\,dx = \int 1\,dx + \int \frac{1}{x - 1}\,dx - \int \frac{1}{x + 1}\,dx$

$$= x + \ln|x - 1| - \ln|x + 1| + \ln A$$

$$= x + \ln \frac{A|x - 1|}{|x + 1|}$$

Even when improper fractions do not need conversion into partial fractions, it is still essential to reduce to proper form before attempting to integrate, i.e.

$$\int \frac{2x + 4}{x + 1}\,dx = \int \frac{2(x + 1) + 2}{x + 1}\,dx = \int 2\,dx + \int \frac{2}{x + 1}\,dx$$

$$= 2x + 2\ln A|x + 1|$$

The reader should not fall into the trap of thinking that, whenever the denominator of a fraction factorises, integration will involve partial fractions.

Careful scrutiny is vital, as fractions requiring quite different integration techniques often *look* very similar. The following example shows this clearly.

Examples 17b (continued)

3. Integrate w.r.t. x,

(a) $\displaystyle\frac{x + 1}{x^2 + 2x - 8}$ (b) $\displaystyle\frac{x + 1}{(x^2 + 2x - 8)^2}$ (c) $\displaystyle\frac{x + 2}{x^2 + 2x - 8}$

(a) This fraction is basically of the form $f'(x)/f(x)$

$$\int \frac{x + 1}{x^2 + 2x - 8}\,dx = \frac{1}{2}\int \frac{2x + 2}{x^2 + 2x - 8}\,dx$$

$$= \tfrac{1}{2}\ln A|x^2 + 2x - 8|$$

(b) This time the numerator is basically the derivative of the function *within* the denominator so we use

$$u = x^2 + 2x - 8 \quad \Rightarrow \quad \ldots du \equiv \ldots (2x + 2)\, dx \equiv \ldots 2(x + 1)\, dx$$

$$\therefore \quad \int \frac{x + 1}{(x^2 + 2x - 8)^2}\, dx = \frac{1}{2} \int \frac{1}{u^2}\, du = -\frac{1}{2u} + K$$

$$= K - \frac{1}{2(x^2 + 2x - 8)}$$

(c) In this fraction the numerator is not related to the derivative of the denominator so, as the denominator factorises, we use partial fractions.

$$\int \frac{x + 2}{x^2 + 2x - 8}\, dx = \int \frac{\frac{1}{3}}{x + 4}\, dx + \int \frac{\frac{2}{3}}{x - 2}\, dx$$

$$= \tfrac{1}{3} \ln|x + 4| + \tfrac{2}{3} \ln|x - 2| + \ln A$$

$$= \ln A \,|(x + 4)^{1/3}(x - 2)^{2/3}|$$

EXERCISE 17b

Integrate each of the following functions w.r.t. x

1. $\dfrac{2}{x(x + 1)}$

2. $\dfrac{4}{(x - 2)(x + 2)}$

3. $\dfrac{x}{(x - 1)(x + 1)}$

4. $\dfrac{x - 1}{x(x + 2)}$

5. $\dfrac{x - 1}{(x - 2)(x - 3)}$

6. $\dfrac{1}{x(x - 1)(x + 1)}$

7. $\dfrac{x}{x + 1}$

8. $\dfrac{x + 4}{x}$

9. $\dfrac{x}{x + 4}$

10. $\dfrac{3x - 4}{x(1 - x)}$

11. $\dfrac{x^2 - 2}{x^2 - 1}$

12. $\dfrac{x^2}{(x + 1)(x + 2)}$

Choose the best method to integrate each function.

13. $\dfrac{x}{x^2 - 1}$

14. $\dfrac{2x}{(x^2 - 1)^2}$

15. $\dfrac{2}{x^2 - 1}$

16. $\dfrac{2x - 5}{x^2 - 5x + 6}$

17. $\dfrac{2x}{x^2 - 5x + 6}$

18. $\dfrac{2x - 3}{x^2 - 5x + 6}$

Evaluate

19. $\displaystyle\int_0^4 \frac{x+2}{x+1}\,dx$ **20.** $\displaystyle\int_{-1}^1 \frac{5}{x^2+x-6}\,dx$ **21.** $\displaystyle\int_1^2 \frac{x+2}{x(x+4)}\,dx$

22. $\displaystyle\int_0^1 \frac{2}{3+2x}\,dx$ **23.** $\displaystyle\int_{1/2}^3 \frac{2}{(3+2x)^2}\,dx$ **24.** $\displaystyle\int_1^2 \frac{2x}{3+2x}\,dx$

The Use of Partial Fractions in Differentiation

Rational functions with two or more factors in the denominator are often easier to differentiate if expressed in partial fractions. This method is an alternative to logarithmic differentiation which can also be used for functions of this type.

However, the use of partial fractions is of particular benefit when a second derivative is required.

Example 17c _____

Find the first and second derivatives of $\dfrac{x}{(x-1)(x+1)}$

Taking $\qquad y = \dfrac{x}{(x-1)(x+1)} = \dfrac{\frac{1}{2}}{(x-1)} + \dfrac{\frac{1}{2}}{(x+1)}$

$$= \tfrac{1}{2}(x-1)^{-1} + \tfrac{1}{2}(x+1)^{-1}$$

gives $\qquad \dfrac{dy}{dx} = -\tfrac{1}{2}(x-1)^{-2} - \tfrac{1}{2}(x+1)^{-2}$

$$= \dfrac{-1}{2(x-1)^2} - \dfrac{1}{2(x+1)^2}$$

and $\qquad \dfrac{d^2y}{dx^2} = (-2)(-\tfrac{1}{2})(x-1)^{-3} - (-2)(\tfrac{1}{2})(x+1)^{-3}$

$$= \dfrac{1}{(x-1)^3} + \dfrac{1}{(x+1)^3}$$

_____

EXERCISE 17c

In each question express the given function in partial fractions and hence find its first and second derivatives.

1. $\dfrac{2}{(x-2)(x-1)}$ **2.** $\dfrac{3x}{(2x-1)(x-3)}$ **3.** $\dfrac{x}{(x+2)(x-4)}$

4. $\dfrac{5}{(x+2)(x-3)}$ **5.** $\dfrac{x}{(2x+3)(x+1)}$ **6.** $\dfrac{3}{(3x-1)(x-1)}$

SPECIAL TECHNIQUES FOR INTEGRATING SOME TRIGONOMETRIC FUNCTIONS

To Integrate a Function Containing an Odd Power of $\sin x$ or $\cos x$

When $\sin x$ or $\cos x$ appear to an odd power other than 1, the identity $\cos^2 x + \sin^2 x \equiv 1$ is often useful in converting the given function to an integrable form,

e.g. $\sin^3 x$ is converted to $(\sin^2 x)(\sin x)$ $\quad\Rightarrow\quad (1 - \cos^2 x)(\sin x)$

$$\Rightarrow\quad \sin x - \cos^2 x \sin x$$

Examples 17d

1. Integrate w.r.t. x, (a) $\cos^5 x$ (b) $\sin^3 x \cos^2 x$

(a)
$$\cos^5 x = (\cos^2 x)^2 \cos x$$
$$= (1 - \sin^2 x)^2 \cos x$$
$$= (1 - 2\sin^2 x + \sin^4 x)\cos x$$

$$\therefore \quad \int \cos^5 x \, dx = \int \cos x \, dx - 2\int \sin^2 x \cos x \, dx + \int \sin^4 x \cos x \, dx$$

For any value of n we know that $\displaystyle\int \sin^n x \cos x \, dx = \frac{1}{n+1}\sin^{n+1} x + K$

$$\therefore \quad \int \cos^5 x \, dx = \sin x - 2(\tfrac{1}{3})\sin^3 x + (\tfrac{1}{5})\sin^5 x + K$$
$$= \sin x - \tfrac{2}{3}\sin^3 x + \tfrac{1}{5}\sin^5 x + K$$

(b)
$$\sin^3 x \cos^2 x = \sin x(1 - \cos^2 x)\cos^2 x$$
$$= \cos^2 x \sin x - \cos^4 x \sin x$$

$$\therefore \quad \int \sin^3 x \cos^2 x \, dx = \int \cos^2 x \sin x \, dx - \int \cos^4 x \sin x \, dx$$

$$= -\tfrac{1}{3}\cos^3 x + \tfrac{1}{5}\cos^5 x + K$$

To Integrate a Function Containing only Even Powers of $\sin x$ or $\cos x$

This time the double angle identities are useful,

e.g. $\cos^4 x$ becomes $(\cos^2 x)^2 = \{\frac{1}{2}(1 + \cos 2x)\}^2$

$$= \tfrac{1}{4}\{1 + 2\cos 2x + \cos^2 2x\}$$

then we can use a double angle identity again

$$= \tfrac{1}{4}(1 + 2\cos 2x) + \tfrac{1}{4}\{\tfrac{1}{2}(1 + \cos 4x)\}$$

$$= \tfrac{3}{8} + \tfrac{1}{2}\cos 2x + \tfrac{1}{4}\cos 4x$$

Now each of these terms can be integrated.

Examples 17d (continued)

2. Integrate w.r.t. x, (a) $\sin^2 x$ (b) $16\sin^4 x \cos^2 x$

(a) $$\int \sin^2 dx = \int \tfrac{1}{2}(1 - \cos 2x)\, dx$$

$$= \int \tfrac{1}{2}\, dx - \tfrac{1}{2}\int \cos 2x\, dx$$

$$= \tfrac{1}{2}x - \tfrac{1}{4}\sin 2x + K$$

(b) $$16\int \sin^4 x \cos^2 x\, dx = 16\int \{\tfrac{1}{2}(1 - \cos 2x)\}^2\{\tfrac{1}{2}(1 + \cos 2x)\}\, dx$$

$$= 2\int (1 - \cos 2x - \cos^2 2x + \cos^3 2x)\, dx$$

$$= 2x - \sin 2x - 2\int \cos^2 2x\, dx + 2\int \cos^3 2x\, dx$$

Now $$2\int \cos^2 2x\, dx = \int (1 + \cos 4x)\, dx = x + \tfrac{1}{4}\sin 4x$$

and $$2\int \cos^3 2x\, dx = 2\int \cos 2x(1 - \sin^2 2x)\, dx = \sin 2x - \tfrac{1}{3}\sin^3 2x$$

$\therefore$ $$16\int \sin^4 x \cos^2 x\, dx = x - \tfrac{1}{4}\sin 4x - \tfrac{1}{3}\sin^3 2x + K$$

Note that for a product with an odd power in one term and an even power in the other, the method for an odd power is usually best (see Example 2(b) on the previous page).

It is important to appreciate that the techniques used in these examples, although they are of most general use, are by no means exhaustive. Because there are so many trig identities there is always the possibility that a particular integral can be dealt with in several different ways,

e.g. to integrate $\sin^2x \cos^2x$ w.r.t. x the best conversion would use $2\sin x \cos x \equiv \sin 2x$, so that

$$\int \sin^2x \cos^2x \, dx = \int (\tfrac{1}{2}\sin 2x)^2 \, dx = \tfrac{1}{4}\int \tfrac{1}{2}(1 - \cos 4x) \, dx$$

So it is always advisable to look for the identity which will make the given function integrable as quickly and simply as possible.

Further, as mentioned earlier, it must be remembered that there are many expressions whose integrals cannot be found as a function at all. (In examination papers, of course, any integral asked for *can* be found!)

To Integrate any Power of tan x

The identity $\tan^2x = \sec^2x - 1$ is useful here,

e.g. $\tan^3x$ becomes $\tan x(\sec^2x - 1) \Rightarrow \sec^2x \tan x - \tan x$

and we know that $\displaystyle\int \sec^2x \tan^n x \, dx = \frac{1}{n+1}\tan^{n+1}x + K$

Examples 17d (continued) _____

3. Integrate w.r.t. x,   (a) $\tan^4x$ (b) $\tan^5x$

(a) $\displaystyle\int \tan^4x \, dx = \int \tan^2x(\sec^2x - 1) \, dx$

$$= \int \sec^2x \tan^2x \, dx - \int \tan^2x \, dx$$

$$= \tfrac{1}{3}\tan^3x - \int (\sec^2x - 1) \, dx$$

$$= \tfrac{1}{3}\tan^3x - \tan x + x + K$$

(b) $\quad \int \tan^5 x \, dx = \int \tan^3 x (\sec^2 x - 1) \, dx$

$$= \int \sec^2 x \tan^3 x - \int \tan x (\sec^2 x - 1) \, dx$$

$$= \int \sec^2 x \tan^3 x \, dx - \int \sec^2 x \tan x \, dx + \int \tan x \, dx$$

$$= \tfrac{1}{4} \tan^4 x - \tfrac{1}{2} \tan^2 x + \ln \sec x + K$$

Note that, when integrating *any* power of $\tan x$, we use the identity $\tan^2 x = \sec^2 x - 1$ to convert $\tan^2 x$ *only, one step at a time,* i.e. it is not a good idea to convert $\tan^4 x$ to $(1 - \sec^2 x)^2$.

Integrals Involving Multiple Angles

To integrate a product such as $\sin 5x \cos 3x$, the appropriate factor formula can be used to express the product as a sum. In this case,

$$\int \sin 5x \cos 3x \, dx = \tfrac{1}{2} \int (\sin 8x + \sin 2x) \, dx$$

and the RHS consists of two standard integrals.

EXERCISE 17d

Integrate each function w.r.t. x

1. $\cos^2 x$ 2. $\cos^3 x$ 3. $\sin^5 x$ 4. $\tan^2 x$

5. $\sin^4 x$ 6. $\tan^3 x$ 7. $\cos x$ 8. $\sin^3 x$

Find

9. $\displaystyle \int \sin^2 \theta \cos^3 \theta \, d\theta$

10. $\displaystyle \int \sin^{10} \theta \cos^3 \theta \, d\theta$

11. $\displaystyle \int \sin^n \theta \cos^3 \theta \, d\theta$

12. $\displaystyle \int \sin^2 \theta \cos^2 \theta \, d\theta$

13. $\displaystyle \int 4 \sin^2 \theta \cos 3\theta \, d\theta$ (*Hint.* Change $\sin^2 \theta$ into the double angle.)

14. $\displaystyle \int \tan^2 \theta \sec^4 \theta \, d\theta$ (*Hint.* Change $\sec^2 \theta$ into $\tan^2 \theta + 1$.)

Integrate w.r.t. t

15. $2 \sin 4t \cos 3t$

16. $2 \cos 2t \cos 5t$

17. $\sin 2t \cos 6t$

18. $\sin t \sin 3t$

19. $2 \sin nt \cos mt$

20. $\cos nt \cos mt$

Evaluate

21. $\displaystyle\int_0^{\pi/2} \sin^3 x \cos^3 x \, dx$

22. $\displaystyle\int_0^{\pi/4} \tan^6 x \, dx$

23. $\displaystyle\int_{\pi/6}^{\pi/3} 2 \sin 3x \cos 2x \, dx$

24. $\displaystyle\int_0^{\pi/3} \cos x \cos 3x \, dx$

CHAPTER 18

SYSTEMATIC INTEGRATION AND DIFFERENTIAL EQUATIONS

At this stage it is possible to classify most of the integrals which the reader is likely to meet.

Once correctly classified, a given expression can be integrated using the method best suited to its category.

The simplest category comprises the quotable results listed below.

STANDARD INTEGRALS

Function	Integral	Function	Integral
x^n	$\dfrac{1}{n+1}x^{n+1}$ $(n \neq -1)$	$-\text{cosec}^2 x$	$\cot x$
		$\tan x$	$\ln\lvert \sec x \rvert$
e^x	e^x	$\sec x$	$\ln\lvert \sec x + \tan x \rvert$
$\dfrac{1}{x}$	$\ln\lvert x \rvert$	$\dfrac{1}{1+x^2}$	$\arctan x$
$\cos x$	$\sin x$		
$\sin x$	$-\cos x$	$\dfrac{1}{\sqrt{(1-x^2)}}$	$\arcsin x$
$\sec^2 x$	$\tan x$		

Each of these should be recognised equally readily when x is replaced by ax or $(ax + b)$, e.g.

for e^{ax+b} the standard integral is $\frac{1}{a}e^{ax+b}$ and for $\cos ax$ the standard integral is $\frac{1}{a}\sin ax$

CLASSIFICATION

When attempting to classify a particular function the following questions should be asked, *in order*, about the form of the integral.

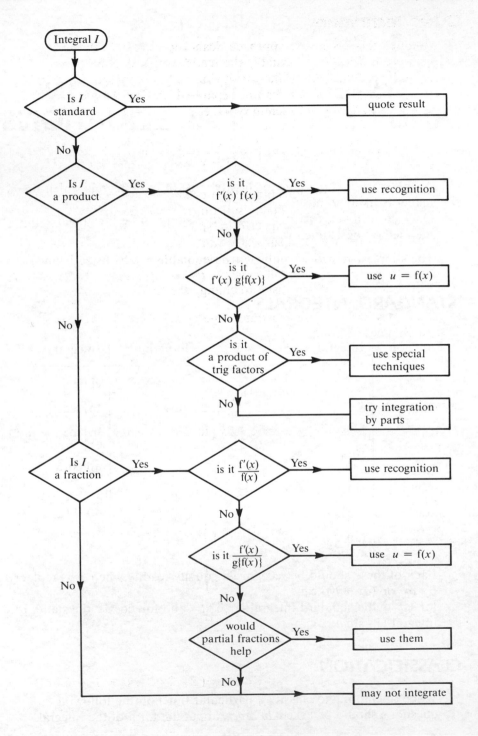

Other Techniques

Although this systematic approach deals successfully with most integrals at this level, inevitably the reader will encounter some integrals for which no method is obvious. A fraction, for instance may be such that its *numerator* can be separated, thus producing *two* (or more) fractions of different types, e.g.

$$\int \frac{x+1}{\sqrt{(1-x^2)}} \, dx = \int \frac{x}{\sqrt{(1-x^2)}} \, dx + \int \frac{1}{\sqrt{(1-x^2)}} \, dx$$

Further, many expressions other than products and fractions can be integrated by making a suitable substitution. Because at this stage the reader cannot always be expected to 'spot' an appropriate change of variable, a substitution is suggested in all but the simplest of cases. The resulting integral must be converted so that is is expressed in terms of the original variable *except in the case of a definite integral* when it is usually much easier to change the limits.

The following examples illustrate some of the integrals which respond to a change of variable.

Examples 18a

1. Use the substitution $u = 1 + 2x$ to find $\int x(1 + 2x)^{11} \, dx$

$$u = 1 + 2x \quad \Rightarrow \quad \ldots \, du \equiv 2 \, dx \quad \Rightarrow \quad \tfrac{1}{2} \, du \equiv \ldots \, dx$$

Hence
$$\int x(1 + 2x)^{11} \, dx = \int \tfrac{1}{2}(u - 1)(u^{11})(\tfrac{1}{2} \, du)$$

$$= \tfrac{1}{4} \int (u^{12} - u^{11}) \, du$$

$$= \tfrac{1}{4}(\tfrac{1}{13} u^{13} - \tfrac{1}{12} u^{12}) + K$$

$$= \tfrac{1}{624} u^{12}(12u - 13) + K$$

i.e.
$$\int x(1 + 2x)^{11} \, dx = \tfrac{1}{624}(1 + 2x)^{12}(24x - 1) + K$$

2. Integrate $\dfrac{1}{\sqrt{(9-16x^2)}}$ w.r.t. x by using $x = \frac{3}{4}\sin\theta$

$$x = \tfrac{3}{4}\sin\theta \quad\Rightarrow\quad \ldots dx \equiv \ldots \tfrac{3}{4}\cos\theta \, d\theta$$

Hence
$$\int \dfrac{1}{\sqrt{(9-16x^2)}}\, dx = \int \dfrac{1}{\sqrt{(9 - 9\sin^2\theta)}}\,(\tfrac{3}{4}\cos\theta)\, d\theta$$

$$= \dfrac{3}{4}\int \dfrac{\cos\theta}{3\sqrt{(1 - \sin^2\theta)}}\, d\theta$$

$$= \dfrac{1}{4}\int \dfrac{\cos\theta}{\cos\theta}\, d\theta = \tfrac{1}{4}\theta + K$$

i.e.
$$\dfrac{1}{\sqrt{(9-16x^2)}}\, dx = \tfrac{1}{4}\arcsin \tfrac{4}{3}x + K$$

---

3. Complete the square in the denominator and then use $u = x + 1$
to find (a) $\displaystyle\int_{-1}^{0} \dfrac{1}{x^2 + 2x + 2}\, dx$ (b) $\displaystyle\int_{-1}^{0} \dfrac{1}{\sqrt{(3 - 2x - x^2)}}\, dx$

$$u = x + 1 \quad\Rightarrow\quad \ldots du \equiv \ldots dx \quad\text{and}\quad \begin{cases} x & -1 & 0 \\ u & 0 & 1 \end{cases}$$

(a) First we deal with the denominator

$$x^2 + 2x + 2 = (x + 1)^2 + 1$$

Hence
$$\int_{-1}^{0} \dfrac{1}{x^2 + 2x + 2}\, dx = \int_{-1}^{0} \dfrac{1}{(x + 1)^2 + 1}\, dx = \int_{0}^{1} \dfrac{1}{u^2 + 1}\, du$$

$$= \Big[\arctan u\Big]_{0}^{1}$$

i.e.
$$\int_{-1}^{0} \dfrac{1}{x^2 + 2x + 2}\, dx = \tfrac{1}{4}\pi$$

(b) $3 - 2x - x^2 = 4 - (x + 1)^2$

Hence
$$\int_{-1}^{0} \dfrac{1}{\sqrt{(3 - 2x - x^2)}}\, dx = \int_{0}^{1} \dfrac{1}{\sqrt{(4 - u^2)}}\, du = \Big[\arcsin \dfrac{u}{2}\Big]_{0}^{1}$$

$$= \tfrac{1}{6}\pi$$

4. Integrate $\sqrt{(1-x^2)}$ w.r.t. x, using the substitution $x = \sin\theta$

$$x = \sin\theta \quad \Rightarrow \quad \ldots dx \equiv \ldots \cos\theta \, d\theta$$

Hence
$$\int\sqrt{(1-x^2)}\,dx = \int\sqrt{(1-\sin^2\theta)}\cos\theta \, d\theta = \int\cos^2\theta \, d\theta$$

$$= \int\tfrac{1}{2}(1+\cos 2\theta)\,d\theta$$

$$= \tfrac{1}{2}(\theta + \tfrac{1}{2}\sin 2\theta) + K$$

This expression must now be given in terms of x, so we use $\theta = \arcsin x$ and
$\sin 2\theta = 2\sin\theta\cos\theta = 2\sin\theta\sqrt{(1-\sin^2\theta)} = 2x\sqrt{(1-x^2)}$

i.e. $\int\sqrt{(1-x^2)}\,dx = \tfrac{1}{2}\{\arcsin x + x\sqrt{(1-x^2)}\} + K$

The fourth example has been included specifically to demonstrate how to convert a term like $\sin 2\theta$ back to the variable x. It is suggested that the reader now re-work this example for the *definite* integral

$$\int_0^1 \sqrt{(1-x^2)}\,dx$$

EXERCISE 18a

Find the following integrals using the suggested substitution.

1. $\int (x+1)(x+3)^5\,dx;$ $x + 3 = u$

2. $\int \dfrac{1}{4+x^2}\,dx;$ $x = 2\tan\theta$

3. $\int \dfrac{x}{\sqrt{(3-x)}}\,dx;$ $3 - x = u^2$

4. $\int x\sqrt{(x+1)}\,dx;$ $x + 1 = u^2$

5. $\int \dfrac{2x+1}{(x-3)^6}\,dx;$ $x - 3 = u$

6. $\int \dfrac{1}{\sqrt{(1+x^2)}}\,dx;$ $x = \tan\theta$

7. $\int 2x\sqrt{(3x-4)}\,dx;$ $3x - 4 = u^2$

8. $\int \dfrac{3}{25+4x^2}\,dx;$ $2x = 5\tan\theta$

9. $\int \dfrac{1}{\sqrt{(x^2+4x+3)}}\,dx;$ $x + 2 = \sec\theta$,   after 'completing the square' in the denominator.

Devise a suitable substitution and hence find:

10. $\int 2x(1-x)^7\,dx$ 11. $\int \dfrac{1}{\sqrt{(1-4x^2)}}\,dx$ 12. $\int \dfrac{x+3}{(4-x)^5}\,dx$

EXERCISE 18b

Use the flow chart to classify each of the following integrals. Hence perform each integration using an appropriate method.

1. $\int e^{2x+3}\,dx$

2. $\int x\sqrt{(2x^2-5)}\,dx$

3. $\int \sin^2 3x\,dx$

4. $\int x\,e^{-x^2}\,dx$

5. $\int \sin 3\theta\cos\theta\,d\theta$

6. $\int u(u+7)^9\,du$

7. $\int \dfrac{x^2}{(x^3+9)^5}\,dx$

8. $\int \dfrac{\sin 2y}{1-\cos 2y}\,dy$

9. $\int \dfrac{1}{2x+7}\,dx$

10. $\int \dfrac{1}{\sqrt{(1-u^2)}}\,du$

11. $\int \sin 3x\sqrt{(1+\cos 3x)}\,dx$

12. $\int x\sin 4x\,dx$

13. $\int \dfrac{x+2}{x^2+4x-5}\,dx$

14. $\int \dfrac{x+1}{x^2+4x-5}\,dx$

15. $\int \dfrac{x+2}{(x^2+4x-5)^3}\,dx$

16. $\int 3y\sqrt{(9-y^2)}\,dy$

17. $\int e^{2x}\cos 3x\,dx$

18. $\int \ln 5x\,dx$

19. $\int \cos^3 2x \, dx$ **20.** $\int \operatorname{cosec}^2 x \, e^{\cot x} \, dx$

21. $\int \dfrac{\sin y}{\sqrt{(7 + \cos y)}} \, dy$ **22.** $\int x^2 \, e^x \, dx$

23. $\int \dfrac{x}{x^2 - 4} \, dx$ **24.** $\int \dfrac{x^2}{x^2 - 4} \, dx$

25. $\int \dfrac{1}{x^2 - 4} \, dx$ **26.** $\int \cos 4x \cos x \, dx$

27. $\int \sin^5 2\theta \, d\theta$ **28.** $\int \cos^2 u \sin^3 u \, du$

29. $\int \tan^2 \theta \, d\theta$ **30.** $\int \dfrac{1 - 2x}{\sqrt{(1 - x^2)}} \, dx$

31. $\int y^2 \cos 3y \, dy$ **32.** $\int \dfrac{\sec^2 x}{1 - \tan x} \, dx$

33. $\int x\sqrt{(7 + x^2)} \, dx$ **34.** $\int \sin\left(5\theta - \tfrac{1}{4}\pi\right) d\theta$

35. $\int \cos \theta \ln \sin \theta \, d\theta$ **36.** $\int \sec^2 u \, e^{\tan u} \, du$

DIFFERENTIAL EQUATIONS

An equation in which at least one term contains $\dfrac{dy}{dx}$, $\dfrac{d^2y}{dx^2}$ etc., is called a *differential equation*. If it contains only $\dfrac{dy}{dx}$ it is of the first order whereas if it contains $\dfrac{d^2y}{dx^2}$ it is of the second order, and so on.

For example, $x + 2\dfrac{dy}{dx} = 3y$ is a first order differential equation and $\dfrac{d^2y}{dx^2} - 5\dfrac{dy}{dx} + 4y = 0$ is a second order differential equation.

Each of these examples is a *linear* differential equation because none of the differential coefficients $\left(\text{i.e. } \dfrac{dy}{dx}, \dfrac{d^2y}{dx^2}\right)$ is raised to a power higher than 1.

A differential equation represents a relationship between two variables. The same relationship can often be expressed in a form which does not contain a differential coefficient,

e.g. $\dfrac{dy}{dx} = 2x$ and $y = x^2 + K$ express the same relationship

between x and y, but $\dfrac{dy}{dx} = 2x$ is a differential equation whereas

$y = x^2 + K$ is not.

Converting a differential equation into a direct one is called *solving the differential equation*. This clearly involves some form of integration. There are many different types of differential equation, each requiring a specific technique for its solution. At this stage however we are going to deal with only one simple type, i.e. linear differential equations of the first order.

FIRST ORDER DIFFERENTIAL EQUATIONS WITH SEPARABLE VARIABLES

Consider the differential equation $\quad 3y\dfrac{dy}{dx} = 5x^2$ [1]

Integrating both sides of the equation gives

$$\int 3y\frac{dy}{dx}\,dx = \int 5x^2\,dx$$

We know that $\quad\ldots\dfrac{dy}{dx}\,dx \equiv \ldots dy$

so $$\int 3y\,dy = \int 5x^2\,dx \qquad [2]$$

Temporarily removing the integral signs from this equation gives

$$3y\,dy = 5x^2\,dx \qquad [3]$$

This can be obtained direct from equation [1] by *separating the variables,* i.e. by *separating* dy *from* dx *and collecting on one side all the terms involving y together with dy, while all the x terms are collected, along with dx on the other side.*

It is vital to appreciate that what is shown in [3] above does not, in itself, have any meaning, and it *should not be written down as a step in the solution.* It simply provides a way of making a quick mental conversion from the differential equation [1] to the form [2] which is ready for two separate integrations.

Now returning to equation [2] and integrating each side we have

$$\tfrac{3}{2}y^2 = \tfrac{5}{3}x^3 + A$$

Note that it is unnecessary to introduce a constant of integration on both sides. It is sufficient to have a constant on one side only.

When solving differential equations, the constant of integration is usually denoted by A, B, etc. and is called the *arbitrary constant*.

The solution of a differential equation including the arbitrary constant is called *the general solution*, or, very occasionally, *the complete primitive*. It represents a family of straight lines or curves, each member of the family corresponding to one value of A.

Example 18c

Find the general solution of the differential equation

$$\frac{1}{x}\frac{dy}{dx} = \frac{2y}{x^2 + 1}$$

$$\frac{1}{x}\frac{dy}{dx} = \frac{2y}{x^2 + 1} \qquad \Rightarrow \qquad \frac{1}{y}\frac{dy}{dx} = \frac{2x}{x^2 + 1}$$

So, after separating the variables we have,

$$\int \frac{1}{y}\,dy = \int \frac{2x}{x^2 + 1}\,dx$$

$$\Rightarrow \qquad \ln|y| = \ln A|x^2 + 1|$$

(Using $\ln A$ as the arbitrary constant.)

i.e. $\qquad\qquad y = A(x^2 + 1)$

Note that, whenever we solve a differential equation, some integration has to be done, so the systematic classification of each integral involved is an essential part of solving differential equations.

EXERCISE 18c

Find the general solution of each differential equation

1. $y\dfrac{dy}{dx} = \sin x$

2. $x^2\dfrac{dy}{dx} = y^2$

3. $\dfrac{1}{x}\dfrac{dy}{dx} = \dfrac{1}{y^2 - 2}$

4. $\tan y\dfrac{dy}{dx} = \dfrac{1}{x}$

5. $\dfrac{dy}{dx} = y^2$

6. $\dfrac{1}{x}\dfrac{dy}{dx} = \dfrac{1}{1-x^2}$

7. $(x-3)\dfrac{dy}{dx} = y$

8. $\tan y\dfrac{dx}{dy} = 4$

9. $u\dfrac{du}{dv} = v+2$

10. $\dfrac{y^2}{x^3}\dfrac{dy}{dx} = \ln x$

11. $e^x\dfrac{dy}{dx} = \dfrac{x}{y}$

12. $\sec x\dfrac{dy}{dx} = e^y$

13. $r\dfrac{dr}{d\theta} = \sin^2\theta$

14. $\dfrac{dv}{du} = \dfrac{v+1}{u+2}$

15. $xy\dfrac{dy}{dx} = \ln x$

16. $y(x+1) = (x^2+2x)\dfrac{dy}{dx}$

17. $v^2\dfrac{dv}{dt} = (2+t)^3$

18. $x\dfrac{dy}{dx} = \dfrac{1}{y}+y$

19. $r\dfrac{d\theta}{dr} = \cos^2\theta$

20. $y\sin^3x\dfrac{dy}{dx} = \cos x$

21. $\dfrac{uv}{u-1} = \dfrac{du}{dv}$

22. $e^x\dfrac{dy}{dx} = e^{y-1}$

23. $\tan x\dfrac{dy}{dx} = 2y^2\sec^2x$

24. $\dfrac{dy}{dx} = \dfrac{x(y^2-1)}{(x^2+1)}$

NATURAL OCCURRENCE OF DIFFERENTIAL EQUATIONS

Differential equations often arise when a physical situation is interpreted mathematically (i.e. when a mathematical model is made of the physical situation).

Consider the following examples.

1) Suppose that a body falls from rest in a medium which causes the velocity to decrease at a rate proportional to the velocity.

Using v for velocity and t for time, the rate of *decrease* of velocity can be written as $-\dfrac{dv}{dt}$.

Thus the motion of the body satisfies the differential equation

$$-\frac{dv}{dt} = kv$$

2) During the initial stages of the growth of yeast cells in a culture, the number of cells present increases in proportion to the number already formed.

Thus n, the number of cells at a particular time t, can be found from the differential equation

$$\frac{dn}{dt} = kn$$

Note. In forming (and subsequently solving) differential equations from naturally occuring data, it is not actually necessary to understand the background of the situation or experiment.

EXERCISE 18d

In each question form, but *do not solve*, the differential equation representing the given data.

1. A body moves with a velocity v which is inversely proportional to its displacement s from a fixed point.

2. The rate at which the height h of a certain plant increases is proportional to the natural logarithm of the difference between its present height and its final height H.

3. Two chemicals, P and Q, are involved in a reaction. The masses of P and Q present at any time t, are p and q respectively. The rate at which p is increasing at time t is k times the product of the two masses. If the masses of P and Q have a constant sum s, find a differential equation expressing $\frac{dp}{dt}$ in terms of p, s and k.

CALCULATION OF THE ARBITRARY CONSTANT

We saw on p. 289 that

$$3y\frac{dy}{dx} = 5x^2 \quad \Longleftrightarrow \quad \tfrac{3}{2}y^2 = \tfrac{5}{3}x^3 + A$$

The equation $\tfrac{3}{2}y^2 = \tfrac{5}{3}x^3 + A$ represents a family of curves with similar characteristics. Each value of A gives one particular member of the family, i.e. a *particular solution*.

The value of A cannot be found from the differential equation alone; further information is needed.

Suppose that we require the equation of a curve which satisfies the differential equation $2\dfrac{dy}{dx} = \dfrac{\cos x}{y}$ and which passes through the point $(0, 2)$.

We want one member of the family of curves represented by the differential equation, i.e. the particular value of the arbitrary constant must be found.

The general solution has to be found first so, separating the variables, we have

$$\int 2y\, dy = \int \cos x\, dx \quad \Rightarrow \quad y^2 = A + \sin x$$

In order to find the required curve we need the value of A such that the general solution is satisfied by

$$x = 0 \text{ and } y = 2$$

i.e. $\qquad\qquad\qquad 4 = A + 0 \quad \Rightarrow \quad A = 4$

Hence the equation of the specified curve is $y^2 = 4 + \sin x$

Examples 18e

1. Describe the family of curves represented by the differential equation $y = x\dfrac{dy}{dx}$ and sketch any three members of this family.

   Find the particular solution for which $y = 2$ when $x = 1$ and sketch this member of the family on the same axes as before.

By separating the variables,

$$y = x\frac{dy}{dx} \quad \Rightarrow \quad \int \frac{1}{y}\, dy = \int \frac{1}{x}\, dx$$

$$\Rightarrow \quad \ln|y| = \ln|x| + \ln A$$

i.e. the general solution is $y = Ax$

This equation represents a family of straight lines through the origin, each line having a gradient A, as shown in the following diagram.

If $y = 2$ when $x = 1$ then $A = 2$ and the corresponding member of the family is the line $y = 2x$

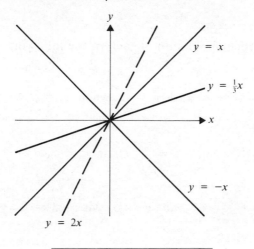

2. A curve is such that the gradient is proportional to the product of the x and y coordinates. If the curve passes through the points $(2, 1)$ and $(4, e^2)$, find its equation.

First find the general solution

$$\frac{dy}{dx} = kxy \text{ where } k \text{ is a constant of proportion.}$$

$$\therefore \qquad \int \frac{1}{y} dy = \int kx \, dx \qquad \Rightarrow \qquad \ln y = \tfrac{1}{2}kx^2 + A$$

There are *two* unknown constants this time so we need two extra pieces of information; these are

(i) $y = 1$ when $x = 2$ $\quad \Rightarrow \quad$ $\ln 1 = 2k + A$

$\qquad\qquad\qquad\qquad\qquad\qquad\quad \ln 1 = 0 \quad$ so $\quad A + 2k = 0$

(ii) $y = e^2$ when $x = 4$ $\quad \Rightarrow \quad$ $\ln e^2 = 8k + A$

$\qquad\qquad\qquad\qquad\qquad\qquad\quad \ln e^2 = 2 \quad$ so $\quad A + 8k = 2$

Solving these equations for A and k we get $k = \tfrac{1}{3}$ and $A = -\tfrac{2}{3}$

$\therefore$ the equation of the specified curve is $\quad \ln y = \tfrac{1}{6}x^2 - \tfrac{2}{3}$

$$= \tfrac{1}{6}(x^2 - 4)$$

$\qquad\qquad\qquad\qquad$ or $\quad y^6 = e^{(x^2 - 4)}$

EXERCISE 18e

Find the particular solution of each of the following differential equations.

1. $y^2 \dfrac{dy}{dx} = x^2 + 1$ and $y = 1$ when $x = 2$

2. $e^t \dfrac{ds}{dt} = \sqrt{s}$ and $s = 4$ when $t = 0$

3. $\dfrac{y}{x} \dfrac{dy}{dx} = \dfrac{y^2 - 1}{x^2 - 1}$ and $y = 3$ when $x = 2$

4. A curve passes through the origin and its gradient function is $2x - 1$. Find the equation of the curve and sketch it.

5. A curve for which $e^{-x} \dfrac{dy}{dx} = 1$, passes through the point $(0, -1)$. Find the equation of the curve.

6. A curve passes through the points $(1, 2)$ and $(\frac{1}{5}, -10)$ and its gradient is inversely proportional to x^2. Find the equation of the curve.

7. If $y = 2$ when $x = 1$, find the coordinates of the point where the curve represented by $\dfrac{2y}{3} \dfrac{dy}{dx} = e^{-3x}$ crosses the y-axis.

8. Find the equation of the curve whose gradient function is $\dfrac{y + 1}{x^2 - 1}$ and which passes through the point $(-3, 1)$.

9. The gradient function of a curve is proportional to $x + 3$. If the curve passes through the origin and the point $(2, 8)$, find its equation.

10. Solve the differential equation $(1 + x^2) \dfrac{dy}{dx} - y(y + 1)x = 0$, given that $y = 1$ when $x = 0$

MIXED EXERCISE 18

Integrate w.r.t. x

1. $x(1 + x^2)^4$

2. xe^{-3x}

3. $\cos 2x \cos 3x$

4. $\dfrac{x + 3}{x + 2}$

5. $\dfrac{x^2}{(x^3 + 1)^2}$

6. $\dfrac{3}{(x - 4)(x - 1)}$

7. $\dfrac{(x + 1)}{x(2x + 1)}$

8. $\dfrac{x - 1}{(x^2 + 1)}$

9. $\dfrac{\sin x}{\sqrt{(\cos x)}}$

Evaluate

10. $\displaystyle\int_{\pi/2}^{\pi} (\sin \tfrac{1}{2}x + \cos 2x)\, dx$

11. $\displaystyle\int_{2}^{5} x\sqrt{(x - 1)}\, dx$

12. $\displaystyle\int_{0}^{\pi/4} \tan^3 x\, dx$

13. $\displaystyle\int_{1}^{2} x\sqrt{(5 - x^2)}\, dx$

14. $\displaystyle\int_{4}^{6} \dfrac{5}{x^2 - x - 6}\, dx$

15. $\displaystyle\int_{2}^{3} \dfrac{1}{x \ln x}\, dx$

16. $\displaystyle\int_{-2}^{-1} \dfrac{2 - x}{x(1 - x)}\, dx$

17. $\displaystyle\int_{0}^{1} \dfrac{x(1 - x)}{2 + x}\, dx$

18. Solve the differential equation $\dfrac{dy}{dx} = 3x^2 y^2$ given that $y = 1$ when $x = 0$

19. If $\dfrac{dy}{dx} = x(y^2 + 1)$ and $y = 0$ when $x = 2$ find the particular solution of the differential equation.

20. Find the equation of the curve which passes through the point $(\tfrac{1}{2}, 1)$ and is defined by the differential equation $ye^{y^2}\dfrac{dy}{dx} = e^{2x}$.
Show that the curve also passes through the point $(2, 2)$ and sketch the curve.

CONSOLIDATION D

SUMMARY

INTEGRATION

Standard Integrals

Function	Integral
x^n	$\dfrac{1}{n+1}x^{n+1}$ $(n \neq -1)$
e^x	e^x
$\dfrac{1}{x}$	$\ln|x|$
$\cos x$	$\sin x$
$\sin x$	$-\cos x$
$\sec^2 x$	$\tan x$
$\operatorname{cosec}^2 x$	$-\cot x$
$\tan x$	$\ln|\sec x|$
$\cot x$	$\ln|\sin x|$
$\dfrac{1}{\sqrt{(1-x^2)}}$	$\arcsin x$
$\dfrac{1}{1+x^2}$	$\arctan x$

The following methods are general guide lines; alternative approaches can be better in individual cases.

Integrating products can be done by

a) recognition:

in particular $\displaystyle\int f'(x)\, e^{f(x)}\, dx = e^{f(x)} + K$

$$\int \sin^p x \cos x \, dx = \frac{1}{p+1}\sin^{p+1}x + K \qquad (p \neq -1)$$

$$\int \cos^p x \sin x \, dx = -\frac{1}{p+1} \cos^{p+1}x + K \quad (p \neq -1)$$

$$\int \tan^p x \sec^2 x \, dx = \frac{1}{p+1} \tan^{p+1}x + K \quad (p \neq -1)$$

b) change of variable: suitable for the type $f'(x)g\{f(x)\}$

c) by parts: $\displaystyle\int v \frac{du}{dx} \, dx = uv - \int u \frac{dv}{dx} \, dx$

Integration by parts can be used also to integrate $\ln x$ and inverse trig functions.

Integrating fractions can be done by

a) recognition: in particular $\displaystyle\int \frac{f'(x)}{f(x)} \, dx = \ln f(x) + K$

b) change of variable: suitable for the type $\dfrac{f'(x)}{g\{f(x)\}}$

c) using partial fractions.

DIFFERENTIAL EQUATIONS

A first order linear differential equation is a relationship between x, y and dy/dx. It can be solved by collecting all the x terms, along with dx, on one side, with all y terms and dy on the other side. Then each side is integrated with respect to its own variable. A constant of integration called an arbitrary constant is introduced on one side only to give a general solution which is a family of lines or curves.

If extra information provides the value of this constant we have a particular solution, i.e. one member of the family.

MULTIPLE CHOICE EXERCISE D

TYPE I

1. e^{x^2} could be the integral w.r.t. x of

   A e^{2x} **C** $\dfrac{e^{x^2}}{2x}$ **E**   none of these

   B $2xe^{x^2}$ **D** $x^2 e^{x^2} - 1$

2. If $\displaystyle\int_1^5 \dfrac{dx}{2x-1} = \ln K$, the value of K is

   A 9   **B** 3   **C** undefined   **D** 81   **E** 8

3. $\displaystyle\int_0^{\pi/6} \sin^n x \cos x \, dx = \tfrac{1}{64};$ n is

   A 6   **B** 5   **C** 4   **D** 3   **E** none of these

4. The differential equation of a curve is $\dfrac{x}{y}\dfrac{dy}{dx} = 1$. The sketch of the curve could be

A
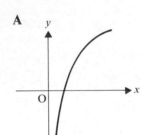

C
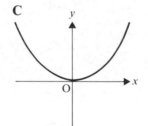

E
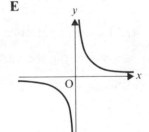

B
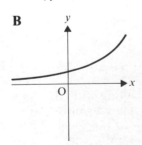

D
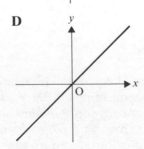

5. The value of $\displaystyle\int_0^2 2e^{2x} \, dx$ is

   A e^4 **B** $e^4 - 1$ **C** ∞ **D** $4e^4$ **E** $\tfrac{1}{2}e^4$

6. Using $x = \sin \theta$ transforms $\int \dfrac{x^2}{\sqrt{(1-x^2)}} = dx$ into

A $\displaystyle\int \frac{\sin^2\theta}{\cos\theta}\, d\theta$

C $-\displaystyle\int \sin^2\theta\, d\theta$

B $\frac{1}{2}\displaystyle\int (1 - \cos 2\theta)\, d\theta$

D $\frac{1}{2}\displaystyle\int (1 + \cos 2\theta)\, d\theta$

TYPE II

7. Integration by parts can be used to find

A $\displaystyle\int x^2 e^x\, dx$

C $\displaystyle\int \ln x\, dx$

B $\displaystyle\int e^x \ln x\, dx$

D $\displaystyle\int (\ln x)(\sin x)\, dx$

8. Which of the following definite integrals can be evaluated?

A $\displaystyle\int_0^1 \frac{1}{x-1}\, dx$

C $\displaystyle\int_1^2 \sqrt{(1-x^2)}\, dx$

B $\displaystyle\int_0^{\pi/2} \sin x\, dx$

D $\displaystyle\int_{-2}^1 \ln x\, dx$

9. Which of the following differential equations can be solved by separating the variables?

A $x\dfrac{dy}{dx} = y + x$

C $e^{x+y} = y\dfrac{dy}{dx}$

B $xy\dfrac{dy}{dx} = x + 1$

D $x + \dfrac{dy}{dx} = \ln y$

10. $\displaystyle\int_1^2 x e^x\, dx$

A is a definite integral

C is equal to $\left[\frac{1}{2}e^{x^2}\right]_1^2$

B is equal to $x e^x - e^x$

D can be integrated by parts.

TYPE III

11. $\displaystyle\int_a^b y\, dx = \left[\frac{1}{2}y^2\right]_a^b$

12. $\displaystyle\int \tan x \, dx = \sec^2 x + K$

13. $\displaystyle\int \sin x \, dx = \cos x + K$

14. $\displaystyle\left[f(x) \right]_0^a = f(a) - 0$

15. A differential equation must contain $\dfrac{dy}{dx}$

MISCELLANEOUS EXERCISE D

1. (a) Find $\displaystyle\int (3x + 4) \, e^{2x} \, dx$

    (b) By using the substitution $x = 2 \tan \theta$, evaluate

$$\int_0^2 \frac{1}{(4 + x^2)^2}$$ (U of L)

2. Find the following integrals

    (a) $\displaystyle\int \cos^2(3x) \, dx$

    (b) $\displaystyle\int \frac{1}{e^x + 4e^{-x}} \, dx$, by means of the substitution $u = e^x$, or otherwise

    (c) $\displaystyle\int x e^{2x} \, dx$ (C)

3. Given that $\dfrac{x}{(1 + x)^2} = \dfrac{A}{(1 + x)^2} + \dfrac{B}{1 + x}$, find the values of the constants A and B.

Hence, or otherwise, evaluate

$$\int_0^1 \frac{x}{(1 + x)^2} \, dx$$ (U of L)

4. By means of the substitution $u = e^x$, or otherwise, evaluate

$$\int_0^1 \frac{e^x}{1 + e^{2x}} \, dx$$

giving the answer correct to two decimal places. (JMB)

5. By using a substitution, or otherwise, find the exact value of

$$\int_0^{\pi/2} \frac{\cos x}{(4 + \sin x)^2}$$ (JMB)

6. Given that $\dfrac{7x - x^2}{(2 - x)(x^2 + 1)} \equiv \dfrac{A}{(2 - x)} + \dfrac{Bx + C}{(x^2 + 1)}$, determine the

values of A, B and C.

A curve has equation $y = \dfrac{7x - x^2}{(2 - x)(x^2 + 1)}$.

Determine the equation of the normal to the curve at the point $(1, 3)$.

Prove that the area of the region bounded by the curve, the x-axis and
the line $x = 1$ is $\frac{7}{2}\ln 2 - \frac{1}{4}\pi$.   (AEB)

7. Given that $x < 4$, find

$$\int \frac{8}{(4 - x)(8 - x)}\, dx$$

A chemical reaction takes place in a solution containing a substance
S. At noon there are two grams of S in the solution and t hours
later there are x grams of S. The rate of the reaction is such that x
satisfies the differential equation

$$8\frac{dx}{dt} = (4 - x)(8 - x)$$

Solve this equation, giving t in terms of x.

Find, to the nearest minute, the time at which there are three grams of
S present.   (JMB)

8. Express $\dfrac{1}{(1 + x)(3 + x)}$ in partial fractions.

Hence find the solution of the differential equation

$$\frac{dy}{dx} = \frac{y}{(1 + x)(3 + x)} \qquad x > -1$$

given that $y = 2$ when $x = 1$

Express your answer in the form $y = f(x)$.   (U of L)

9. Given that $y = x \cos 3x$, find

(a) $\dfrac{dy}{dx}$ (b) $\displaystyle\int y\, dx$ (JMB)

10. Evaluate

$$\int_1^e (2x + 1) \ln x \, dx$$

giving the answer in a simplified form in terms of e. (JMB)

11. Evaluate

$$\int_0^2 \frac{11 + 5x}{(3 - x)(4 + x^2)} \, dx$$

expressing your answer in the form $\ln p + \dfrac{\pi}{q}$, where p and q are integers. (JMB)

12. Showing your method of integration clearly in each case, evaluate

(a) $\displaystyle\int_0^{\pi/2} \sin^2 x \cos^5 x \, dx$ (b) $\displaystyle\int_1^{\sqrt3} \frac{x^2}{1 + x^2} \, dx$ (AEB)

13. Find y explicitly in terms of x given that

$$x \frac{dy}{dx} = 1 - y$$

and $y = 0$ when $x = 2$ (AEB)

14. Evaluate the following integrals

(a) $\displaystyle\int_0^\pi x \cos 3x \, dx$ (b) $\displaystyle\int_0^{\pi/2} \sin^3 x \, dx$ (AEB)

15. Find the general solution of the differential equation

$$(4 + t^2) \frac{ds}{dt} = 1$$

Given that $s = 0$ when $t = 2$, express s in terms of t. (U of L)

16. The amount of product, x units, formed in a chemical reaction is such that $\dfrac{dx}{dt} = 12e^{-4t}$, where t is the time in minutes for x units to form. Given that x tends to 3 as t tends to infinity, show that $x = c(1 - e^{-4t})$ and state the value of c. (U of L)

17. Express y in terms of x, given that $\dfrac{dy}{dx} = -2y^2$, and that $y = 10$
 when $x = 0$ (O/C, SU & C)

18. Given that $f(x) = \dfrac{2x - 3}{(2x - 1)(x - 1)}$, express $f(x)$ in partial fractions.
 Hence, show that $\displaystyle\int_2^3 f(x)\,dx = \ln\frac{25}{18}$. (AEB)

19. In an experiment, two chemicals A and B react yielding another
 chemical C. The rate of formation of C at time t seconds is k times
 the product of the concentrations of A and B present at that time,
 where k is a positive constant. At time t seconds, the concentration
 of each of A and B is $(p - x)$ units, where x is the concentration of
 C at that time and p is a positive constant.
 (a) Find a differential equation satisfied by x.
 (b) Given that $x = 0$ when $t = 0$, obtain x in terms of k, t and p.
 (U of L)

20. Differentiate with respect to x
 (a) $\sec x \tan x$,
 (b) $\ln(\sec x + \tan x)$.
 Hence show that

$$\int \sec^3 x\,dx = \tfrac{1}{2}\sec x \tan x + \tfrac{1}{2}\ln(\sec x + \tan x) + C,$$

 where C is an arbitrary constant.
 Find the general solution of the differential equation

$$\frac{dy}{dx} = \sec^3 2x \cot^2 y$$

 Given that $y = \tfrac{1}{4}\pi$ when $x = 0$, show that when $x = \tfrac{1}{6}\pi$

$$4\tan y - 4y + \pi = 4 + 2\sqrt{3} + \ln(2 + \sqrt{3})$$ (AEB)

21. Find the solution of the differential equation

$$(x^2 - 5)^{1/2}\frac{dy}{dx} = 2xy^{1/2}, \quad x^2 > 5$$

 for which $y = 4$ when $x = 3$, expressing y in terms of x. (JMB)

22. The integral I is defined by

$$I = \int_0^{\pi/2} \cos^2 x \, dx \qquad [1]$$

Use the substitution $x = \frac{1}{2}\pi - y$ to show that

$$I = \int_0^{\pi/2} \sin^2 x \, dx \qquad [2]$$

Hence, by considering the sum of [1] and [2], deduce the value of I.

(WJEC)

23. (a) Show that $\dfrac{1}{1-p} + \dfrac{1}{p} = \dfrac{1}{p(1-p)}$

A disease spreads through an urban population. At time t, p is the proportion of the population who have the disease. The rate of change of p is proportional to the product of p and the proportion $(1-p)$ who do not have the disease.

(b) Form a differential equation in terms of p and t.

(c) Given that when $p = \dfrac{1}{4}$, $\dfrac{dp}{dt} = \dfrac{3}{8}$, show that $\dfrac{dp}{dt} = 2p(1-p)$.

(d) Solve the differential equation in (c) given that $p = \frac{1}{2}$ at $t = 0$.

(U of L)

24. Use the binomial expansion to show that, for x sufficiently small,

$$\frac{x^2}{\sqrt{(4-x^2)}} \approx \frac{1}{2}x^2 + \frac{1}{16}x^4$$

Use this approximation to show that the value of the integral

$$I = \int_0^1 \frac{x^2 \, dx}{\sqrt{(4-x^2)}}$$

is approximately $\frac{43}{240}$.

Putting $x = 2 \sin \theta$, show that

$$I = 4 \int_0^{\pi/6} \sin^2 \theta \, d\theta$$

Hence show that the value of I is $\frac{1}{3}\pi - \frac{1}{2}\sqrt{3}$

Using integration by parts, or otherwise, evaluate the integral

$$I = \int_0^1 \frac{x^4 \, dx}{(4-x^2)^{3/2}}$$

giving your answer correct to three decimal places.   (WJEC)

CHAPTER 19

WORKING WITH VECTORS

VECTORS

Although we usually assume that two and two make four, this is not always the case.

If, for example, a point B is 2 cm from a point A and C is 2 cm from B then, in general, C is *not* 4 cm from A.

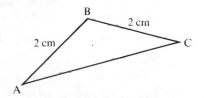

AB, BC and AC are displacements. Each of them has a magnitude and is related to a definite direction in space and so is called a *vector*.

> A vector is a quantity which has both magnitude and a specific direction in space.

A scalar quantity is one that is fully defined by magnitude alone. Length, for example, is a scalar quantity as the length of a piece of string does not depend on its direction when it is measured.

Vector Representation

A vector can be represented by a section of a straight line, whose length represents the magnitude of the vector and whose direction, indicated by an arrow, represents the direction of the vector.
Such vectors can be denoted by a letter in bold type, e.g. **a** or, when hand-written, by a̲.

Alternatively we can represent a vector by the magnitude and direction of a line joining A to B. When we denote the vector by $\overrightarrow{AB}$ or **AB**, the vector in the opposite direction, i.e. from B to A, is written $\overrightarrow{BA}$ or **BA**.

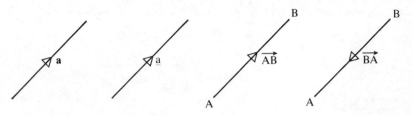

Equivalent Displacements

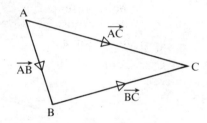

The displacement from A to B, followed by the displacement from B to C, is equivalent to the displacement from A to C.

This is written as the vector equation

$$\overrightarrow{AB} + \overrightarrow{BC} = \overrightarrow{AC}$$

Note that, in vector equations like the one above,
    + means 'together with'   and   = means 'is equivalent to'

Note also that $\overrightarrow{AB}$, $\overrightarrow{BC}$ and $\overrightarrow{AC}$ are called *displacement vectors*.

Many quantities other than displacements behave in the same way and all of them are vectors.

PROPERTIES OF VECTORS

The Modulus of a Vector

The *modulus* of a vector **a** is its magnitude and is written $|\mathbf{a}|$ or a i.e. $|\mathbf{a}|$ is the length of the line representing **a**

Equal Vectors

Two vectors with the same magnitude and the same direction are equal.

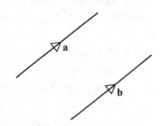

i.e. $\mathbf{a} = \mathbf{b}$ $\Longleftrightarrow$ $\begin{cases} |\mathbf{a}| = |\mathbf{b}| \quad \text{and} \\ \text{the directions of } \mathbf{a} \text{ and } \mathbf{b} \text{ are the same.} \end{cases}$

It follows that a vector can be represented by *any* line of the right length and direction, regardless of position, i.e. each of the lines in the diagram below represents the vector **a**

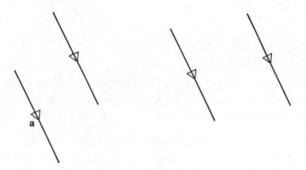

Negative Vectors

If two vectors, **a** and **b**, have the same magnitude but opposite directions we say that

$$\mathbf{b} = -\mathbf{a}$$

i.e. $-\mathbf{a}$ is a vector of magnitude $|\mathbf{a}|$ and in the direction opposite to that of **a**

We also say that **a** and **b** are *equal and opposite* vectors.

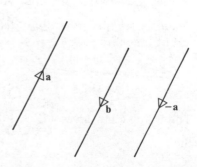

Multiplication of a Vector by a Scalar

If λ is a positive real number, then $\lambda\mathbf{a}$ is a vector in the same direction as $\mathbf{a}$ and of magnitude $\lambda|\mathbf{a}|$

It follows that $-\lambda\mathbf{a}$ is a vector in the opposite direction, with magnitude $\lambda|\mathbf{a}|$

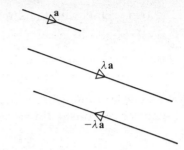

Addition of Vectors

If the sides AB and BC of a triangle ABC represent the vectors $\mathbf{p}$ and $\mathbf{q}$ then the third side AC represents the vector sum, or resultant, of $\mathbf{p}$ and $\mathbf{q}$, which is denoted by $\mathbf{p} + \mathbf{q}$

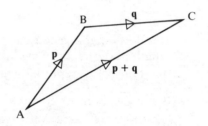

(This property was demonstrated for displacement vectors at the beginning of the chapter.)

Note that $\mathbf{p}$ and $\mathbf{q}$ follow each other round the triangle (in this case in the clockwise sense), whereas the resultant, $\mathbf{p} + \mathbf{q}$, goes the opposite way round (anticlockwise in the diagram).

This is known as *the triangle law* for addition of vectors. It can be extended to cover the addition of more than two vectors.
Let $\overrightarrow{AB}$, $\overrightarrow{BC}$, $\overrightarrow{CD}$ and $\overrightarrow{DE}$ represent the vectors $\mathbf{a}$, $\mathbf{b}$, $\mathbf{c}$ and $\mathbf{d}$ respectively.

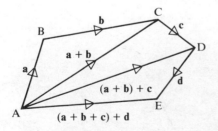 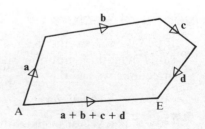

The triangle law gives $\qquad \overrightarrow{AB} + \overrightarrow{BC} = \mathbf{a} + \mathbf{b} = \overrightarrow{AC}$

then $\qquad\qquad\qquad \overrightarrow{AC} + \overrightarrow{CD} = (\mathbf{a} + \mathbf{b}) + \mathbf{c} = \overrightarrow{AD}$

and $\qquad\qquad\qquad \overrightarrow{AD} + \overrightarrow{DE} = (\mathbf{a} + \mathbf{b} + \mathbf{c}) + \mathbf{d} = \overrightarrow{AE}$

Now AE completes the polygon of which AB, BC, CD and DE are four sides taken in order, (i.e. they follow each other round the polygon in the *same sense*). Note that, again, the side representing the resultant closes the polygon in the *opposite* sense.

Note that the vectors **a**, **b**, **c** and **d** are not necessarily coplanar so the polygon may not be a plane figure.

The order in which the addition is performed does not matter as we can see by considering a parallelogram ABCD

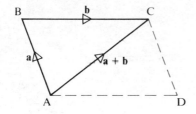 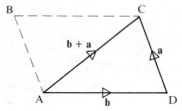

Because the opposite sides of a parallelogram are equal and parallel, $\overrightarrow{AB}$ and $\overrightarrow{DC}$ both represent **a**   and $\overrightarrow{BC}$ and $\overrightarrow{AD}$ both represent **b**

In $\triangle ABC$ $\overrightarrow{AC} = \mathbf{a} + \mathbf{b}$ and   in $\triangle ADC$ $\overrightarrow{AC} = \mathbf{b} + \mathbf{a}$

Therefore $\mathbf{a} + \mathbf{b} = \mathbf{b} + \mathbf{a}$

The Angle between Two Vectors

There are two angles between two lines
i.e. α and $180° - \alpha$

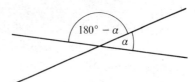

The angle between two vectors,
however, is defined uniquely.
It is the angle between their directions
when the lines representing them
both converge or *both diverge*
(see diagrams i and ii).

In some cases one of the lines
may have to be produced in order
to mark the correct angle
(see diagram iii).

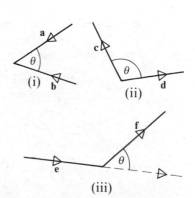

Examples 19a _____

1. Two vectors, **a** and **b**, are such that $|\mathbf{a}| = 3$, $|\mathbf{b}| = 5$ and the angle between **a** and **b** is $\frac{1}{3}\pi$. If the line OP represents the vector $3\mathbf{a} - 2\mathbf{b}$, find, correct to 1 d.p., the angle between $\overrightarrow{OP}$ and **a**.

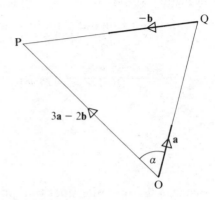

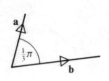

The line OP is found by drawing OQ parallel to **a** such that $\overrightarrow{OQ} = 3\mathbf{a}$, followed by QP parallel to **b** such that $\overrightarrow{QP} = -2\mathbf{b}$

Thus $\overrightarrow{OP} = 3\mathbf{a} - 2\mathbf{b}$

Now $OQ = 3|\mathbf{a}| = 9$

and $QP = 2|\mathbf{b}| = 10$

The angle between $\overrightarrow{OP}$ and **a** is α

where $\dfrac{\sin \alpha}{10} = \dfrac{\sin \frac{1}{3}\pi}{OP}$

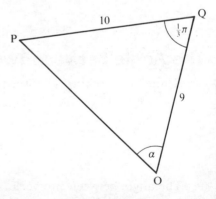

So first we must find OP

Using the cosine formula in OPQ gives

$$OP^2 = 81 + 100 - (2)(9)(10)\cos \tfrac{1}{3}\pi = 91$$

$\Rightarrow \qquad OP = \sqrt{91}$

Then $\dfrac{\sin \alpha}{10} = \dfrac{\sin \frac{1}{3}\pi}{\sqrt{91}} \qquad \Rightarrow \qquad \sin \alpha = 0.9078$

$\therefore$ OP is included at 1.2 rad to **a**

_____

2. In a triangle ABC, $\overrightarrow{AB}$ represents **a** and $\overrightarrow{BC}$ represents **b**. If D is the midpoint of AB express in terms of **a** and **b** the vectors $\overrightarrow{CA}$ and $\overrightarrow{DC}$.

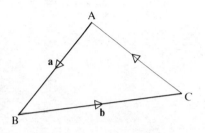

$$\overrightarrow{CA} = \overrightarrow{CB} + \overrightarrow{BA}$$

Now $\overrightarrow{CB} = -\overrightarrow{BC} = -\mathbf{b}$
and $\overrightarrow{BA} = -\overrightarrow{AB} = -\mathbf{a}$

$$\therefore \quad \overrightarrow{CA} = -\mathbf{b} - \mathbf{a} = -(\mathbf{a} + \mathbf{b})$$

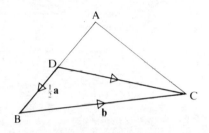

$$\overrightarrow{DC} = \overrightarrow{DB} + \overrightarrow{BC}$$

Now $DB = \tfrac{1}{2}AB$
$\Rightarrow \quad \overrightarrow{DB} = \tfrac{1}{2}\overrightarrow{AB} = \tfrac{1}{2}\mathbf{a}$

$$\therefore \quad \overrightarrow{DC} = \tfrac{1}{2}\mathbf{a} + \mathbf{b}$$

---

3. If D is the midpoint of the side BC of a triangle ABC, show that
$$\overrightarrow{AB} + \overrightarrow{AC} = 2\overrightarrow{AD}$$

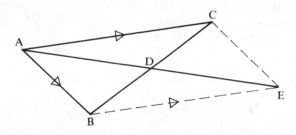

Completing the parallelogram ABEC we see that $\overrightarrow{BE} = \overrightarrow{AC}$

Therefore $\qquad\qquad \overrightarrow{AB} + \overrightarrow{AC} = \overrightarrow{AB} + \overrightarrow{BE} = \overrightarrow{AE}$

The diagonals of a parallelogram bisect each other.

Therefore $\qquad\qquad\qquad\qquad\qquad \overrightarrow{AE} = 2\overrightarrow{AD}$

$\Rightarrow \qquad\qquad\qquad\qquad \overrightarrow{AB} + \overrightarrow{AC} = 2\overrightarrow{AD}$

---

4. Four points O, A, B and C are such that $\overrightarrow{OA} = 10\mathbf{a}$, $\overrightarrow{OB} = 5\mathbf{b}$ and $\overrightarrow{OC} = 4\mathbf{a} + 3\mathbf{b}$. Show that A, B and C are collinear.

If A, B and C are collinear then AB and BC have the same direction so this is what we must show.

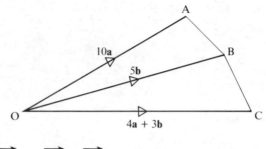

$$\overrightarrow{AB} = \overrightarrow{AO} + \overrightarrow{OB} = -10\mathbf{a} + 5\mathbf{b} = 5(\mathbf{b} - 2\mathbf{a})$$

$$\overrightarrow{BC} = \overrightarrow{BO} + \overrightarrow{OC} = -5\mathbf{b} + 4\mathbf{a} + 3\mathbf{b}$$

$$= 4\mathbf{a} - 2\mathbf{b} = -2(\mathbf{b} - 2\mathbf{a})$$

AB and BC both have a direction given by $\lambda(\mathbf{b} - 2\mathbf{a})$ so they are parallel. Hence, since C is a common point, A, B and C are collinear.

Note that $\overrightarrow{BC} = -\frac{2}{5}\overrightarrow{AB}$ so, although $\overrightarrow{AB}$ and $\overrightarrow{BC}$ are parallel, they are in opposite directions, showing that the diagram really looks like this.

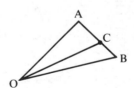

EXERCISE 19a

1. ABCD is a quadrilateral. Find the single vector which is equivalent to

  (a) $\overrightarrow{AB} + \overrightarrow{BC}$ (b) $\overrightarrow{BC} + \overrightarrow{CD}$

  (c) $\overrightarrow{AB} + \overrightarrow{BC} + \overrightarrow{CD}$ (d) $\overrightarrow{AB} + \overrightarrow{DA}$

2. ABCDEF is a regular hexagon in which $\overrightarrow{BC}$ represents $\mathbf{b}$ and $\overrightarrow{FC}$ represents $2\mathbf{a}$.
   Express the vectors $\overrightarrow{AB}$, $\overrightarrow{CD}$ and $\overrightarrow{BE}$ in terms of $\mathbf{a}$ and $\mathbf{b}$

3. Draw diagrams representing the following vector equations.

(a) $\vec{AB} - \vec{CB} = \vec{AC}$ (b) $\vec{AB} = 2\vec{PQ}$

(c) $\vec{AB} + \vec{BC} = 3\vec{AD}$ (d) $2\vec{AB} + \vec{PQ} = 0$

4. If A, B, C, D are four points such that $\vec{AB} = \vec{DC}$ and $\vec{BC} + \vec{DA} = 0$ prove that ABCD is a parallelogram.

5. O, A, B, C, D are five points such that $\vec{OA} = a$, $\vec{OB} = b$, $\vec{OC} = a + 2b$, $\vec{OD} = 2a - b$
Express $\vec{AB}$, $\vec{BC}$, $\vec{CD}$, $\vec{AC}$, $\vec{BD}$ in terms of **a** and **b**

6.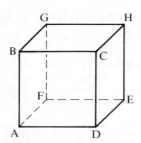

If **a**, **b**, **c** are represented by the edges $\vec{AB}$, $\vec{AD}$, $\vec{AF}$ of the cube in the diagram, find, in terms of **a**, **b** and **c**, the vectors represented by the remaining edges.

7. If O, A, B, C are four points such that $\vec{OA} = a$, $\vec{OB} = 2a - b$, $\vec{OC} = b$ show that A, B and C are collinear.

8.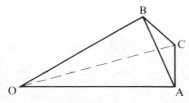

If OABC is a tetrahedron and $\vec{OA} = a$, $\vec{OB} = b$, $\vec{OC} = c$, find $\vec{AC}$, $\vec{AB}$, $\vec{CB}$ in terms of **a**, **b**, **c**

9. For the cube defined in Question 6, find, in terms of **a**, **b** and **c**, the vectors $\vec{BE}$, $\vec{GD}$, $\vec{AH}$, $\vec{FC}$.

10. If **a** and **b** are vectors such that $|a| = 2$, $|b| = 4$ and the angle between **a** and **b** is $\frac{1}{3}\pi$, find the angle between

(a) **a** and **a** − **b**   (b) **b** and **a** + **b**   (c) 3**a** − **b** and **b**

POSITION VECTORS

In general a vector has no specific location in space and is called a *free vector*. Some vectors, however, are constrained to a specific position, e.g. the vector $\overrightarrow{OA}$ where O is a fixed origin.

$\overrightarrow{OA}$ is called the position vector of A relative to O.

This displacement is unique and *cannot* be represented by any other line of equal length and direction.

Vectors such as $\overrightarrow{OA}$, representing quantities that have a specific location, are called *position* vectors or *tied* vectors.

The Position Vector of a Point Dividing a Given Line in a Given Ratio

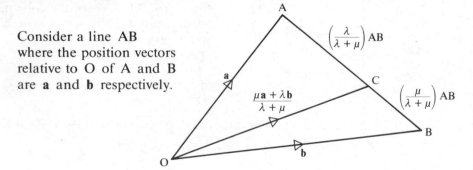

Consider a line AB where the position vectors relative to O of A and B are **a** and **b** respectively.

If C divides AB in the ratio $\lambda : \mu$ then

$$\overrightarrow{AC} = \frac{\lambda}{\lambda + \mu} \overrightarrow{AB} = \frac{\lambda}{\lambda + \mu}(\overrightarrow{OB} - \overrightarrow{OA})$$

i.e. $$\overrightarrow{AC} = \frac{\lambda}{\lambda + \mu}(\mathbf{b} - \mathbf{a})$$

Now $$\overrightarrow{OC} = \overrightarrow{OA} + \overrightarrow{AC} = \mathbf{a} + \frac{\lambda}{\lambda + \mu}(\mathbf{b} - \mathbf{a})$$

i.e.   if A and B are points with position vectors **a** and **b**, and C divides AB in the ratio $\lambda : \mu$,

then   the position vector of C is $\dfrac{\mu\mathbf{a} + \lambda\mathbf{b}}{\lambda + \mu}$

A special case of this quotable result arises when $\lambda = \mu$ i.e. when C is the midpoint of AB.

Then the position vector of C is

$$\tfrac{1}{2}(a + b)$$

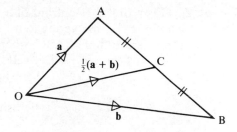

Examples 19b

1. In a triangle ABC, P is the midpoint of AB and Q divides BC in the ratio $2:1$. Given that $\overrightarrow{OA} = a$, $\overrightarrow{OB} = b$ and $\overrightarrow{OC} = c$, find in terms of a, b and c, the position vectors of P, Q and the midpoint of PQ.

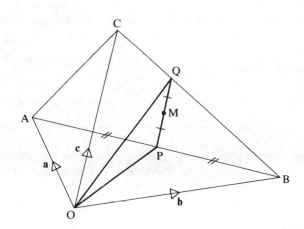

$AP:PB = 1:1$

∴ the position vector of P is $\tfrac{1}{2}(a + b)$

$BQ:QC = 2:1$

∴ the position vector of Q is $\dfrac{(1)(b) + (2)(c)}{2 + 1} = \tfrac{1}{3}(b + 2c)$

If M is the midpoint of PQ then the position vector of M is

$$\tfrac{1}{2}(\overrightarrow{OP} + \overrightarrow{OQ}) = \tfrac{1}{2}\{\tfrac{1}{2}(a + b) + \tfrac{1}{3}(b + 2c)\}$$

$$= \tfrac{1}{4}a + \tfrac{5}{12}b + \tfrac{1}{3}c$$

2. Given that the centroid of a triangle divides each median in the ratio $2:1$ from vertex to opposite side, find the position vector of the centroid of $\triangle ABC$ where the position vectors of A, B and C are $\mathbf{a}$, $\mathbf{b}$ and $\mathbf{c}$ respectively

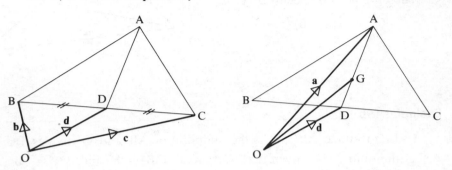

Considering the median AD we first find the position vector of D, the midpoint of BC.

The position vector of D is $\mathbf{d}$ where $\mathbf{d} = \frac{1}{2}(\mathbf{b} + \mathbf{c})$
The centroid, G, divides AD in the ratio $2:1$

So the position vector of G is $\dfrac{(1)(\mathbf{a}) + (2)(\mathbf{d})}{2 + 1} = \dfrac{\mathbf{a} + 2\{\frac{1}{2}(\mathbf{b} + \mathbf{c})\}}{3}$

$$= \tfrac{1}{3}(\mathbf{a} + \mathbf{b} + \mathbf{c})$$

Note that this result shows that the position vectors of the centroid of a triangle is the average of the respective position vectors of the three vertices.

EXERCISE 19b

In this exercise the position vectors, relative to O, of A, B, C and D are $\mathbf{a}$, $\mathbf{b}$, $\mathbf{c}$ and $\mathbf{d}$ respectively. P, Q and R are the midpoints of AB, BC and CD respectively.

In Questions 1 to 4 find the position vector of each given point.

1. (a) The midpoint of AC   (b) The midpoint of BD.

2. The point L which divides AD in the ratio $1:3$

3. The point M which divides BC in the ratio $4:3$

4. (a) The midpoint of PQ   (b) The midpoint of QR.

5. Show that PQ is parallel to AC.

6. Find the vector represented by LD.

7. Find the vector represented by AR.

8. Find the position vector of the point which

   (a) divides CP in the ratio $2:1$

   (b) divides AQ in the ratio $2:1$

   Say what you notice about your answers to (a) and (b) and explain this relationship.

THE LOCATION OF A POINT IN SPACE

We saw in Module A that any point P in a plane can be located by giving its distances from a fixed point O, in each of two perpendicular directions. These distances are the Cartesian coordinates of the point.

Now we consider locating a point in three-dimensional space.

If we have a fixed point, O, then any other point can be located by giving its distances from O in each of *three* mutually perpendicular directions, i.e. we need *three* coordinates to locate a point in space. So we use the familiar x and y axes, together with a third axis Oz. Then any point has coordinates (x, y, z) relative to the origin O.

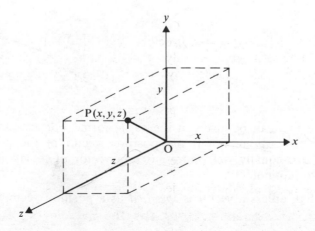

Cartesian Unit Vectors

A unit vector is a vector whose magnitude is one unit.

Now if

i is a unit vector in the direction of Ox
j is a unit vector in the direction of Oy
k is a unit vector in the direction of Oz

then the position vector, relative to O, of any point P can be given in terms of **i**, **j** and **k**.

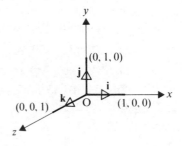

e.g. the point distant

   3 units from O in the direction Ox
   4 units from O in the direction Oy
   5 units from O in the direction Oz

has coordinates $(3, 4, 5)$

and $\overrightarrow{OP} = 3\mathbf{i} + 4\mathbf{j} + 5\mathbf{k}$

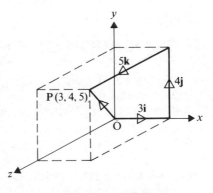

In general, if P is a point, (x, y, z) and $\overrightarrow{OP} = \mathbf{r}$, then

$$\mathbf{r} = x\mathbf{i} + y\mathbf{j} + z\mathbf{k}$$

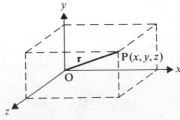

Free vectors can be given in the same form. For example, the vector $3\mathbf{i} + 4\mathbf{j} + 5\mathbf{k}$ *can* represent the position vector of the point $P(3, 4, 5)$ but it can equally well represent *any* vector of length and direction equal to those of OP.

Note that, unless a vector is *specified* as a position vector it is taken to be free.

OPERATIONS ON CARTESIAN VECTORS

Addition and Subtraction

To add or subtract vectors given in $\mathbf{i\,j\,k}$ form, the coefficients of $\mathbf{i}$, $\mathbf{j}$ and $\mathbf{k}$ are collected separately,

e.g.   if $\mathbf{v}_1 = 3\mathbf{i} + 2\mathbf{j} + 2\mathbf{k}$ and $\mathbf{v}_2 = \mathbf{i} + 2\mathbf{j} - 3\mathbf{k}$

then

$$\mathbf{v}_1 + \mathbf{v}_2 = (3\mathbf{i} + 2\mathbf{j} + 2\mathbf{k}) + (\mathbf{i} + 2\mathbf{j} - 3\mathbf{k})$$
$$= (3 + 1)\mathbf{i} + (2 + 2)\mathbf{j} + (2 - 3)\mathbf{k}$$
$$= 4\mathbf{i} + 4\mathbf{j} - \mathbf{k}$$

and

$$\mathbf{v}_1 - \mathbf{v}_2 = (3 - 1)\mathbf{i} + (2 - 2)\mathbf{j} + (2 - \{-3\})\mathbf{k}$$
$$= 2\mathbf{i} + 5\mathbf{k}$$

Modulus

The modulus of $\mathbf{v}$, where $\mathbf{v} = 12\mathbf{i} - 3\mathbf{j} + 4\mathbf{k}$, is the length of OP where P is the point $(12, -3, 4)$.

Using Pythagoras twice we have

$$OB^2 = OA^2 + AB^2 = 12^2 + 4^2$$

$$OP^2 = OB^2 + BP^2 = (12^2 + 4^2) + (-3)^2$$

$$\therefore \quad OP = \sqrt{(12^2 + 4^2 + 3^2)} = 13$$

In general, if $\mathbf{v} = a\mathbf{i} + b\mathbf{j} + c\mathbf{k}$.
$$|\mathbf{v}| = \sqrt{(a^2 + b^2 + c^2)}$$

Parallel Vectors

Two vectors $\mathbf{v}_1$ and $\mathbf{v}_2$ are parallel if $\mathbf{v}_1 = \lambda \mathbf{v}_2$

e.g. $2\mathbf{i} - 3\mathbf{j} - \mathbf{k}$ is parallel to $4\mathbf{i} - 6\mathbf{j} - 2\mathbf{k}$ $(\lambda = 2)$
and $\mathbf{i} + \mathbf{j} + \mathbf{k}$ is parallel to $-3\mathbf{i} - 3\mathbf{j} - 3\mathbf{k}$ $(\lambda = -3)$

Equal Vectors

If two vectors $\mathbf{v}_1 = a_1\mathbf{i} + b_1\mathbf{j} + c_1\mathbf{k}$ and $\mathbf{v}_2 = a_2\mathbf{i} + b_2\mathbf{j} + c_2\mathbf{k}$ are equal then

$$a_1 = a_2 \text{ and } b_1 = b_2 \text{ and } c_1 = c_2$$

UNIT VECTORS

We know that the vectors **i**, **j** and **k** are unit vectors in the directions of the x, y and z axes but these are not the only unit vectors.

> **Any vector whose magnitude is 1 is a unit vector.**

Suppose that the vector $\mathbf{a} = 3\mathbf{i} + 2\mathbf{j} + 6\mathbf{k}$ is represented by $\overrightarrow{OP}$ where P is the point $(3, 2, 6)$.

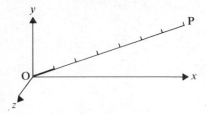 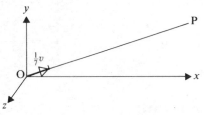

Now $|\mathbf{a}| = \sqrt{(3^2 + 2^2 + 6^2)} = 7$, so OP is 7 units long.

Therefore $\frac{1}{7}\mathbf{a}$ is a vector of unit magnitude.

i.e. a unit vector in the direction of $\mathbf{a}$ is $\dfrac{3\mathbf{i} + 2\mathbf{j} + 6\mathbf{k}}{7}$

> **In general a unit vector in the direction of $\mathbf{v}$ is given by $\dfrac{\mathbf{v}}{|\mathbf{v}|}$**

For example, a unit vector in the direction of $2\mathbf{i} - \mathbf{j} - 2\mathbf{k}$ is

$$\frac{2\mathbf{i} - \mathbf{j} - 2\mathbf{k}}{\sqrt{(2^2 + \{-1\}^2 + \{-2\}^2)}} = \frac{2\mathbf{i} - \mathbf{j} - 2\mathbf{k}}{3} = \tfrac{2}{3}\mathbf{i} - \tfrac{1}{3}\mathbf{j} - \tfrac{2}{3}\mathbf{k}$$

Examples 19c

1. Given the vector **v** where $\mathbf{v} = 5\mathbf{i} - 2\mathbf{j} + 4\mathbf{k}$, state whether each of the following vectors is parallel to **v**, equal to **v** or neither.

   (a) $10\mathbf{i} - 4\mathbf{j} + 8\mathbf{k}$ (b) $-\tfrac{1}{2}(-10\mathbf{i} + 4\mathbf{j} - 8\mathbf{k})$
   (c) $-5\mathbf{i} + 2\mathbf{j} - 4\mathbf{k}$ (d) $4\mathbf{i} - 2\mathbf{j} + 5\mathbf{k}$

(a) $10\mathbf{i} - 4\mathbf{j} + 8\mathbf{k} = 2(5\mathbf{i} - 2\mathbf{j} + 4\mathbf{k})$ $(\lambda = 2)$

                       $\therefore\quad 10\mathbf{i} - 4\mathbf{j} + 8\mathbf{k}$ is parallel to **v**

(b) $-\tfrac{1}{2}(-10\mathbf{i} + 4\mathbf{j} - 8\mathbf{k}) = 5\mathbf{i} - 2\mathbf{j} + 4\mathbf{k}$

                       $\therefore\quad -\tfrac{1}{2}(-10\mathbf{i} + 4\mathbf{j} - 8\mathbf{k})$ is equal to **v**

(c) $\qquad -5\mathbf{i} + 2\mathbf{j} - 4\mathbf{k} = -(5\mathbf{i} - 2\mathbf{j} + 4\mathbf{k})$ $\qquad\qquad$ ($\lambda = -1$)

$\qquad\therefore\quad -5\mathbf{i} + 2\mathbf{j} - 4\mathbf{k}$ is parallel to $\mathbf{v}$

(d) $\qquad 4\mathbf{i} - 2\mathbf{j} + 5\mathbf{k}$ is not a multiple of $5\mathbf{i} - 2\mathbf{j} + 4\mathbf{k}$

$\qquad\therefore\quad 4\mathbf{i} - 2\mathbf{j} + 5\mathbf{k}$ is not parallel to $\mathbf{v}$

---

2. A triangle ABC has its vertices at the points $A(2, -1, 4)$, $B(3, -2, 5)$ and $C(-1, 6, 2)$. Find, in the form $a\mathbf{i} + b\mathbf{j} + c\mathbf{k}$, the vectors $\overrightarrow{AB}$, $\overrightarrow{BC}$ and $\overrightarrow{CA}$ and hence find the lengths of the sides of the triangle.

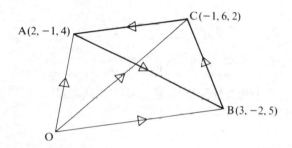

The coordinate axes are not drawn in this diagram as they tend to cause confusion when two or more points are illustrated. The origin should always be included however as it provides a reference point.

$$\overrightarrow{AB} = \overrightarrow{OB} - \overrightarrow{OA}$$
$$= (3\mathbf{i} - 2\mathbf{j} + 5\mathbf{k}) - (2\mathbf{i} - \mathbf{j} + 4\mathbf{k}) = \mathbf{i} - \mathbf{j} + \mathbf{k}$$
$$\overrightarrow{BC} = \overrightarrow{OC} - \overrightarrow{OB}$$
$$= (-\mathbf{i} + 6\mathbf{j} + 2\mathbf{k}) - (3\mathbf{i} - 2\mathbf{j} + 5\mathbf{k}) = -4\mathbf{i} + 8\mathbf{j} - 3\mathbf{k}$$
$$\overrightarrow{CA} = \overrightarrow{OA} - \overrightarrow{OC}$$
$$= (2\mathbf{i} - \mathbf{j} + 4\mathbf{k}) - (-\mathbf{i} + 6\mathbf{j} + 2\mathbf{k}) = 3\mathbf{i} - 7\mathbf{j} + 2\mathbf{k}$$

Hence $\quad AB = |\overrightarrow{AB}| = \sqrt{\{(1)^2 + (-1)^2 + (1)^2\}} = \sqrt{3}$

$\qquad\quad BC = |\overrightarrow{BC}| = \sqrt{\{(-4)^2 + (8)^2 + (-3)^2\}} = \sqrt{89}$

$\qquad\quad CA = |\overrightarrow{CA}| = \sqrt{\{(3)^2 + (-7)^2 + (2)^2\}} = \sqrt{62}$

Two-dimensional problems can be solved by using the same principles as for three-dimensional cases but the working tends to be easier because it involves fewer terms.

---

3. Given that $\mathbf{p} = \mathbf{i} + 3\mathbf{j}$, $\mathbf{q} = 4\mathbf{i} - 2\mathbf{j}$, $\mathbf{OA} = 2\mathbf{p}$ and $\mathbf{OB} = 3\mathbf{q}$, find   (a) $|\mathbf{OA}|$ (b) $|\mathbf{OB}|$ (c) $|\mathbf{AB}|$

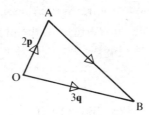

(a) $|\mathbf{OA}| = 2|\mathbf{i} + 3\mathbf{j}| = 2\sqrt{(1^2 + 3^2)} = 2\sqrt{10}$

(b) $|\mathbf{OB}| = 3|4\mathbf{i} - 2\mathbf{j}| = 3\sqrt{\{4^2 + (-2)^2\}} = 6\sqrt{5}$

(c) $\mathbf{AB} = \mathbf{OB} - \mathbf{OA} = 3(4\mathbf{i} - 2\mathbf{j}) - 2(\mathbf{i} + 3\mathbf{j}) = 10\mathbf{i} - 12\mathbf{k}$

$|\mathbf{AB}| = \sqrt{(10^2 + 12^2)} = 2\sqrt{61}$

---

4. If P is a point with position vector $(\cos\theta)\mathbf{i} + (\sin\theta)\mathbf{j}$, find the Cartesian equation of the curve on which P lies.

If P is the point (x, y) and $\mathbf{OP} = (\cos\theta)\mathbf{i} + (\sin\theta)\mathbf{j}$ then

$$x = \cos\theta \quad \text{and} \quad y = \sin\theta$$

These are the parametric equations of the locus of P. We can find, the Cartesian equation by eliminating θ between them.

Using $\cos^2\theta + \sin^2\theta = 1$

gives $x^2 + y^2 = 1$

Therefore P lies on the curve $x^2 + y^2 = 1$ which can be recognised as a circle with radius 1 and centre at O.

EXERCISE 19c

1. Write down, in the form $a\mathbf{i} + b\mathbf{j} + c\mathbf{k}$, the vector represented by $\overrightarrow{OP}$ if P is a point with coordinates
   (a) $(3, 6, 4)$ (b) $(1, -2, -7)$
   (c) $(1, 0, -3)$ (d) $(0, -4, 0)$

2. $\overrightarrow{OP}$ represents a vector $\mathbf{r}$. Write down the coordinates of P if
   (a) $\mathbf{r} = 5\mathbf{i} - 7\mathbf{j} + 2\mathbf{k}$ (b) $\mathbf{r} = \mathbf{i} + 4\mathbf{j}$
   (c) $\mathbf{r} = \mathbf{j} - \mathbf{k}$ (d) $\mathbf{r} = 2\mathbf{i} - 3\mathbf{k}$

3. Find the length of the line OP if P is the point
   (a) $(2, -1, 4)$ (b) $(3, 0, 4)$
   (c) $(-2, -2, 1)$ (d) $(2, -6, 3)$

4. Find the modulus of the vector $\mathbf{V}$ if
   (a) $\mathbf{V} = 2\mathbf{i} - 4\mathbf{j} + 4\mathbf{k}$ (b) $\mathbf{V} = 6\mathbf{i} + 2\mathbf{j} - 3\mathbf{k}$
   (c) $\mathbf{V} = 11\mathbf{i} - 7\mathbf{j} - 6\mathbf{k}$ (d) $\mathbf{V} = 4\mathbf{i} + \mathbf{j} - 8\mathbf{k}$

5. Find a unit vector in the direction of $\mathbf{v}$ if $\mathbf{v}$ is given by
   (a) $3\mathbf{i} + 4\mathbf{k}$ (b) $\mathbf{i} + 8\mathbf{j} - 4\mathbf{k}$
   (c) $4\mathbf{i} - 4\mathbf{j} + 2\mathbf{k}$ (d) $6\mathbf{i} - 3\mathbf{j} - 2\mathbf{k}$

6. If $\mathbf{a} = \mathbf{i} + \mathbf{j} + \mathbf{k}$, $\mathbf{b} = 2\mathbf{i} - \mathbf{j} + 3\mathbf{k}$, $\mathbf{c} = -\mathbf{i} + 3\mathbf{j} - \mathbf{k}$ find
   (a) $\mathbf{a} + \mathbf{b}$ (b) $\mathbf{a} - \mathbf{c}$
   (c) $\mathbf{a} + \mathbf{b} + \mathbf{c}$ (d) $\mathbf{a} - 2\mathbf{b} + 3\mathbf{c}$

7. A, B, C and D are the points $(0, 0, 2)$, $(-1, 3, 2)$, $(1, 0, 4)$ and $(-1, 2, -2)$ respectively. Find the vectors representing $\overrightarrow{AB}$, $\overrightarrow{BD}$, $\overrightarrow{CD}$, $\overrightarrow{AD}$

In Questions 8 to 10, $\overrightarrow{OA} = \mathbf{a} = 4\mathbf{i} - 12\mathbf{j}$ and $\overrightarrow{OB} = \mathbf{b} = \mathbf{i} + 6\mathbf{j}$.

8. Which of the following vectors are parallel to $\mathbf{a}$?

    (a) $\mathbf{i} + 3\mathbf{j}$ (b) $-\mathbf{i} + 3\mathbf{j}$ (c) $12\mathbf{i} - 4\mathbf{j}$ (d) $-4\mathbf{i} + 12\mathbf{j}$ (e) $\mathbf{i} - 3\mathbf{j}$

9. Which of the following vectors are equal to $\mathbf{b}$?

    (a) $2\mathbf{i} + 12\mathbf{j}$ (b) $-\mathbf{i} - 6\mathbf{j}$

    (c) $\overrightarrow{AE}$ if E is $(5, -6)$ (d) $\overrightarrow{AF}$ if F is $(6, 0)$

10. If $\overrightarrow{OD} = \lambda\overrightarrow{OA}$, find the value of λ for which $\overrightarrow{OD} + \overrightarrow{OB}$ is parallel to the x-axis

11. Which of the following vectors are parallel to $3\mathbf{i} - \mathbf{j} - 2\mathbf{k}$?
    (a) $6\mathbf{i} - 3\mathbf{j} - 4\mathbf{k}$ (b) $-9\mathbf{i} + 3\mathbf{j} + 6\mathbf{k}$
    (c) $-3\mathbf{i} - \mathbf{j} - 2\mathbf{k}$ (d) $-2(3\mathbf{i} + \mathbf{j} + 2\mathbf{k})$
    (e) $\frac{3}{2}\mathbf{i} - \frac{1}{2}\mathbf{j} - \mathbf{k}$ (f) $-\mathbf{i} + \frac{1}{3}\mathbf{j} + \frac{2}{3}\mathbf{k}$

12. Given that $\mathbf{a} = 4\mathbf{i} + \mathbf{j} - 6\mathbf{k}$, state whether each of the following vectors is parallel or equal to $\mathbf{a}$ or neither.
    (a) $8\mathbf{i} + 2\mathbf{j} - 10\mathbf{k}$ (b) $-4\mathbf{i} - \mathbf{j} + 6\mathbf{k}$
    (c) $2(2\mathbf{i} + \frac{1}{2}\mathbf{j} - 3\mathbf{k})$ (d) $6\mathbf{i} + \mathbf{j} + 4\mathbf{k}$

13. The triangle ABC has its vertices at the points $A(-1, 3, 0)$, $B(-3, 0, 7)$, $C(-1, 2, 3)$,. Find in the form $a\mathbf{i} + b\mathbf{j} + c\mathbf{k}$ the vectors representing

    (a) $\overrightarrow{AB}$ (b) $\overrightarrow{AC}$ (c) $\overrightarrow{CB}$

14. Find the lengths of the sides of the triangle described in Question 13.

15. Find $|\mathbf{a} - \mathbf{b}|$ where $\mathbf{a} = \mathbf{i} - \mathbf{j} + 2\mathbf{k}$, $\mathbf{b} = 2\mathbf{i} - \mathbf{j}$

16. If the position vector of P is $t^2\mathbf{i} + 2t\mathbf{j}$, find the Cartesian equation of the locus of P and name this curve.

17. Show that, for all values of t, the point whose equation vector is $\mathbf{r} = t\mathbf{i} + (2t - 1)\mathbf{j}$ lies on the line $y = 2x - 1$

PROPERTIES OF A LINE JOINING TWO POINTS

Consider the line joining the points $A(x_1, y_1, z_1)$ and $B(x_2, y_2, z_2)$

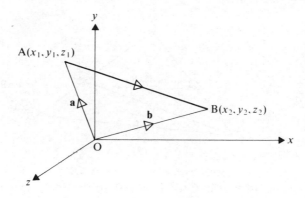

$$\overrightarrow{OA} = x_1\mathbf{i} + y_1\mathbf{j} + z_1\mathbf{k} \quad \text{and} \quad \overrightarrow{OB} = x_2\mathbf{i} + y_2\mathbf{j} + z_2\mathbf{k}$$

and $\quad \overrightarrow{AB} = \overrightarrow{AO} + \overrightarrow{OB} = \overrightarrow{OB} - \overrightarrow{OA}$

hence $\quad \overrightarrow{AB} = (x_2 - x_1)\mathbf{i} + (y_2 - y_1)\mathbf{j} + (z_2 - z_1)\mathbf{k}$

The Length of AB

$$AB = |(x_2 - x_1)\mathbf{i} + (y_2 - y_1)\mathbf{j} + (z_2 - z_1)\mathbf{k}|$$

so the length of the line joining (x_1, y_1, z_1) and (x_2, y_2, z_2) is

$$\sqrt{[(x_2 - x_1)^2 + (y_2 - y_1)^2 + (z_2 - z_1)^2]}$$

A Direction Vector for the Line through A and B

A vector which specifies the direction of a line can be called a *direction vector*.

The direction of the line through A and B is given by $\overrightarrow{AB}$ where

$$\overrightarrow{AB} = (x_2 - x_1)\mathbf{i} + (y_2 - y_1)\mathbf{j} + (z_2 - z_1)\mathbf{k}$$

Therefore $(x_2 - x_1)\mathbf{i} + (y_2 - y_1)\mathbf{j} + (z_2 - z_1)\mathbf{k}$ is a direction vector for this line.

The Position Vector of a Point Dividing AB in the Ratio $\lambda : \mu$

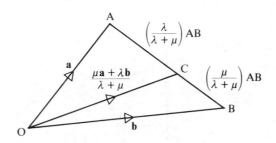

We saw on p. 314 that if C divides AB in the ratio $\lambda : \mu$ then the position vector of C is $\dfrac{\mu \overrightarrow{OA} + \lambda \overrightarrow{OB}}{\lambda + \mu}$

If $\overrightarrow{OA} = x_1 \mathbf{i} + y_1 \mathbf{j} + z_1 \mathbf{k}$ and $\overrightarrow{OB} = x_2 \mathbf{i} + y_2 \mathbf{j} + z_2 \mathbf{k}$

Then $\overrightarrow{OC} = \dfrac{1}{\lambda + \mu} \{ (\mu x_1 + \lambda x_2)\mathbf{i} + (\mu y_1 + \lambda y_2)\mathbf{j} + (\mu z_1 + \lambda z_2)\mathbf{k} \}$

Therefore the coordinates of C are

$$\frac{\mu x_1 + \lambda x_2}{\lambda + \mu}, \quad \frac{\mu y_1 + \lambda y_2}{\lambda + \mu}, \quad \frac{\mu z_1 + \lambda z_2}{\lambda + \mu}$$

e.g. if A and B are the points $(5, -1, 6)$ and $(1, 7, -10)$ and C divides AB in the ratio $1:3$, then the coordinates of C are

$$\frac{3(5) + 1(1)}{4}, \quad \frac{3(-1) + 1(7)}{4}, \quad \frac{3(6) + 1(-10)}{4}, \quad \text{i.e. } (4, 1, 2)$$

In particular, if $\lambda = \mu$ so that C bisects AB we see that the coordinates of the midpoint of AB are

$$\tfrac{1}{2}(x_1 + x_2), \tfrac{1}{2}(y_1 + y_2), \tfrac{1}{2}(z_1 + z_2)$$

i.e.   the coordinates of the midpoint are the averages of the respective coordinates of the end points

Examples 19d

1. Find the length of the median through O of the triangle OAB, where A is the point $(2, 7, -1)$ and B is the point $(4, 1, 2)$

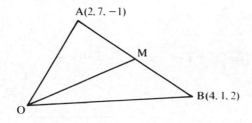

A(2, 7, -1)

M

B(4, 1, 2)

O

The coordinates of M, the midpoint of AB, are

$$(\tfrac{1}{2}\{2 + 4\}, \tfrac{1}{2}\{7 + 1\}, \tfrac{1}{2}\{-1 + 2\})$$

i.e. $(3, 4, \tfrac{1}{2})$

So the length of OM is $\sqrt{(3^2 + 4^2 + \{\tfrac{1}{2}\}^2)} = \tfrac{1}{2}\sqrt{101}$

2. The points A, B and C have coordinates $(3, -1, 5)$, $(7, 1, 3)$ and $(-5, 9, -1)$ respectively. If L is the midpoint of AB and M is the midpoint of BC, find the length of LM and a direction vector for LM.

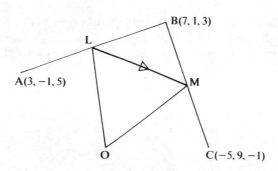

B(7, 1, 3)

L

A(3, -1, 5)

M

O

C(-5, 9, -1)

L is the point $(\tfrac{1}{2}\{3 + 7\}, \tfrac{1}{2}\{-1 + 1\}, \tfrac{1}{2}\{5 + 3\})$, i.e. $(5, 0, 4)$

M is the point $(\tfrac{1}{2}\{7 - 5\}, \tfrac{1}{2}\{1 + 9\}, \tfrac{1}{2}\{3 - 1\})$, i.e. $(1, 5, 1)$

∴ $LM = \sqrt{(\{1 - 5\}^2 + \{5 - 0\}^2 + \{1 - 4\}^2)} = 5\sqrt{2}$

Now $\overrightarrow{LM} = (1 - 5)\mathbf{i} + (5 - 0)\mathbf{j} + (1 - 4)\mathbf{k}$

So $-4\mathbf{i} + 5\mathbf{j} - 3\mathbf{k}$ is a direction vector for LM.

3. A and B are two points with position vectors $3\mathbf{i} + \mathbf{j} - 2\mathbf{k}$ and $\mathbf{i} - 3\mathbf{j} - \mathbf{k}$ respectively. Find the position vectors of the points P and Q if

(a) P divides AB internally in the ratio $3:1$
(b) Q divides AB externally in the ratio $3:1$

(a) $AP:PB = 3:1$

$\therefore \quad \overrightarrow{OP} = \dfrac{1(3\mathbf{i} + \mathbf{j} - 2\mathbf{k}) + 3(\mathbf{i} - 3\mathbf{j} - \mathbf{k})}{3 + 1}$

$\qquad = \tfrac{3}{2}\mathbf{i} - 2\mathbf{j} - \tfrac{5}{4}\mathbf{k}$

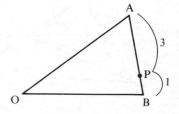

(b) $AQ:QB = 3:-1$

$\therefore \quad \overrightarrow{OQ} = \dfrac{-1(3\mathbf{i} + \mathbf{j} - 2\mathbf{k}) + 3(\mathbf{i} - 3\mathbf{j} - \mathbf{k})}{3 - 1}$

$\qquad = -5\mathbf{j} - \tfrac{1}{2}\mathbf{k}$

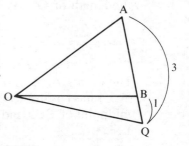

---

4. A, B and C are the points with position vectors $2\mathbf{i} - \mathbf{j} + 5\mathbf{k}$, $\mathbf{i} - 2\mathbf{j} + \mathbf{k}$ and $3\mathbf{i} + \mathbf{j} - 2\mathbf{k}$ respectively. If D and E are the respective midpoints of BC and AC, show that DE is parallel to AB.

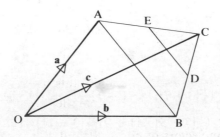

Using $\mathbf{a} = 2\mathbf{i} - \mathbf{j} + 5\mathbf{k}$, $\mathbf{b} = \mathbf{i} - 2\mathbf{j} + \mathbf{k}$ and $\mathbf{c} = 3\mathbf{i} + \mathbf{j} - 2\mathbf{k}$ we have

$$\overrightarrow{OD} = \tfrac{1}{2}(\mathbf{b} + \mathbf{c}) = \tfrac{1}{2}(4\mathbf{i} - \mathbf{j} - \mathbf{k})$$

and
$$\overrightarrow{OE} = \tfrac{1}{2}(\mathbf{a} + \mathbf{c}) = \tfrac{1}{2}(5\mathbf{i} + 3\mathbf{k})$$

$$\therefore \qquad \overrightarrow{DE} = \overrightarrow{OE} - \overrightarrow{OD} = \tfrac{1}{2}(i + j + 4k)$$

Also
$$\overrightarrow{AB} = b - a$$
$$= -i - j - 4k$$
$$= -2\overrightarrow{DE}$$

$\therefore$ AB and DE are parallel.

EXERCISE 19d

In Questions 1 to 7 A, B, C and D are the points with position vectors $i + j - k$, $i - j + 2k$, $j + k$ and $2i + j$ respectively.

1. Find $|AB|$ and $|BD|$.

2. If P divides BD in the ratio $1:2$, find the position vector of P.

3. Find the position vector of the point which
   (a) divides BC internally in the ratio $3:2$
   (b) divides AC externally in the ratio $3:2$

4. Determine whether any of the following pairs of lines are parallel.
   (a) AB and CD   (b) AC and BD   (c) AD and BC.

5. If L and M are the position vectors of the midpoints of AD and BD respectively, show that $\overrightarrow{LM}$ is parallel to $\overrightarrow{AB}$.

6. If H and K are the midpoints of AC and CD respectively show that $\overrightarrow{HK} = \tfrac{1}{2}\overrightarrow{AD}$.

7. If L, M, N and P are the midpoints of AD, BD, BC and AC respectively, show that $\overrightarrow{LM}$ is parallel to $\overrightarrow{NP}$.

8. The position vectors of points P and R are $2i - 3j + 7k$ and $4i + 5j + 3k$ respectively. Given that R divides PQ in the ratio $2:1$, find the position vector of Q if
   (a) R divides PQ internally
   (b) R divides PQ externally.
   (Try using $xi + yi + zk$ as the position vector of Q.)

THE EQUATION OF A STRAIGHT LINE

A straight line is located uniquely in space if
either it passes through a known fixed point and has a known direction,
or it passes through two known fixed points.

In each of these cases we can find equations which describe the set of
points on the line.

A Line with Known Direction and Passing through a Known Fixed Point

Consider a line for which **d** is a direction vector and which passes
through a fixed point A with position vector **a**

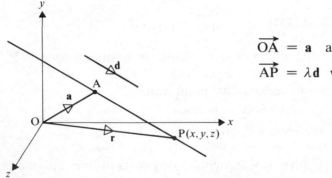

$$\overrightarrow{OA} = \mathbf{a} \quad \text{and}$$

$$\overrightarrow{AP} = \lambda\mathbf{d} \quad \text{where } \lambda \text{ is scalar}$$

If $P(x, y, z)$ is *any* point on the line and **r** is the position vector of P

then $\qquad \overrightarrow{OP} = \overrightarrow{OA} + \overrightarrow{AP} \quad \Rightarrow \quad \mathbf{r} = \mathbf{a} + \lambda\mathbf{d}$

For example, the vector equation of the line which passes through the
point $(5, -2, 4)$ and is parallel to the vector $2\mathbf{i} - \mathbf{j} + 3\mathbf{k}$, is

$$\mathbf{r} = 5\mathbf{i} - 2\mathbf{j} + 4\mathbf{k} + \lambda(2\mathbf{i} - \mathbf{j} + 3\mathbf{k}) \qquad [1]$$

$\mathbf{r} = \mathbf{a} + \lambda\mathbf{d}$ is called a *vector equation* of the line.

Each value of the parameter λ gives the position vector of one point
on the line, e.g. taking $\lambda = 1$ in equation [1] above gives the
position vector $7\mathbf{i} - 3\mathbf{j} + 7\mathbf{k}$, so $(7, -3, 7)$ is a point on the given line.

Note that the lines $\mathbf{r} = \mathbf{a}_1 + \lambda\mathbf{d}_1$ and $\mathbf{r} = \mathbf{a}_2 + \mu\mathbf{d}_2$ are parallel if
$\mathbf{d}_1$ is a multiple of $\mathbf{d}_2$.

Note also that a line does not have a unique vector equation as **a** can
be the position vector of *any* point on the line.

A Line Passing through Two Fixed Points

If a line passes through the points $A(x_1, y_1, z_1)$ and $B(x_2, y_2, z_2)$ then a direction vector for the line is $\mathbf{d}$ where

$$\mathbf{d} = (x_2 - x_1)\mathbf{i} + (y_2 - y_1)\mathbf{j} + (z_2 - z_1)\mathbf{k}$$

This fact, together with A or B as a fixed point on the line, allow the equations of the line to be written down in the form

$$\mathbf{r} = \mathbf{a} + \lambda\mathbf{d} \quad \text{or} \quad \mathbf{r} = \mathbf{b} + \lambda\mathbf{d}$$

Examples 19e

1. A line passing through the point $(2, -1, 4)$ is in the direction of $\mathbf{i} + \mathbf{j} - 2\mathbf{k}$. Find a vector equation for the line.

The position vector relative to O of the point $(2, -1, 4)$ is

$$2\mathbf{i} - \mathbf{j} + 4\mathbf{k}$$

Therefore the vector equation of the line is

$$\mathbf{r} = 2\mathbf{i} - \mathbf{j} + 4\mathbf{k} + \lambda(\mathbf{i} + \mathbf{j} - 2\mathbf{k})$$

Note that this equation can also be given in the form

$$\mathbf{r} = (2 + \lambda)\mathbf{i} + (-1 + \lambda)\mathbf{j} + (4 - 2\lambda)\mathbf{k}$$

2. Find a vector equation for the line through the points $A(3, 4, -7)$ and $B(1, -1, 6)$.

$$\overrightarrow{OA} = \mathbf{a} = 3\mathbf{i} + 4\mathbf{j} - 7\mathbf{k}$$

$$\overrightarrow{OB} = \mathbf{b} = \mathbf{i} - \mathbf{j} + 6\mathbf{k}$$

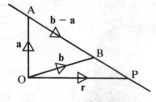

A direction vector of $\overrightarrow{AB}$ is $\mathbf{b} - \mathbf{a}$

For any point P on the line, $\overrightarrow{OP} = \mathbf{r}$

so

$$\mathbf{r} = \mathbf{a} + \lambda(\mathbf{b} - \mathbf{a})$$

$$= 3\mathbf{i} + 4\mathbf{j} - 7\mathbf{k} + \lambda(-2\mathbf{i} - 5\mathbf{j} + 13\mathbf{k})$$

3. Show that the line through the points A and B with position vectors $i + j - 3k$ and $4i + 7j + k$ is parallel to the line l with equation $r = i - j + \lambda(\frac{3}{2}i + 3j + 2k)$.

For $\overrightarrow{AB}$ a direction vector is $(4i + 7j + k) - (i + j - 3k)$

i.e. $3i + 6j + 4k$

For l a direction vector is $\frac{3}{2}i + 3j + 2k$

i.e. $\frac{1}{2}(3i + 6j + 4k)$

Therefore the two lines are parallel.

---

4. A line L passes through the point with position vector $i - 3j + 2k$ and is parallel to the line $r = 4i - 5j + k + \lambda(5i + 2j - k)$. Write down a vector equation for L and hence find the coordinates of the points where L crosses the xy plane and the yz plane.

For the line $r = 4i - 5j + k + \lambda(5i + 2j - k)$ a direction vector is

$$5i + 2j - k$$

This line is parallel to L so

$5i + 2j - k$ is also a direction vector for L

Therefore a vector equation of L is

$$r = i - 3j + 2k + \lambda(5i + 2j - k)$$
$$= (1 + 5\lambda)i + (-3 + 2\lambda)j + (2 - \lambda)k$$

At the point where L crosses the xy plane the z-coordinate is zero, i.e. the coefficient of k is zero.

i.e. $2 - \lambda = 0 \quad \Rightarrow \quad \lambda = 2$

When $\lambda = 2$, $r = 11i + j$

So L crosses the xy plane at the point $(11, 1, 0)$.

Similarly, L crosses the yz plane where $x = 0 \quad \Rightarrow \quad 1 + 5\lambda = 0$

i.e. $\lambda = -\frac{1}{5} \quad \Rightarrow \quad r = -\frac{17}{5}j + \frac{11}{5}k$.

So L crosses the yz plane at the point $(0, -\frac{17}{5}, \frac{11}{5})$.

EXERCISE 19e

1. Write down a direction vector for each line
   (a) $r = 2i + 3j - k + \lambda(i + j + k)$
   (b) $r = 4j + \lambda(3i + 5k)$
   (c) $r = \lambda(2i + 3j + 4k)$

2. Write down a vector equation for the line through a point A with position vector **a** and with a direction vector **b** if
   (a) $a = i - 3j + 2k$ $b = 5i + 4j - k$
   (b) $a = 2i + j$ $b = 3j - k$
   (c) A is the origin $b = i - j - k$

3. State whether or not the following pairs of lines are parallel.
   (a) $r = i + j - k + \lambda(2i - 3j + k)$ and
   $r = 2i - 4j + 5k + \lambda(i + j - k)$
   (b) $r = \lambda(3i - 3j + 6k)$ and $r = 4j + \lambda(-i + j - 2k)$
   (c) $r = 3i + k + \lambda(i - j - 2k)$ and $r = 3i + k - \lambda(i + j + 2k)$

4. The points A(4, 5, 10), B(2, 3, 4) and C(1, 2, −1) are three vertices of a parallelogram ABCD. Find
   (a) the coordinates of the midpoint of AC
   (b) the coordinates of D
   (c) vector equations for the lines AB, BC and AC.

5. Write down a vector equation for the line through A and B if
   (a) $\overrightarrow{OA}$ is $3i + j - 4k$ and $\overrightarrow{OB}$ is $i + 7j + 8k$
   (b) A and B have coordinates (1, 1, 7) and (3, 4, 1).
   Find, in each case, the coordinates of the points where the line crosses the *xy* plane, the *yz* plane and the *zx* plane.

6. A line has a vector equation $r = i - 2j + 4k + \lambda(3i + 4j + 5k)$. Find a vector equation for a parallel line passing through the point with position vector $5i - 2j - 4k$ and find the coordinates of the point on this line where $y = 0$

7. The vector equation of a line is $r = 2i - 5j + 7k + \lambda(6i + 3j + 4k)$ Write down the vector equation of the line through (2, −1, −1) which is parallel to the given line.

8. Given four points, $A(1, 2, 4)$, $B(2, 4, 2)$, $C(6, 7, 1)$ and $D(-3, 4, -2)$, find

(a) a vector equation for the line, L, through A and B,

(b) the coordinates of the point P where L cuts the xy plane.

Show that P divides CD in the ratio $1:2$

PAIRS OF LINES

The location of two lines in space may be such that the lines

either   are parallel
or   are not parallel and intersect
or   are not parallel and do not intersect (such lines are *skew*).

Parallel Lines

We know how to recognise parallel vectors. So if two lines are parallel, this property can be observed from their direction vectors.

Non-parallel Lines

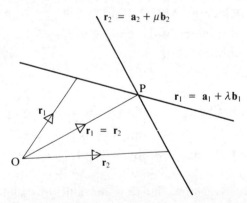

Consider two lines whose vector equations are

$$\mathbf{r}_1 = \mathbf{a}_1 + \lambda\mathbf{b}_1 \quad \text{and} \quad \mathbf{r}_2 = \mathbf{a}_2 + \mu\mathbf{b}_2$$

If these lines intersect it will be at a point where $\mathbf{r}_1 = \mathbf{r}_2$

This is possible only if there are unique values for λ and μ for which

$$\mathbf{a}_1 + \lambda\mathbf{b}_1 = \mathbf{a}_2 + \mu\mathbf{b}_2$$

If no such values can be found the lines do not intersect.

Example 19f

Find out whether the following pairs of lines are parallel, non-parallel and intersecting, or non-parallel and non-intersecting:

(a) $\mathbf{r}_1 = \mathbf{i} + \mathbf{j} + 2\mathbf{k} + \lambda(3\mathbf{i} - 2\mathbf{j} + 4\mathbf{k})$
    $\mathbf{r}_2 = 2\mathbf{i} - \mathbf{j} + 3\mathbf{k} + \mu(-6\mathbf{i} + 4\mathbf{j} - 8\mathbf{k})$,

(b) $\mathbf{r}_1 = \mathbf{i} - \mathbf{j} + 3\mathbf{k} + \lambda(\mathbf{i} - \mathbf{j} + \mathbf{k})$
    $\mathbf{r}_2 = 2\mathbf{i} + 4\mathbf{j} + 6\mathbf{k} + \mu(2\mathbf{i} + \mathbf{j} + 3\mathbf{k})$,

(c) $\mathbf{r}_1 = \mathbf{i} + \mathbf{k} + \lambda(\mathbf{i} + 3\mathbf{j} + 4\mathbf{k})$
    $\mathbf{r}_2 = 2\mathbf{i} + 3\mathbf{j} + \mu(4\mathbf{i} - \mathbf{j} + \mathbf{k})$.

(a) Checking first whether the lines are parallel we compare the direction vectors of the two lines.

The direction vector given for the first line is $3\mathbf{i} - 2\mathbf{j} + 4\mathbf{k}$

The direction vector given for the second line is $-6\mathbf{i} + 4\mathbf{j} - 8\mathbf{k}$

$$= -2(3\mathbf{i} - 2\mathbf{j} + 4\mathbf{k})$$

Therefore these two lines are parallel.

(b) Comparing the two direction vectors shows that these two lines are not parallel.

Now if the lines intersect it will be at a point where $\mathbf{r}_1 = \mathbf{r}_2$, i.e. where

$$(1 + \lambda)\mathbf{i} - (1 + \lambda)\mathbf{j} + (3 + \lambda)\mathbf{k} = 2(1 + \mu)\mathbf{i} + (4 + \mu)\mathbf{j} + (6 + 3\mu)\mathbf{k}$$

Equating the coefficients of $\mathbf{i}$ and $\mathbf{j}$, we have

$$1 + \lambda = 2(1 + \mu)$$

and $$-(1 + \lambda) = 4 + \mu$$

Hence $$\mu = -2, \quad \lambda = -3$$

With these values for λ and μ, the coefficients of $\mathbf{k}$ become

for the first line $\qquad\qquad 3 + \lambda = 0$
for the second line $\qquad\quad 6 + 3\mu = 0$ $\Big\}$ i.e. equal values.

So $\mathbf{r}_1 = \mathbf{r}_2$ when $\lambda = -3$ and $\mu = -2$

Therefore the lines *do* intersect at the point with position vector

$$(1 - 3)\mathbf{i} - (1 - 3)\mathbf{j} + (3 - 3)\mathbf{k} \qquad (\lambda = -3 \text{ in } \mathbf{r}_1)$$

i.e. $$-2\mathbf{i} + 2\mathbf{j}$$

(c) Comparing the direction vectors of these two lines shows that the lines are not parallel.

If the lines intersect it will be where $r_1 = r_2$,

i.e. where $\quad (1 + \lambda)i + 3\lambda j + (1 + 4\lambda)k = (2 + 4\mu)i + (3 - \mu)j + \mu k$

Equating the coefficients of i and j we have

$$\left. \begin{array}{r} 1 + \lambda = 2 + 4\mu \\ \text{and} \qquad 3\lambda = 3 - \mu \end{array} \right\} \quad \Rightarrow \quad \mu = 0, \ \lambda = 1$$

With these values of λ and μ, the coefficients of k become

for the first line $\quad 1 + 4\lambda = 5$ and for the second line $\quad \mu = 0$

As these are unequal there are no values of λ and μ for which $r_1 = r_2$

Therefore these two lines do not intersect and are skew.

EXERCISE 19f

In Questions 1 to 3 find whether the given two lines are parallel, intersecting or skew. If they intersect, state the position vector of the common point.

1. $r_1 = i - j + k + \lambda(3i - 4j + k)$ and $r_2 = \mu(-9i + 12j - 3k)$

2. $r_1 = 4i + 8j + 3k + \lambda(i + 2j + k)$ and
$r_2 = 7i + 6j + 5k + \mu(6i + 4j + 5k)$

3. $r_1 = i + 3k + \lambda(2i + j + k)$ and $r_2 = 2i - j + k + \mu(i - 2j)$

4. Two lines have equations $r_1 = 2i + 9j + 13k + \lambda(i + 2j + 3k)$ and
$r_2 = ai + 7j - 2k + \mu(-i + 2j - 3k)$.
Given that they intersect, find the value of a and the position vector of the point of intersection.

5. If the line $r_1 = 5j + tk + \lambda(-i + 3j + 4k)$ intersects the line
$r_2 = -4i + j + k + \mu(5i + j + k)$, find the value of t and the coordinates of the point of intersection.

6. A line L has equation $r_1 = -3i + j + 4k + \lambda(4i + pj - 3k)$
Find the value of p if

(a) the line $r_2 = 5i - j - 2k + \mu(4i - 2j - 3k)$ is parallel to L,

(b) the line $r_3 = -i + 3j - 3k + \mu(i - j + 2k)$ intersects L.

THE SCALAR PRODUCT

We are now going to look at an operation involving two vectors and the angle between them. This operation is called a product but, because it involves vectors, it is in no way related to the product of real numbers.

The Definition of the Scalar Product

The scalar product of two vectors **a** and **b** is denoted by **a.b** and defined as $ab \cos \theta$ where θ is the angle between **a** and **b**

i.e. $\mathbf{a.b} = ab \cos \theta$

PROPERTIES OF THE SCALAR PRODUCT

Parallel Vectors

If **a** and **b** are parallel then

either $\mathbf{a.b} = ab \cos 0$ or $\mathbf{a.b} = ab \cos \pi$

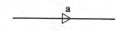

i.e.   for like parallel vectors $\mathbf{a.b} = ab$

and   for unlike parallel vectors $\mathbf{a.b} = -ab$

In the special case when $\mathbf{a} = \mathbf{b}$

$$\mathbf{a.b} = \mathbf{a.a} = a^2 \quad \text{(sometimes } \mathbf{a.a} \text{ is written } \mathbf{a}^2)$$

In particular, for the Cartesian unit vectors **i**, **j** and **k**

$$\mathbf{i.i} = \mathbf{j.j} = \mathbf{k.k} = 1$$

Perpendicular Vectors

If **a** and **b** are perpendicular then $\theta = \frac{1}{2}\pi$, $\Rightarrow$ **a.b** $= ab \cos\frac{1}{2}\pi = 0$

i.e.   for perpendicular vectors **a.b** $= 0$

For the unit vectors **i, j** and **k** we have

$$\mathbf{i.j} = \mathbf{j.k} = \mathbf{k.i} = 0$$

The Scalar Product is Commutative

This means that **a.b** $=$ **b.a** and this property is easy to prove as follows.

From the definition we have **a.b** $= ab\cos\theta$ and **b.a** $= ba\cos\theta$

Now $ab\cos\theta = ba\cos\theta$ therefore **a.b** $=$ **b.a**

The Scalar Product is Distributive for Addition

This means that **a.(b + c)** $=$ **a.b** $+$ **a.c**

It is not necessary to be able to prove this property but a proof is given below for readers who would like to see it.

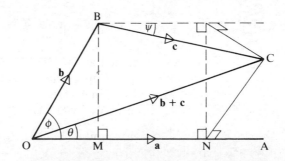

$$\mathbf{a.(b + c)} = |\mathbf{a}||\mathbf{b + c}|\cos\theta$$

$$= (OA)(OC)\cos\theta$$

$$= (OA)(ON)$$

$$= (OA)(OM + MN)$$

$$= (OA)(OB\cos\phi) + (OA)(BC\cos\psi)$$

$$= \mathbf{a.b} + \mathbf{a.c}$$

CALCULATING **a.b** IN CARTESIAN FORM

If $\mathbf{a} = x_1\mathbf{i} + y_1\mathbf{j} + z_1\mathbf{k}$ and $\mathbf{b} = x_2\mathbf{i} + y_2\mathbf{j} + z_2\mathbf{k}$ then we can find **a.b** by using the properties given above.

$$\mathbf{a.b} = (x_1\mathbf{i} + y_1\mathbf{j} + z_1\mathbf{k}).(x_2\mathbf{i} + y_2\mathbf{j} + z_2\mathbf{k})$$

$$= (x_1x_2\mathbf{i.i} + y_1y_2\mathbf{j.j} + z_1z_2\mathbf{k.k})$$

$$+ (x_1y_2\mathbf{i.j} + y_1z_2\mathbf{j.k} + z_1x_2\mathbf{k.i})$$

$$+ (y_1x_2\mathbf{j.i} + z_1y_2\mathbf{k.j} + x_1z_2\mathbf{i.k})$$

$$= (x_1x_2 + y_1y_2 + z_1z_2) + (0) + (0),$$

i.e.

$$(x_1\mathbf{i} + y_1\mathbf{j} + z_1\mathbf{k}).(x_2\mathbf{i} + y_2\mathbf{j} + z_2\mathbf{k}) = x_1x_2 + y_1y_2 + z_1z_2$$

For example,

$$(2\mathbf{i} - 3\mathbf{j} + 4\mathbf{k}).(\mathbf{i} + 3\mathbf{j} - 2\mathbf{k}) = (2)(1) + (-3)(3) + (4)(-2) = -15$$

Using the Scalar Product to Find the Angle Between Two Lines

The lines with equations $\mathbf{r}_1 = \mathbf{a}_1 + \lambda\mathbf{d}_1$ and $\mathbf{r}_2 = \mathbf{a}_2 + \mu\mathbf{d}_2$ are in the directions of $\mathbf{d}_1$ and $\mathbf{d}_2$ respectively.
The angle between two lines is defined as the angle between their direction vectors and does not depend upon their positions. It does not even depend on whether the lines intersect, so we are looking for the angle between the vectors $\mathbf{d}_1$ and $\mathbf{d}_2$ in any convenient position.

Drawing $\mathbf{d}_1$ and $\mathbf{d}_2$ from a common point, the angle θ between them is given by $\mathbf{d}_1.\mathbf{d}_2 = d_1d_2 \cos\theta$

$$\therefore \quad \cos\theta = \frac{\mathbf{d}_1.\mathbf{d}_2}{d_1d_2}$$

This confirms that, for perpendicular lines, $\mathbf{d}_1.\mathbf{d}_2 = 0$

Examples 19g

1. Simplify $(\mathbf{a} - \mathbf{b}).(\mathbf{a} + \mathbf{b})$ and $(\mathbf{a} + \mathbf{b}).\mathbf{c} - (\mathbf{a} + \mathbf{c}).\mathbf{b}$

$$(\mathbf{a} - \mathbf{b}).(\mathbf{a} + \mathbf{b}) = \mathbf{a}.\mathbf{a} - \mathbf{b}.\mathbf{a} + \mathbf{a}.\mathbf{b} - \mathbf{b}.\mathbf{b}$$

but $\mathbf{a}.\mathbf{b} = \mathbf{b}.\mathbf{a}$ hence $\mathbf{a}.\mathbf{b} - \mathbf{b}.\mathbf{a} = 0$

also $\mathbf{a}.\mathbf{a} = a^2$ and $\mathbf{b}.\mathbf{b} = b^2$

Therefore $(\mathbf{a} - \mathbf{b}).(\mathbf{a} + \mathbf{b}) = a^2 - b^2$

Also $(\mathbf{a} + \mathbf{b}).\mathbf{c} - (\mathbf{a} + \mathbf{c}).\mathbf{b} = \mathbf{a}.\mathbf{c} + \mathbf{b}.\mathbf{c} - \mathbf{a}.\mathbf{b} - \mathbf{c}.\mathbf{b}$

$$= \mathbf{a}.\mathbf{c} - \mathbf{a}.\mathbf{b}$$

$$= \mathbf{a}.(\mathbf{c} - \mathbf{b})$$

---

2. Find the scalar product of $\mathbf{a} = 2\mathbf{i} - 3\mathbf{j} + 5\mathbf{k}$ and $\mathbf{b} = \mathbf{i} - 3\mathbf{j} + \mathbf{k}$
   and hence find the cosine of the angle between $\mathbf{a}$ and $\mathbf{b}$.

$$\mathbf{a}.\mathbf{b} = (2)(1) + (-3)(-3) + (5)(1) = 16$$

But $\mathbf{a}.\mathbf{b} = |\mathbf{a}||\mathbf{b}|\cos\theta$

$$|\mathbf{a}| = \sqrt{(4 + 9 + 25)} = \sqrt{38}$$

$$|\mathbf{b}| = \sqrt{(1 + 9 + 1)} = \sqrt{11}$$

Hence $\cos\theta = \dfrac{\mathbf{a}.\mathbf{b}}{|\mathbf{a}||\mathbf{b}|} = \dfrac{16}{\sqrt{11}\sqrt{38}} = \dfrac{16}{\sqrt{418}}$

---

3. If $\mathbf{a} = 10\mathbf{i} - 3\mathbf{j} + 5\mathbf{k}$, $\mathbf{b} = 2\mathbf{i} + 6\mathbf{j} - 3\mathbf{k}$ and $\mathbf{c} = \mathbf{i} + 10\mathbf{j} - 2\mathbf{k}$,
   verify that $\mathbf{a}.\mathbf{b} + \mathbf{a}.\mathbf{c} = \mathbf{a}.(\mathbf{b} + \mathbf{c})$

$$\mathbf{a}.\mathbf{b} = (10)(2) + (-3)(6) + (5)(-3) = -13$$

$$\mathbf{a}.\mathbf{c} = (10)(1) + (-3)(10) + (5)(-2) = -30$$

$$\mathbf{b} + \mathbf{c} = 3\mathbf{i} + 16\mathbf{j} - 5\mathbf{k}$$

Hence $\mathbf{a}.(\mathbf{b} + \mathbf{c}) = (10)(3) + (-3)(16) + (5)(-5) = -43$

But $\mathbf{a}.\mathbf{b} + \mathbf{a}.\mathbf{c} = -13 - 30 = -43$

Therefore $\mathbf{a}.\mathbf{b} + \mathbf{a}.\mathbf{c} = \mathbf{a}.(\mathbf{b} + \mathbf{c})$

4. Find the angle between the lines

$$\mathbf{r}_1 = \mathbf{i} - 2\mathbf{j} + 3\mathbf{k} + \lambda(2\mathbf{i} - 3\mathbf{j} + 6\mathbf{k}) \qquad [1]$$

$$\mathbf{r}_2 = 2\mathbf{i} - 7\mathbf{j} + 10\mathbf{k} + \mu(\mathbf{i} + 2\mathbf{j} + 2\mathbf{k}) \qquad [2]$$

The angle between the lines depends only upon their directions.

Line [1] has a direction vector $\mathbf{d}_1 = 2\mathbf{i} - 3\mathbf{j} + 6\mathbf{k}$

Line [2] has a direction vector $\mathbf{d}_2 = \mathbf{i} + 2\mathbf{j} + 2\mathbf{k}$

The angle θ between the lines is given by

$$\cos\theta = \frac{\mathbf{d}_1 . \mathbf{d}_2}{d_1 d_2} = \frac{2 - 6 + 12}{(7)(3)}$$

$$\Rightarrow \qquad \theta = \arccos \tfrac{8}{21}$$

---

5. Find a vector which is perpendicular to AB and AC if
$\overrightarrow{AB} = \mathbf{i} + 2\mathbf{j} + 3\mathbf{k}$ and $\overrightarrow{AC} = 4\mathbf{i} - \mathbf{j} + 2\mathbf{k}$

Let $a\mathbf{i} + b\mathbf{j} + c\mathbf{k}$ be a vector perpendicular to both AB and AC.

It is perpendicular to AB so $(a\mathbf{i} + b\mathbf{j} + c\mathbf{k}).(\mathbf{i} + 2\mathbf{j} + 3\mathbf{k}) = 0$

It is perpendicular to AC so $(a\mathbf{i} + b\mathbf{j} + c\mathbf{k}).(4\mathbf{i} - \mathbf{j} + 2\mathbf{k}) = 0$

Therefore $\begin{cases} a + 2b + 3c = 0 \\ 4a - b + 2c = 0 \end{cases}$

Eliminating b gives $a = -\tfrac{7}{9}c$

Eliminating a gives $b = -\tfrac{10}{9}c$

Hence $a\mathbf{i} + b\mathbf{j} + c\mathbf{k} = -\tfrac{7}{9}c\mathbf{i} - \tfrac{10}{9}c\mathbf{j} + c\mathbf{k}$

$$= \tfrac{1}{9}c(-7\mathbf{i} - 10\mathbf{j} + 9\mathbf{k})$$

This vector is perpendicular to both AB and AC for any value of c.

Thus $-7\mathbf{i} - 10\mathbf{j} + 9\mathbf{k}$ is a vector perpendicular to both AB and AC.

EXERCISE 19g

1. Calculate **a.b** if
   (a) $\mathbf{a} = 2\mathbf{i} - 4\mathbf{j} + 5\mathbf{k}, \quad \mathbf{b} = \mathbf{i} + 3\mathbf{j} + 8\mathbf{k}$
   (b) $\mathbf{a} = 3\mathbf{i} - 7\mathbf{j} + 2\mathbf{k}, \quad \mathbf{b} = 5\mathbf{i} + \mathbf{j} - 4\mathbf{k}$
   (c) $\mathbf{a} = 2\mathbf{i} - 3\mathbf{j} + 6\mathbf{k}, \quad \mathbf{b} = \mathbf{i} + \mathbf{j}$
   What conclusion can you draw in (b)?

2. Find **p.q** and the cosine of the angle between **p** and **q** if
   (a) $\mathbf{p} = 2\mathbf{i} + 4\mathbf{j} + \mathbf{k}, \quad \mathbf{q} = \mathbf{i} + \mathbf{j} + \mathbf{k}$
   (b) $\mathbf{p} = -\mathbf{i} + 3\mathbf{j} - 2\mathbf{k}, \quad \mathbf{q} = \mathbf{i} + \mathbf{j} - 6\mathbf{k}$
   (c) $\mathbf{p} = -2\mathbf{i} + 5\mathbf{j}, \quad \mathbf{q} = \mathbf{i} + \mathbf{j}$
   (d) $\mathbf{p} = 2\mathbf{i} + \mathbf{j}, \quad \mathbf{q} = \mathbf{j} - 2\mathbf{k}$

3. Simplify
   (a) $(\mathbf{a} - \mathbf{b}).\mathbf{a}$ (b) $(\mathbf{a} - \mathbf{b}).(\mathbf{a} - \mathbf{b})$
   (c) $(\mathbf{a} + \mathbf{b}).\mathbf{b} - (\mathbf{a} + \mathbf{b}).\mathbf{a}$ (d) $(\mathbf{a} + \mathbf{b}).\mathbf{c} - (\mathbf{a} - \mathbf{b}).\mathbf{c}$

4. Given that **a** and **b** are perpendicular, simplify
   (a) $\mathbf{a}.\mathbf{b}$ (b) $(\mathbf{a} - \mathbf{b}).\mathbf{b}$ (c) $(\mathbf{a} + \mathbf{b}).\mathbf{a}$ (d) $(\mathbf{a} - \mathbf{b}).(2\mathbf{a} + \mathbf{b})$

5. The angle between two vectors $\mathbf{v}_1$ and $\mathbf{v}_2$ is $\arccos \frac{4}{21}$
   If $\mathbf{v}_1 = 6\mathbf{i} + 3\mathbf{j} - 2\mathbf{k}$ and $\mathbf{v}_2 = -2\mathbf{i} + \lambda\mathbf{j} - 4\mathbf{k}$, find the positive value of λ

6. If $\mathbf{a} = 3\mathbf{i} + 4\mathbf{j} - \mathbf{k}$, $\mathbf{b} = \mathbf{i} - \mathbf{j} + 3\mathbf{k}$ and $\mathbf{c} = 2\mathbf{i} + \mathbf{j} - 5\mathbf{k}$, find
   (a) $\mathbf{a}.\mathbf{b}$ (b) $\mathbf{a}.\mathbf{c}$ (c) $\mathbf{a}.(\mathbf{b} + \mathbf{c})$ (d) $(2\mathbf{a} + 3\mathbf{b}).\mathbf{c}$ (e) $(\mathbf{a} - \mathbf{b}).\mathbf{c}$

7. In a triangle ABC, $\overrightarrow{AB} = \mathbf{i} + 2\mathbf{j} + 3\mathbf{k}$ and $\overrightarrow{BC} = -\mathbf{i} + 4\mathbf{j}$.
   Find the cosine of angle ABC.
   Find the vector $\overrightarrow{AC}$ and use it to calculate the angle BAC.

8. A, B and C are points with position vectors **a**, **b** and **c** respectively, relative to the origin O. AB is perpendicular to OC and BC is perpendicular to OA.
   Show that AC is perpendicular to OB.

9. Given two vectors **a** and **b** (**a** ≠ 0, **b** ≠ 0), show that
   (a) if **a** + **b** and **a** − **b** are perpendicular then |**a**| = |**b**|,
   (b) if |**a** + **b**| = |**a** − **b**| then **a** and **b** are perpendicular.

10. Three vectors **a**, **b** and **c** are such that **a** ≠ **b** ≠ **c** ≠ 0.
    (a) If **a**.(**b** + **c**) = **b**.(**a** − **c**) prove that **c**.(**a** + **b**) = 0
    (b) If (**a**.**b**)**c** = (**b**.**c**)**a** show that **c** and **a** are parallel.

11. Find the angle between each of the following pairs of lines.
    (a) $\mathbf{r}_1 = 3\mathbf{i} + 2\mathbf{j} - 4\mathbf{k} + \lambda(\mathbf{i} + 2\mathbf{j} + 2\mathbf{k})$ and
        $\mathbf{r}_2 = 5\mathbf{j} - 2\mathbf{k} - \mu(3\mathbf{i} + 2\mathbf{j} + 6\mathbf{k})$
    (b) $\mathbf{r} = 4\mathbf{i} - \mathbf{j} + \lambda(\mathbf{i} + 2\mathbf{j} - 2\mathbf{k})$ and
        $\mathbf{r} = \mathbf{i} - \mathbf{j} + 2\mathbf{k} - \mu(2\mathbf{i} + 4\mathbf{j} - 4\mathbf{k})$

12. Show that $\mathbf{i} + 7\mathbf{j} + 3\mathbf{k}$ is perpendicular to both $\mathbf{i} - \mathbf{j} + 2\mathbf{k}$ and
    $2\mathbf{i} + \mathbf{j} - 3\mathbf{k}$

13. Show that $13\mathbf{i} + 23\mathbf{j} + 7\mathbf{k}$ is perpendicular to both $2\mathbf{i} + \mathbf{j} - 7\mathbf{k}$
    and $3\mathbf{i} - 2\mathbf{j} + \mathbf{k}$

MIXED EXERCISE 19

1. In the diagram, $\overrightarrow{AB} = \mathbf{a}$, $\overrightarrow{AC} = \mathbf{c}$, $\overrightarrow{AE} = \mathbf{e}$ and ED is equal
   and parallel to AB. Find, in terms of **a**, **c** and **e**, the vectors
   represented by $\overrightarrow{DE}$, $\overrightarrow{BC}$, $\overrightarrow{CE}$, $\overrightarrow{CD}$ and $\overrightarrow{AD}$.

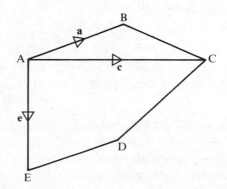

2. ABCD is a quadrilateral; AC and BD meet at P and P divides AC in the ratio $1:2$. Show that

   (a) $\overrightarrow{AB} + \overrightarrow{CD} - \overrightarrow{PD} = \overrightarrow{AP} + \overrightarrow{CB}$

   (b) $\frac{1}{3}\overrightarrow{DP} - \frac{2}{3}\overrightarrow{DA} + \frac{1}{3}\overrightarrow{DC} = 0$

3. Given the vectors, $\mathbf{a} = 2\mathbf{i} + \mathbf{j}$, $\mathbf{b} = \mathbf{i} - 3\mathbf{j}$ and $\mathbf{c} = -\mathbf{i} + 2\mathbf{j}$ represent on one diagram the vectors

   (a) $-\mathbf{a}$ (b) $\mathbf{a} - \mathbf{b}$ (c) $\mathbf{b} + \mathbf{c}$ (d) $\mathbf{a} + \mathbf{b} - \mathbf{c}$ (e) $\mathbf{a} + 2\mathbf{c}$

In Questions 4 and 5, ABC is a triangle in which D, E and F are the respective midpoints of BC, CA and AB, and AD meets BE at G.

4. Use vector methods to prove that

   (a) DE is parallel to AB

   (b) the length of DE is half the length of AB.

5. By taking $BG:GE = \lambda:1$ and $AG:GD = \mu:1$ find $\overrightarrow{CG}$, first in terms of λ, then in terms of μ. Hence prove that G divides both AD and BE in the ratio of $2:1$

For Questions 6 to 9 the position vectors of points A, B, C and D are given by:

$\overrightarrow{OA} = \mathbf{i} - 2\mathbf{j} + 2\mathbf{k}$, $\overrightarrow{OB} = -\mathbf{i} - 4\mathbf{k}$, $\overrightarrow{OC} = 6\mathbf{i} + 3\mathbf{j} - 2\mathbf{k}$,

$\overrightarrow{OD} = 8\mathbf{i} + \mathbf{j} + 4\mathbf{k}$

6. Write down a vector equation for the line

   (a) through A and B

   (b) through D and parallel to BC.

7. Find the position vector of the point where each of the lines in Question 6 crosses the yz plane.

8. (a) Write down the position vector of the point which divides BD externally in the ratio $2:1$

   (b) Find the position vector of the point E such that B is the midpoint of AE.

9. State which sides of quadrilateral ABCD are parallel.

10. Calculate $\mathbf{a} \cdot \mathbf{b}$ given that
    (a) $\mathbf{a} = 2\mathbf{j} - 5\mathbf{k}$ and $\mathbf{b} = \mathbf{i} + \mathbf{j} - 3\mathbf{k}$
    (b) $\mathbf{a} = 4\mathbf{i} - 3\mathbf{j} + 6\mathbf{k}$ and $\mathbf{b} = 3\mathbf{i} + \mathbf{j} - 2\mathbf{k}$.

11. Determine whether the following pairs of lines are parallel, perpendicular or neither *and* whether they intersect or are skew. If they intersect give the position vector of the point of intersection.
    (a) $\mathbf{r}_1 = 2\mathbf{i} - \mathbf{j} + t(3\mathbf{i} - 2\mathbf{j} + \mathbf{k})$
        $\mathbf{r}_2 = 2\mathbf{i} - \mathbf{j} + s(\mathbf{i} - \mathbf{j} - 5\mathbf{k})$
    (b) $\mathbf{r}_1 = t(3\mathbf{i} - \mathbf{j} - \mathbf{k})$
        $\mathbf{r}_2 = \mathbf{i} - 2\mathbf{j} - 4\mathbf{k} + s(2\mathbf{i} + \mathbf{j} + 3\mathbf{k})$.

12. Find, to the nearest degree, the acute angle between the pair of lines given in Question 11(b).

CHAPTER 20

APPLICATIONS OF CALCULUS

RATE OF INCREASE

When the variation of y depends upon another variable x, then

$\dfrac{dy}{dx}$ gives the rate at which y increases compared with x

This fact forms the basis of methods which can be used to analyse practical situations in which two variables are related.

SMALL INCREMENTS

Consider two variables, x and y, related by the equation $y = f(x)$

If $\qquad$ x increases by a small increment δx

then $\qquad$ y increases by a corresponding small amount δy

Now $$\lim_{\delta x \to 0} \frac{\delta y}{\delta x} = \frac{dy}{dx}$$

so,

provided that δx is small, $\dfrac{\delta y}{\delta x} \approx \dfrac{dy}{dx}$ $\quad \Rightarrow \quad$ $\delta y \approx \dfrac{dy}{dx}(\delta x)$

or, alternatively, $\qquad\qquad$ $\delta\{f(x)\} \approx f'(x)\,\delta x$

This approximation can be used to estimate the value of a function close to a known value, i.e. $y + \delta y$ can be estimated if y is known at a particular value of x

346

For example, knowing that $\ln 1 = 0$, an approximate value for $\ln 1.1$ can be found from $y = \ln x$ as follows

$$y = \ln x \quad \Rightarrow \quad \frac{dy}{dx} = \frac{1}{x}$$

so

$$\delta y \approx \frac{dy}{dx}(\delta x) = \frac{1}{x}(\delta x)$$

Now x increases from 1 to 1.1, i.e. $\delta x = 0.1$

$\therefore$

$$\delta y \approx \tfrac{1}{1}(0.1)$$

Hence

$$\ln 1.1 = y + \delta y \approx (\ln 1) + 0.1$$

but $\ln 1 = 0$

$\therefore$

$$\ln 1.1 \approx 0.1$$

Examples 20a

1. Using $y = \sqrt{x}$, estimate the value of $\sqrt{101}$

$$y = \sqrt{x} \quad \Rightarrow \quad \frac{dy}{dx} = \tfrac{1}{2}x^{-1/2}$$

$$\delta y \approx \frac{dy}{dx}\delta x \quad \text{gives} \quad \delta y \approx \frac{1}{2\sqrt{x}}\delta x$$

So that the value of y can be written down, the value we take for x must be a number with a known square root.

Taking $x = 100$, $y = \sqrt{100}$ and $\delta x = 1$ gives

$$\delta y \approx \frac{1}{2\sqrt{100}}(1) = \frac{1}{20}$$

Then

$$\sqrt{101} = y + \delta y \approx \sqrt{100} + \tfrac{1}{20}$$

i.e. $\sqrt{101} \approx 10.05$

2. Given that $1° = 0.0175$ rad and that $\cos 30° = 0.8660$, use $f(\theta) = \sin \theta$ to find an approximate value for

(a) $\sin 31°$ (b) $\sin 29°$

Using $\delta\{f(\theta)\} \approx f'(\theta)\,\delta\theta$ gives $\delta(\sin \theta) \approx (\cos \theta)\,\delta\theta$

(a) Taking $\theta = \frac{1}{6}\pi$, $\sin \theta = \frac{1}{2}$ and $\delta\theta = 0.0175$ gives

$\delta(\sin \theta) \approx (\cos \frac{1}{6}\pi)(0.0175)$ and $\sin 31° \approx \sin 30° + \delta(\sin \theta)$

Hence $\sin 31° \approx 0.5 + (0.8660)(0.0175)$

i.e. $\sin 31° \approx 0.515$

(b) Again $\theta = \frac{1}{6}\pi$ and $\sin \theta = \frac{1}{2}$, but this time, because the angle *decreases* from 30° to 29°, $\delta\theta = -0.0175$

$\delta(\sin \theta) \approx (\cos \frac{1}{6}\pi)(-0.0175)$

Hence $\sin 29° \approx \sin 30° + \delta(\sin \theta) = 0.5 + (0.8660)(-0.0175)$

i.e. $\sin 29° \approx 0.485$

Small Percentage Increases

In order to adapt the method used above to estimate the percentage change in a dependent variable caused by a small change in the independent variable we use the additional fact that

$$\text{if } x \text{ increases by } r\% \text{ then } \delta x = \frac{r}{100}(x)$$

and the corresponding percentage increase in y is $\dfrac{\delta y}{y} \times 100$

Examples 20a (continued)

3. The period, T, of a simple pendulum is calculated from the formula $T = 2\pi\sqrt{(l/g)}$, where l is the length of the pendulum and g is the constant gravitational acceleration. Find the percentage change in the period caused by lengthening the pendulum by 2 per cent.

$$T = 2\pi\sqrt{\left(\frac{l}{g}\right)} \quad\Rightarrow\quad \frac{dT}{dl} = \left(\frac{2\pi}{\sqrt{g}}\right)\left(\tfrac{1}{2}l^{-1/2}\right) = \frac{\pi}{\sqrt{(lg)}}$$

The length *increases* so the small increment in the length is positive and is given by

$$\delta l = \left(\tfrac{2}{100}\right)(l) = \tfrac{1}{50}l$$

Using $\delta T \approx \dfrac{dT}{dl}\delta l$ gives

$$\delta T \approx \left(\tfrac{1}{50}l\right)\{\pi/\sqrt{(lg)}\} = \left(\tfrac{1}{50}\right)\pi\sqrt{(l/g)}$$

The percentage change in the period is given by $\dfrac{\delta T}{T} \times 100$

i.e. $\tfrac{1}{50}\{\pi\sqrt{(l/g)}\} \div \{2\pi\sqrt{(l/g)}\} \times 100\% = 1\%$

This is a positive change, so we see that the period *increases* by 1%

EXERCISE 20a

1. Using $y = \sqrt[3]{x}$, find, *without using a calculator*, an approximate value for

   (a) $\sqrt[3]{1001}$ (b) $\sqrt[3]{9}$ (c) $\sqrt[3]{63}$

   Work to 6 d.p.
   Now use a calculator to find the accuracy of each approximation.

2. Given that $1° = 0.0175$ rad, $\sin 60° = 0.8660$ and $\sin 45° = 0.7071$, use $f(\theta) = \cos\theta$ to find an approximate value for

   (a) $\cos 31°$ (b) $\cos 59°$ (c) $\cos 44°$

3. If $f(x) = x\ln(1 + x)$ find an approximation for the increase in $f(x)$ when x increases by δx
   Hence estimate the value of $\ln(2.1)$ given that $\ln 2 = 0.6931$

4. If $y = \tan x$ find an approximation for δy when x is increased by δx and use it to estimate, in terms of π, the value of $\tan\tfrac{9}{32}\pi$

5. Use $f(x) = \sqrt[5]{x}$ to find the approximate value of $\sqrt[5]{33}$

6. Given that $y = \sqrt{\left(\dfrac{x-2}{x-1}\right)}$ determine the value of $\dfrac{dy}{dx}$ when $x = 3$ Deduce the approximate increase in the value of y when x increases from 3 to $3 + \alpha$ where α is small.

COMPARATIVE RATES OF CHANGE

Some problems involving the rate of change of one variable compared with another do not provide a direct relationship between these two variables. Instead, each of them is related to a third variable.

The identity $\dfrac{dy}{dx} = \dfrac{dy}{dt} \times \dfrac{dt}{dx}$ is useful in solving problems of this type.

Suppose, for instance, that the radius, r, of a circle is increasing at a rate of 1 mm per second. This means that $\dfrac{dr}{dt} = 1$. The rate at which the area, A, of the circle is increasing is $\dfrac{dA}{dt}$

We do not know A as a function of t but we do know that $A = \pi r^2$ and that

$$\frac{dA}{dt} = \frac{dA}{dr} \times \frac{dr}{dt}$$

Then $\dfrac{dA}{dt}$ can be calculated, as $\dfrac{dr}{dt}$ is given and $\dfrac{dA}{dr}$ can be found from $A = \pi r^2$

In some cases, more than three variables may be involved but the same approach is used with a relationship of the form

$$\frac{dy}{dx} = \frac{dy}{dp} \times \frac{dp}{dq} \times \frac{dq}{dx}$$

Examples 20b

1. A spherical balloon is being blown up so that its volume increases at a constant rate of $1.5 \text{ cm}^3/\text{s}$. Find the rate of increase of the radius when the volume of the balloon is 56 cm^3.

If, at time t, the radius of the balloon is r and the volume is V then

$$V = \tfrac{4}{3}\pi r^3 \quad \Rightarrow \quad \frac{dV}{dr} = 4\pi r^2$$

We are looking for $\dfrac{dr}{dt}$ and we are given $\dfrac{dV}{dt} = 1.5$ so we use

$$\dfrac{dr}{dt} = \dfrac{dr}{dV} \times \dfrac{dV}{dt} = \dfrac{dV}{dt} \div \dfrac{dV}{dr} \quad\Rightarrow\quad \dfrac{dr}{dt} = \dfrac{1.5}{4\pi r^2} = \dfrac{3}{8\pi r^2}$$

Now substituting $V = 56$ in $V = \frac{4}{3}\pi r^3$ gives $r = 2.373$ to 4 s.f.

Therefore, when $V = 56$, $\quad \dfrac{dr}{dt} = \dfrac{3}{8\pi(2.373)^2} = 0.021\,20$

i.e. the radius is increasing at a rate of 0.0212 cm/s (correct to 3 s.f.)

---

2. A vessel containing liquid is in the form of an inverted hollow cone
   with a semi-vertical angle of 30°. The liquid is running out of a
   small hole at the vertex of the cone, at a constant rate of 3 cm³/s.
   Find the rate at which the surface area which is in contact with the
   liquid is changing, at the instant when the volume of liquid left in
   the vessel is 81π cm³.

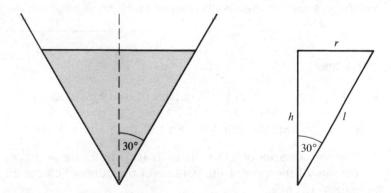

At any time, t, the volume, V, of liquid in the vessel is given by
$$V = \tfrac{1}{3}\pi r^2 h$$

In this equation, V, r and h are all variables so it cannot yet be differentiated.

Using $r = l\sin 30° = \tfrac{1}{2}l$ and $h = l\cos 30° = \tfrac{1}{2}l\sqrt{3}$ gives

$$V = \tfrac{1}{24}\pi l^3\sqrt{3} \quad\Rightarrow\quad \dfrac{dV}{dl} = \tfrac{1}{8}\pi l^2\sqrt{3}$$

The surface area, S, in contact with the liquid at any time is given by

$$S = \pi r l = \tfrac{1}{2}\pi l^2 \quad\Rightarrow\quad \frac{\mathrm{d}S}{\mathrm{d}l} = \pi l$$

Now $\dfrac{\mathrm{d}S}{\mathrm{d}t}$ is required and $\dfrac{\mathrm{d}V}{\mathrm{d}t}$ is given as -3 (negative because the volume is decreasing). There is no equation from which $\dfrac{\mathrm{d}S}{\mathrm{d}V}$ can be obtained so this time we use a three-step link,

i.e. $\quad \dfrac{\mathrm{d}S}{\mathrm{d}t} = \dfrac{\mathrm{d}S}{\mathrm{d}l} \times \dfrac{\mathrm{d}l}{\mathrm{d}V} \times \dfrac{\mathrm{d}V}{\mathrm{d}t} = \dfrac{\mathrm{d}S}{\mathrm{d}l} \div \dfrac{\mathrm{d}V}{\mathrm{d}l} \times \dfrac{\mathrm{d}V}{\mathrm{d}t}$

$$= (\pi l)\left(\frac{8}{\pi l^2 \sqrt{3}}\right)(-3) = \frac{-8\sqrt{3}}{l}$$

At the instant that the value of $\dfrac{\mathrm{d}S}{\mathrm{d}t}$ is required, $V = 81\pi$

i.e. $\qquad\qquad \tfrac{1}{24}\pi l^3 \sqrt{3} = 81\pi \quad\Rightarrow\quad l = 6\sqrt{3}$

At this instant, $\qquad\qquad \dfrac{\mathrm{d}S}{\mathrm{d}t} = \dfrac{-8\sqrt{3}}{6\sqrt{3}} = -\dfrac{4}{3}$

i.e. the wet surface area is decreasing at $1\tfrac{1}{3}$ cm^2/s.

EXERCISE 20b

1. Ink is dropped on to blotting paper forming a circular stain which increases in area at a rate of 2.5 cm^2/s. Find the rate at which the radius is changing when the area of the stain is 16π cm^2.

2. The surface area of a cube is increasing at a rate of 10 cm^2/s. Find the rate of increase of the volume of the cube when the edge is of length 12 cm.

3. The circumference of a circular patch of oil on the surface of a pond is increasing at 2 m/s. When the radius is 4 m, at what rate is the area of the oil changing?

4. A container in the form of a right circular cone of height 16 cm and base radius 4 cm is held vertex downward and filled with liquid. If the liquid leaks out from the vertex at a rate of 4 cm^3/s, find the rate of change of the depth of the liquid in the cone when half of the liquid has leaked out.

5. A right circular cone has a constant volume. The height h and the base radius r can both vary. Find the rate at which h is changing with respect to r at the instant when r and h are equal.

6. The radius of a hemispherical bowl is a cm. The bowl is being filled with water at a steady rate of $3\pi a^3$ cm^3 per minute. Find, in terms of a, the rate at which the water is rising when the depth of water in the bowl is $\frac{1}{2}a$ cm.

(The volume of the shaded part of this hemisphere is $\frac{1}{3}\pi h^2(3a - h)$)

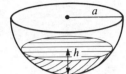

AREA FOUND BY INTEGRATION

The area bounded by part of a curve $y = f(x)$, the x-axis and the lines $x = a$ and $x = b$ is found by summing the areas of vertical strips and taking the limiting value of that sum.

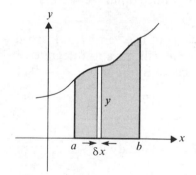

$$\text{Shaded area} = \lim_{\delta x \to 0} \sum_a^b y\,\delta x = \int_a^b y\,dx$$

Similarly the area bounded by a curve $x = f(y)$, the y-axis and the lines $y = a$ and $y = b$ is found by summing the areas of horizontal strips. The limit of this sum is $\displaystyle\int_a^b x\,dy$.

In Module A, curves with algebraic equations only were used as boundaries of required areas. The methods used there apply equally well to the graphs of other functions and to areas where *two* of the boundaries are curves, provided that an element can be found,

1) which has the same format throughout, i.e. the ends of all the elements are on the same boundaries;

2) whose length and width are measured parallel to the x and y axes.

Examples 20c

1. A plane region is defined by the line $y = 4$, the x and y axes and part of the curve $y = \ln x$. Find the area of the region.

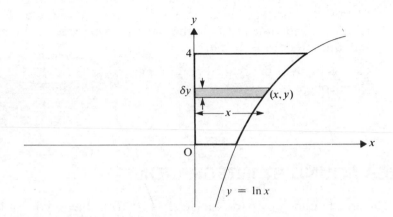

A vertical element is unsuitable in this case as the top and bottom are not always on the same boundaries, but a horizontal element is satisfactory.

The area, δA, of a typical horizontal element is given by $\delta A \approx x \, \delta y$. Because the width of our element is δy we will have to integrate w.r.t y, so we need the equation of the curve in the form $x = f(y)$

$$y = \ln x \quad \Rightarrow \quad x = e^y$$

$\therefore \quad \delta A \approx e^y \, \delta y$

$$\Rightarrow \qquad\qquad A = \lim_{\delta y \to 0} \sum_{y=0}^{y=4} e^y \, \delta y = \int_0^4 e^y \, dy$$

$$= \left[e^y \right]_0^4 = e^4 - e^0$$

The defined area is $(e^4 - 1)$ square units

Note that this area can also be found by subtracting the shaded region from the area of the rectangle OABC but this alternative is not always suitable.

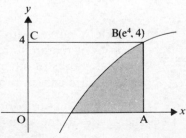

2. Find the area between the curve $y = x^2$ and the line $y = 3x$

The line and curve meet where $x^2 = 3x$,
i.e. where $x = 0$ and $x = 3$

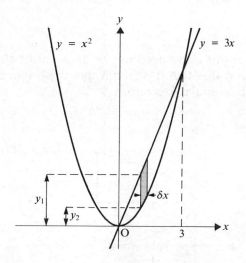

A vertical strip always has its top on the line and its foot on the curve so it is a suitable element. It is approximately a rectangle whose width is δx and whose height is the vertical distance between the line and the curve.

The area of the element, δA, is given by

$$\delta A \approx (y_1 - y_2)\delta x = (3x - x^2)\delta x$$

$$\therefore \quad A = \lim_{\delta x \to 0} \sum_{x=0}^{x=3} (3x - x^2)\delta x = \int_0^3 (3x - x^2)\,dx$$

$$= \left[\tfrac{3}{2}x^2 - \tfrac{1}{3}x^3 \right]_0^3$$

$$= 4\tfrac{1}{2}$$

The required area is $4\tfrac{1}{2}$ square units.

The area bounded partly by a curve whose equation is given parametrically can also be found by summing the areas of suitable elements.

---

3. Find the area bounded in the first quadrant by the x-axis, the line $x = 2$ and part of the curve with parametric equations $x = 2t^2$, $y = 4t$

The sketch of this curve need not be an accurate shape but it is important to realise that it goes through the origin because when $t = 0$, both x and y are zero.

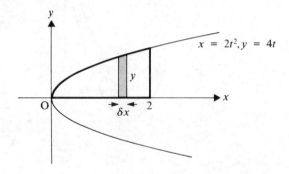

A suitable element is a vertical strip which is approximately a rectangle of height, y, width δx and area δA where $\delta A \approx y\,\delta x$

Considering $x = 2t^2$ gives

$$\delta x \approx \left(\frac{\mathrm{d}x}{\mathrm{d}t}\right)(\delta t) = 4t\,\delta t \qquad \text{and} \qquad \delta t \to 0 \quad \text{when} \quad \delta x \to 0$$

it also shows that, when $x = 2$, $t = 1$

$t \neq -1$, as it gives $y = -4$ which is not in the first quadrant.

$\therefore$
$$\delta A \approx (4t)(4t\,\delta t)$$

Then
$$A = \lim_{\delta x \to 0} \sum_{x=0}^{x=2} 16t^2\,\delta t = \lim_{\delta t \to 0} \sum_{t=0}^{t=1} 16t^2\,\delta t$$

$\Rightarrow$
$$A = \int_0^1 16t^2\,\mathrm{d}t = \left[\frac{16}{3}t^3\right]_0^1 = \frac{16}{3}$$

The required area is $\frac{16}{3}$ square units.

EXERCISE 20c

1. Find, by integration, the area bounded by
    (a) the x-axis, the line $x = 2$ and the curve $y = x^2$
    (b) the y-axis, the line $y = 4$ and the curve $y = x^2$
    Sketch these two areas on the same diagram and hence check the
    sum of the answers to (a) and (b).

2. Calculate the area bounded by the curve $y = \sqrt{x}$, the y-axis
    and the line $y = 3$

3. Calculate the area in the first quadrant between the curve $y^2 = x$
    and the line $x = 9$

4. Find the area between the y-axis and the curve $y^2 = 1 - x$

5. Find the area of the region whose boundaries are the y-axis, the
    line $y = \frac{1}{2}\pi$ and the curve $y = \arcsin x$

6. A region in the xy plane is bounded by the lines $y = 1$ and
    $x = 1$, and the curve $y = e^x$. Find its area.

7. Find the area bounded by the inequalities $y \leqslant 1 - x^2$ and
    $y \geqslant 1 - x$

8. Calculate the area bounded by the curve $y = \sin x$ and the
    lines $y = \frac{1}{2}$ and $x = \frac{1}{2}\pi$

9. Find the area of the region of the xy plane defined by
    (a) $y \geqslant e^x$, $x \geqslant 0$, $y \leqslant e$ (b) $1 \geqslant y \geqslant 1/(x + 1)$, $x \leqslant 2$

10. Find the area bounded by the x-axis and the ordinates at $x = 1$,
    $x = 4$ and the curve with equations $x = 2t$, $y = 2/t$.

11. Find, by integration the area bounded in the first quadrant of the
    xy plane by the x and y axes and the curve whose equations are

    $$x = a\cos\theta, \quad y = a\sin\theta$$

    Sketch this curve and hence check your answer.

12. (a) By taking values of t from 0 to 3, sketch part of the curve
        with parametric equations $x = t^2$, $y = t^3$.
    (b) Find the area bounded in the first quadrant of the xy plane
        by the x-axis, the curve whose equations are $x = at^2$,
        $y = at^3$, and the line $x = 9a$, giving your answer in terms
        of a^2.

13. Evaluate the area between the line $y = x - 1$ and the curve
    (a) $y = x(1 - x)$ (b) $y = (2x + 1)(x - 1)$

14. Calculate the area of the region of the xy plane defined by the
    inequalities $y \geqslant (x + 1)(x - 2)$ and $y \leqslant x$

THE APPROXIMATE VALUE OF A DEFINITE INTEGRAL

We know that the definite integral $\displaystyle\int_a^b f(x)\,dx$ can be used to evaluate
the area between the curve $y = f(x)$, the x-axis and the ordinates at
$x = a$ and $x = b$. It is not always possible, however, to find a
function whose derivative is $f(x)$. In such cases the definite integral,
and hence the exact value of the specified area, cannot be found.

If, on the other hand, we divide the area into a *finite* number of strips
then the sum of their areas gives an approximate value for the required
area and hence an approximate value of the definite integral. When
using this approach it is convenient to choose strips whose widths are
all the same.

The Trapezium Rule

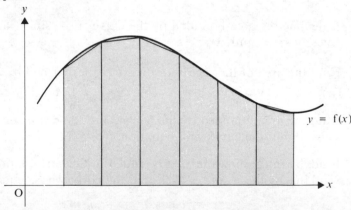

When the area shown in the diagram is divided into vertical strips,
each strip is approximately a trapezium.

If the width of the strip and its two vertical sides are known, the area
of the strip can be found using the formula

$$\text{area} = \tfrac{1}{2}(\text{sum of } \| \text{ sides}) \times \text{width}$$

The sum of the areas of all the strips then gives an approximate value
for the area under the curve.

Now suppose that there are n strips, *all with the same width*, d say, and that the vertical edges of the strips (i.e. the ordinates) are labelled $y_0, y_1, y_2, \ldots, y_{n-1}, y_n$

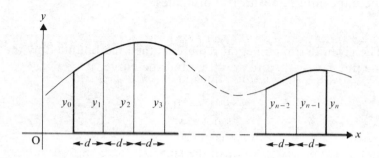

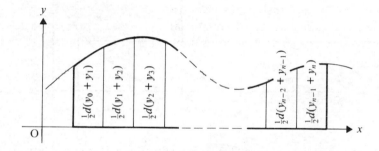

The sum of the areas of all the strips can be written down as follows:

$$\tfrac{1}{2}(y_0 + y_1)(d) + \tfrac{1}{2}(y_1 + y_2)(d) + \tfrac{1}{2}(y_2 + y_3)(d) + \ldots$$

$$\ldots + \tfrac{1}{2}(y_{n-2} + y_{n-1})(d) + \tfrac{1}{2}(y_{n-1} + y_n)(d)$$

Therefore the area, A, under the curve is given approximately by

$$A \approx \tfrac{1}{2}(d)[y_0 + 2y_1 + 2y_2 + \ldots + 2y_{n-1} + y_n]$$

This formula is known as the *Trapezium Rule*

An easy way to remember the formula in terms of ordinates is

half width of strip × (first + last + twice all the others)

Example 20d

Find an approximate value for the definite integral $\displaystyle\int_1^5 x^3\,dx$ using the trapezium rule with five ordinates.

The given definite integral represents the area bounded by the x-axis, the lines $x = 1$ and $x = 5$, and the curve $y = x^3$

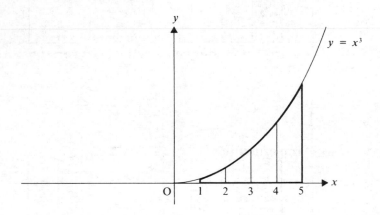

When five ordinates are used there are four strips and their widths must all be the same. From $x = 1$ to $x = 5$ there are four units so the width of each strip must be 1 unit. Hence the five ordinates are where $x = 1$, $x = 2$, $x = 3$, $x = 4$ and $x = 5$

$$\therefore\quad y_0 = 1^3 = 1,\ y_1 = 2^3 = 8,\ y_2 = 3^3 = 27,\ y_3 = 64,\ y_4 = 125$$

Using the trapezium rule the required area, A, is given by

$$A \approx \tfrac{1}{2}(1)[1 + 125 + 2\{8 + 27 + 64\}] = 162$$

The required area is 162 square units.

The degree of accuracy of an answer given by the Trapezium Rule clearly depends upon the number of strips into which the required area is divided because, the narrower the strip the nearer its shape becomes to a trapezium.

EXERCISE 20d

In Questions 1 to 4 estimate the value of each definite integral, using the trapezium rule with

(a) 3 ordinates   (b) 5 ordinates.

1. $\displaystyle\int_0^4 x^2\,dx$

2. $\displaystyle\int_1^3 \frac{1}{x^2}\,dx$

3. $\displaystyle\int_0^{2\pi/3} \sqrt{\sin x}\,dx$

4. $\displaystyle\int_1^3 \ln x\,dx$

5. Find the true value of the definite integrals given in Questions 1 and 2. Complete the table below and note the comparative accuracy of the results.

Value using	$\displaystyle\int_0^4 x^2\,dx$	$\displaystyle\int_1^3 \frac{1}{x^2}\,dx$
Trapezium rule with 3 ordinates		
Trapezium rule with 5 ordinates		
Definite integration		

6. Models fired through a wind tunnel are brought to rest by a set of buffers.

The diagram shows the velocity–time graph of the free end of the buffers from the time the model hits the buffers until it is brought to rest.

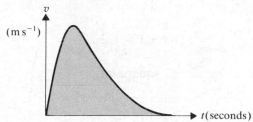

Estimate the distance moved by the model in this time by finding an approximate value for the shaded area using the trapezium rule with 5 ordinates.

Take the equation of the curve as $v = 40e^{-5t}\cos(2t - \tfrac{1}{2}\pi)$.

VOLUME OF REVOLUTION

If an area is rotated about a straight line, the three-dimensional object so formed is called a *solid of revolution*, and its volume is a *volume of revolution*.

The line about which rotation takes place is always an axis of symmetry for the solid of revolution. Also, any cross-section of the solid which is perpendicular to the axis of rotation, is circular.

Consider the solid of revolution formed when the area shown in the diagram is rotated about the x-axis.

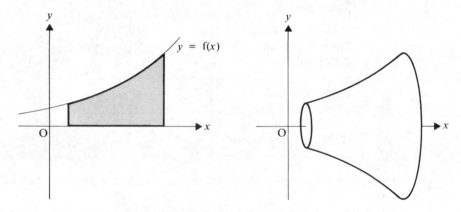

To calculate the volume of this solid we can divide it into 'slices' by making cuts perpendicular to the axis of rotation.

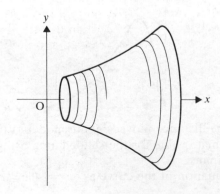

If the cuts are reasonably close together, each slice is approximately cylindrical and the approximate volume of the solid can be found by summing the volumes of these cylinders.

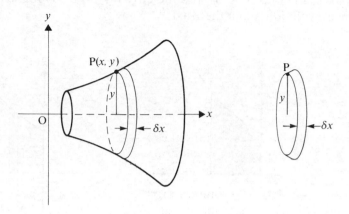

Consider an element formed by one cut through the point P(x, y) and the other cut distant δx from the first.

The volume, δV, of this element is approximately that of a cylinder of radius y and 'height' δx

i.e. $$\delta V \approx \pi y^2 \, \delta x$$

Then the total volume of the solid is V, where

$$V \approx \sum \pi y^2 \, \delta x$$

The smaller δx is, the closer is this approximation to V,

i.e. $$V = \lim_{\delta x \to 0} \sum \pi y^2 \, \delta x = \int \pi y^2 \, dx$$

Now if the equation of the rotated curve is given, this integral can be evaluated and the volume of the solid of revolution found,
e.g. to find the volume generated when the area between part of the curve $y = e^x$ and the x-axis is rotated about the x-axis, we use

$$\int \pi (e^x)^2 \, dx = \pi \int e^{2x} \, dx$$

When an area rotates about the y-axis we can use a similar method based on slices perpendicular to the y-axis, giving

$$V = \int \pi x^2 \, dy$$

Examples 20e _____

1. Find the volume generated when the area bounded by the x and y axes, the line $x = 1$ and the curve $y = e^x$ is rotated through one revolution about the x-axis.

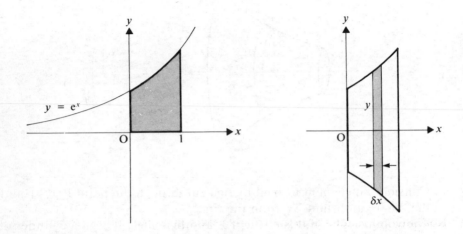

The volume, δV, of the element shown is approximately that of a cylinder of radius y and thickness δx, therefore

$$\delta V \approx \pi y^2 \, \delta x$$

$\therefore$ the total volume is V, where $V \approx \sum_{x=0}^{x=1} \pi y^2 \, \delta x$

$\Rightarrow$
$$V = \lim_{\delta x \to 0} \sum_{x=0}^{x=1} \pi y^2 \, \delta x = \int_0^1 \pi y^2 \, dx$$

$$= \pi \int_0^1 (e^x)^2 \, dx$$

$$= \pi \int_0^1 e^{2x} \, dx$$

$$= \pi \left[\tfrac{1}{2} e^{2x} \right]_0^1$$

$$= \tfrac{1}{2}\pi (e^2 - e^0)$$

i.e. the specified volume of revolution is $\tfrac{1}{2}\pi(e^2 - 1)$ cubic units.

_____

2. The area defined by the inequalities $y \geqslant x^2 + 1$, $x \geqslant 0$, $y \leqslant 2$, is rotated completely about the y-axis. Find the volume of the solid generated.

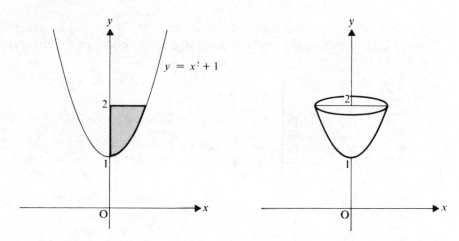

Rotating the shaded area about the y-axis gives the solid shown.

This time we use horizontal cuts to form elements which are approximately cylinders with radius x and thickness δy

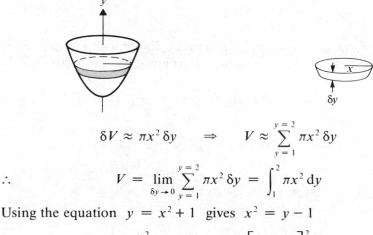

$$\delta V \approx \pi x^2 \, \delta y \qquad \Rightarrow \qquad V \approx \sum_{y=1}^{y=2} \pi x^2 \, \delta y$$

$$\therefore \qquad V = \lim_{\delta y \to 0} \sum_{y=1}^{y=2} \pi x^2 \, \delta y = \int_1^2 \pi x^2 \, dy$$

Using the equation $y = x^2 + 1$ gives $x^2 = y - 1$

$$\therefore \qquad V = \pi \int_1^2 (y - 1) \, dy = \pi \left[\tfrac{1}{2} y^2 - y \right]_1^2$$

$$= \pi \{ (2 - 2) - (\tfrac{1}{2} - 1) \}$$

i.e. the volume of the specified solid is $\tfrac{1}{2}\pi$ cubic units.

3. The area enclosed by the curve $y = 4x - x^2$ and the line $y = 3$ is rotated about the line $y = 3$. Find the volume of the solid generated.

The line $y = 3$ meets the curve $y = 4x - x^2$ at the points $(1, 3)$ and $(3, 3)$, therefore the volume generated is as shown in the diagram.

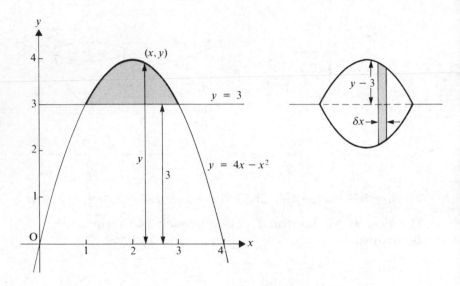

The element shown is approximately a cylinder with radius $(y - 3)$ and thickness δx, so its volume, δV, is given by $\delta V \approx \pi(y - 3)^2 \delta x$

i.e.
$$V = \lim_{\delta x \to 0} \sum_{x=1}^{x=3} \pi(y - 3)^2 \delta x = \pi \int_1^3 (y - 3)^2 \, dx$$

$\Rightarrow$
$$V = \pi \int_1^3 (4x - x^2)^2 \, dx$$

$$= \pi \int_1^3 (16x^2 - 8x^3 + x^4) \, dx$$

$$= \pi \left[\tfrac{16}{3}x^3 - 2x^4 + \tfrac{1}{5}x^5 \right]_1^3$$

$$= \tfrac{406}{15}\pi$$

$\therefore$ the required volume is $\tfrac{406}{15}\pi$ cubic units.

4. Find the volume generated when the area between the curve $y^2 = x$
and the line $y = x$ is rotated completely about the x-axis.

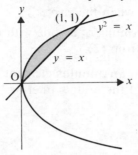

The defined area is shown in the diagram.

When this area rotates about Ox, the solid generated is bowl-shaped
on the outside, with a conical hole inside.
The cross-section this time is not a simple circle but is an annulus, i.e.
the area between two concentric circles.

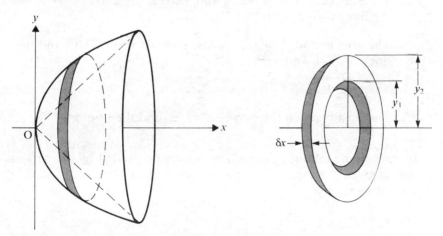

For a typical element the area of cross-section is $\pi y_1^2 - \pi y_2^2$

Therefore the volume of an element is given by $\delta V \approx \pi\{y_1^2 - y_2^2\}\,\delta x$

$$\therefore \qquad V \approx \sum_{x=0}^{x=1} \pi\{y_1^2 - y_2^2\}\,\delta x = \pi \int_0^1 (y_1^2 - y_2^2)\,dx$$

Now $y_1 = \sqrt{x}$ and $y_2 = x$

$$\therefore \qquad V = \pi \int_0^1 (x - x^2)\,dx = \pi\left[\tfrac{1}{2}x^2 - \tfrac{1}{3}x^3\right]_0^1 = \tfrac{1}{6}\pi$$

The volume generated is $\tfrac{1}{6}\pi$ cubic units.

Note that the volume specified in Example 4 could be found by calculating separately

1) the volume given when the curve $y^2 = x$ rotates about the x-axis;

2) the volume, by formula, of a cone with base radius 1 and height 1 and subtracting **(1)** from **(2)**.

The method in which an annulus element is used, however, applies whatever the shape of the hollow interior.

EXERCISE 20e

In each of the following questions, find the volume generated when the area defined is rotated completely about the x-axis.

1. The area between the curve $y = x(4 - x)$ and the x-axis.

2. The area bounded by the x and y axes, the curve $y = e^x$ and the line $x = 3$

3. The area bounded by the x-axis, the curve $y = 1/x$ and the lines $x = 1$ and $x = 2$

4. The area defined by the inequalities $0 \leqslant y \leqslant x^2$ and $-2 \leqslant x \leqslant 2$

5. The area between the curve $y^2 = x$ and the line $x = 2$

In each of the following questions, the area bounded by the curve and line(s) given is rotated about the y-axis to form a solid. Find the volume generated.

6. $y = x^2$, $y = 4$

7. $y = 4 - x^2$, $y = 0$

8. $y = x^3$, $y = 1$, $y = 2$, for $x \geqslant 0$

9. $y = \ln x$, $x = 0$, $y = 0$, $y = 1$

10. Find the volume generated when the area enclosed between $y^2 = x$ and $x = 1$ is rotated about the line $x = 1$

11. The area defined by the inequalities

$$y \geqslant x^2 - 2x + 4, \ y \leqslant 4$$

is rotated about the line $y = 4$. Find the volume generated.

12. The area enclosed by $y = \sin x$ and the x-axis for $0 \leqslant x \leqslant \pi$ is rotated about the x-axis. Find the volume generated.

13. An area is bounded by the line $y = 1$, the x-axis and parts of the curve $y = 3 - x^2$. Find the volume generated when this area rotates completely about the y-axis.

14. The area enclosed between the curves $y = x^2$ and $y^2 = x$ is rotated about the x-axis. Find the volume generated.

MIXED EXERCISE 20

1. The surface area of an expanding sphere is increasing at a constant rate of 0.02 cm^2 per second. Find the rate of increase of the volume of the sphere when the radius is 15 cm.

2. A region of the xy plane is defined by the inequalities $0 \leqslant x \leqslant 4$ and $0 \leqslant y \leqslant e^x$. Find the area of the region.

3. Find the area of the region in the first quadrant bounded by the y-axis, the line $y = 6$ and the curve $y = x^2 + 2$
   If this area is rotated completely about Oy to form a solid, find the volume of the solid.

4. (a) Find the area enclosed between the x-axis and the curve $y = x^2 - 3$
   (b) Find the volume generated when this area is rotated completely about
      (i) the x-axis   (ii) the y-axis.

5. A region of the xy plane is bounded by the curve $y = e^x$, the x-axis and the lines $x = -1$ and $x = 1$
   (a) Find the area of the region.
   (b) Find the volume generated when this area is rotated completely about the x-axis.

6. Without using a calculator find, to 5 s.f. an approximation for $f(x)$ if
   (a) $f(x) = \sqrt[3]{x}$ and $x = 65$
   (b) $f(x) = \sqrt{x}$ and $x = 120$

7. A plane region is bounded by the curve $y = 6 - x^2$ and the line $y = 2$. Find

    (a) the area of the region

    (b) the volume generated when this area rotates through $360°$ about the line $y = 2$

8. The equation of a curve is $y = \sin x$. Find the area bounded by this curve and the x-axis between $x = 0$ and $x = \pi$
   Find also the volume generated when this area rotates through one revolution about the x-axis.

9. (a) Find an approximate value for the area between the x-axis and the curve $y = (x - 1)(x - 4)$, using the trapezium rule with 4 ordinates.

    (b) Evaluate $\displaystyle\int_{1}^{4} (x - 1)(x - 4)\, dx$,

10. (a) Use the trapezium rule with 3 ordinates to estimate the value of $\displaystyle\int_{0}^{5} (3 + x)\, dx$,

    (b) Find the value of $\displaystyle\int_{0}^{5} (3 + x)\, dx$,

    (c) Explain the connection between the results of (a) and (b).

CHAPTER 21

COMPLEX NUMBERS

IMAGINARY NUMBERS

The numbers we have worked with up to now have been such that, when squared, the result is either positive or zero, i.e. for a number k, $k^2 \geqslant 0$

Such numbers are *real* numbers.

However, the roots of an equation such as $x^2 = -1$ are clearly not real since they give -1 when squared.
If we are to work with equations of this type we need another category of numbers, i.e. the set of numbers whose squares are negative real numbers.

Members of this set are called *imaginary* numbers and some examples are

$$\sqrt{-1}, \quad \sqrt{-7}, \quad \sqrt{-20}$$

In general any imaginary number can be represented by $\sqrt{(-n^2)}$ where n is real.

Now $\qquad \sqrt{(-n^2)} = \sqrt{[(n^2)(-1)]} = \sqrt{(n^2)} \times \sqrt{(-1)}$

i.e. $\qquad \sqrt{(-n^2)} = n\mathrm{i} \quad$ where $\quad \mathrm{i} = \sqrt{(-1)}$

So any imaginary number can be written $n\mathrm{i}$ where n is real and

$$\mathrm{i} = \sqrt{(-1)}$$

e.g. $\qquad \sqrt{-16} = 4\mathrm{i} \quad$ and $\quad \sqrt{-3} = \mathrm{i}\sqrt{3}$

Usually the real number is placed before i but we write $\mathrm{i}\sqrt{3}$ rather than $\sqrt{3}\mathrm{i}$ in order to avoid ambiguity.
Note that j is sometimes used instead of i for $\sqrt{-1}$ but in this book we use i.

Imaginary numbers can be added to, or subtracted from, each other.

For example $3i + 9i = 12i$ and $i\sqrt{7} - 2i = (\sqrt{7} - 2)i$

The product of two imaginary numbers is always real.

For example $2i \times 5i = 10i^2 = 10(\sqrt{-1})^2 = 10(-1)$

i.e. $2i \times 5i = -10$

The quotient of two imaginary numbers is always real.

For example $\dfrac{3i}{7i} = \dfrac{3}{7}$

Powers of i can be simplified.

For example $i^3 = (i^2)(i) = -i$ $i^4 = (i^2)^2 = (-1)^2 = 1$

$i^5 = (i^4)(i) = i$ $i^{-1} = i/i^2 = i/-1 = -i$

COMPLEX NUMBERS

When a real number and an imaginary number are added or subtracted, the expression which is obtained cannot be simplified and is called a *complex number*,
e.g. $2 + 3i$, $4 - 7i$, and $-1 + 4i$ are complex numbers.

A general complex number can be represented by $a + bi$ where a and b can have any real value including zero.

If $a = 0$ we have numbers of the form bi, i.e. imaginary numbers.
If $b = 0$ we have numbers of the form a, i.e. real numbers.

Therefore

> the set of complex numbers includes all real numbers and all imaginary numbers.

OPERATIONS ON COMPLEX NUMBERS

Addition and Subtraction

Real terms and imaginary terms are collected separately in two groups,

e.g. $(2 + 3i) + (4 - i) = (2 + 4) + (3i - i) = 6 + 2i$

and $(4 - 2i) - (3 + 5i) = (4 - 3) - (2i + 5i) = 1 - 7i$

Multiplication

The distributive law of multiplication applied to two complex numbers gives their product,

e.g.
$$i(5 - 2i) = 5i - 2i^2$$
$$= 5i - 2(-1)$$
$$= 2 + 5i$$

and
$$(2 + 3i)(4 - i) = 8 - 2i + 12i - 3i^2$$
$$= 8 + 10i - 3(-1)$$
$$= 11 + 10i$$

and
$$(2 + 3i)(2 - 3i) = 4 - 6i + 6i - 9i^2$$
$$= 4 + 9$$
$$= 13$$

Conjugate Complex Numbers

Notice that the product in the last example above is a real number. This is because of the special form of the given complex numbers, $2 \pm 3i$, which are the factors of a 'difference of two squares'

Any pair of complex numbers of the form $a \pm bi$ have a product which is real, since

$$(a + bi)(a - bi) = a^2 - abi + abi - b^2i^2$$
$$= a^2 + b^2$$

Such complex numbers are said to be *conjugate* and each is the conjugate of the other. Thus $4 + 5i$ and $4 - 5i$ are conjugate complex numbers and $4 + 5i$ is the conjugate of $4 - 5i$

If $a + bi$ is denoted by z then its conjugate, $a - bi$, is denoted by $\bar{z}$ or z^*

Division

Division by a complex number cannot be carried out directly because the denominator is made up of two independent terms. This problem can be overcome by making the denominator real, a process called 'realising the denominator'. This is done by using the property that the product of conjugate complex numbers is real.

e.g. if we wish to divide $2 + 9i$ by $5 - 2i$ we multiply both numerator and denominator by the conjugate of the denominator, which is $5 + 2i$ in this case, giving

$$\frac{2 + 9i}{5 - 2i} = \frac{(2 + 9i)(5 + 2i)}{(5 - 2i)(5 + 2i)}$$

$$= \frac{10 + 49i + 18i^2}{25 - 4i^2}$$

$$= \frac{-8 + 49i}{29} = -\frac{8}{29} + \frac{49}{29}i$$

Note that the real term is given first, even when it is negative.

THE ZERO COMPLEX NUMBER

A complex number is zero if, and only if, the real term and the imaginary term are each zero.

i.e. $a + bi = 0$ $\Longleftrightarrow$ $a = 0$ *and* $b = 0$

EQUAL COMPLEX NUMBERS

If $a + bi = c + di$ then $(a + bi) - (c + di) = 0$

$\Rightarrow$ $(a - c) + (b - d)i = 0$

Hence $a - c = 0$ and $b - d = 0$

$\Rightarrow$ $a = c$ and $b = d$

i.e.   two complex numbers are equal if, and only if, the real terms are equal and the imaginary terms are equal.

Denoting the real part of $a + bi$ by $\text{Re}(a + bi)$ and the imaginary part by $\text{Im}(a + bi)$ we have

$$a + bi = c + di \quad \Longleftrightarrow \quad \begin{cases} \text{Re}(a + bi) = \text{Re}(c + di) \\ \text{Im}(a + bi) = \text{Im}(c + di) \end{cases}$$

A complex equation is therefore equivalent to two separate equations.

This property provides an alternative (but not better) method for division by a complex number.

e.g. to divide $3 - 2i$ by $5 + i$ we can represent the quotient by $p + qi$, where p and q are real,

i.e.
$$\frac{3 - 2i}{5 + i} = p + qi$$

Hence
$$3 - 2i = (p + qi)(5 + i)$$
$$= 5p + pi + 5qi + qi^2$$
$$= (5p - q) + (p + 5q)i$$

Equating real and imaginary parts gives

$$3 = 5p - q \quad \text{and} \quad -2 = p + 5q$$

Solving these two equations gives $p = \frac{1}{2}$ and $q = -\frac{1}{2}$

$$\therefore \quad (3 - 2i) \div (5 + i) = \tfrac{1}{2} - \tfrac{1}{2}i$$

FINDING THE SQUARE ROOTS OF A COMPLEX NUMBER

Equating the real parts and the imaginary parts of a complex equation provides one method for determining the square roots of a complex number.

For example, if we assume that a square root of a complex number is itself a complex number then, using $\sqrt{(15 + 8i)}$ as an example we can say

$$\sqrt{(15 + 8i)} = a + bi \qquad \text{where } a \text{ and } b \text{ are real}$$

$$\Rightarrow \quad 15 + 8i = (a + bi)^2$$

$$= a^2 - b^2 + 2abi$$

Equating real and imaginary parts gives

$$a^2 - b^2 = 15 \qquad\qquad [1]$$

and
$$2ab = 8 \qquad\qquad [2]$$

Using $b = \dfrac{4}{a}$ in [1] gives $\qquad a^2 - \dfrac{16}{a^2} = 15$

$\Rightarrow \qquad\qquad\qquad\qquad a^4 - 15a^2 - 16 = 0$

$\Rightarrow \qquad\qquad\qquad\qquad (a^2 - 16)(a^2 + 1) = 0$

Thus $\qquad\qquad\qquad\qquad a^2 - 16 = 0 \quad$ or $\quad a^2 + 1 = 0$

Now a is real, so $a^2 + 1 = 0$ gives no suitable values

therefore the only values for a are $a = \pm 4$

Then from equation [2] we have

$$a = 4 \quad \Rightarrow \quad b = 1$$

and $\qquad\qquad\qquad\qquad a = -4 \; \Rightarrow \; b = -1$

Note. It is not correct to say $a = \pm 4$ therefore $b = \pm 1$ as this offers four different pairs of values for a and b (i.e. 4, 1; 4, -1; -4, 1; -4, -1) two of which are invalid.

Hence $\qquad\qquad\qquad \sqrt{(15 + 8i)} = 4 + i \quad$ or $\quad -4 - i$

$$= \pm(4 + i)$$

This result justifies our original assumption that the square root of a complex number is another complex number.

Sometimes it is possible to find the square roots of a complex number simply by observation.

In the example above, equation [2] shows that the product of a and b is half the coefficient of i in the given complex number.

Suitable integral values for a and b can then be checked quite quickly, e.g. to find $\sqrt{(8 - 6i)}$ we note that $ab = -3$ so possible values for a and b are: 1, -3; 3, -1

Checking: $\qquad (1 - 3i)^2 = -8 - 6i \quad$ which is not correct

$$(3 - i)^2 = 8 - 6i \qquad \text{which is correct}$$

Hence one square root of $8 - 6i$ is $3 - i$ and the other is $-(3 - i)$

i.e. $\qquad\qquad\qquad \sqrt{(8 - 6i)} = \pm(3 - i)$

Note. Unless a and b are integers, this method is unlikely to be useful.

EXERCISE 21a

1. Simplify: i^7, i^{-3}, i^9, i^{-5}, i^{4n}, i^{4n+1}

2. Add the following pairs of complex numbers.
   (a) $3 + 5i$ and $7 - i$ (b) $4 - i$ and $3 + 3i$
   (c) $2 + 7i$ and $4 - 9i$ (d) $a + bi$ and $c + di$

3. Subtract the second number from the first in each part of Question 2.

4. Simplify
   (a) $(2 + i)(3 - 4i)$ (b) $(5 + 4i)(7 - i)$ (c) $(3 - i)(4 - i)$
   (d) $(3 + 4i)(3 - 4i)$ (e) $(2 - i)^2$ (f) $(1 + i)^3$
   (g) $i(3 + 4i)$ (h) $(x + yi)(x - yi)$ (i) $i(1 + i)(2 + i)$
   (j) $(a + bi)^2$

5. Realise the denominator of each of the following fractions and hence express each in the form $a + bi$
   (a) $\dfrac{2}{1 - i}$ (b) $\dfrac{3 + i}{4 - 3i}$ (c) $\dfrac{4i}{4 + i}$ (d) $\dfrac{1 + i}{1 - i}$

   (e) $\dfrac{7 - i}{1 + 7i}$ (f) $\dfrac{x + yi}{x - yi}$ (g) $\dfrac{3 + i}{i}$ (h) $\dfrac{-2 + 3i}{-i}$

6. Solve the following equations for x and y
   (a) $x + yi = (3 + i)(2 - 3i)$ (b) $\dfrac{2 + 5i}{1 - i} = x + yi$

   (c) $3 + 4i = (x + yi)(1 + i)$ (d) $x + yi = 2$
   (e) $x + yi = (3 + 2i)(3 - 2i)$ (f) $x + yi = (4 + i)^2$
   (g) $\dfrac{x + yi}{2 + i} = 5 - i$ (h) $(x + yi)^2 = 3 + 4i$

7. Find the real and imaginary parts of
   (a) $(2 - i)(3 + i)$ (b) $(1 - i)^3$ (c) $(3 + 4i)(3 - 4i)$
   (d) $\dfrac{3 + 2i}{4 - i}$ (e) $\dfrac{2}{3 + i} + \dfrac{3}{2 + i}$ (f) $\dfrac{1}{x + yi} - \dfrac{1}{x - yi}$

8. Find the square roots of
   (a) $3 - 4i$ (b) $21 - 20i$ (c) $2i$ (d) $15 + 8i$ (e) $-24 + 10i$

COMPLEX ROOTS OF QUADRATIC EQUATIONS

Consider the quadratic equation $x^2 + 2x + 2 = 0$

The formula $x = \dfrac{-b \pm \sqrt{(b^2 - 4ac)}}{2a}$ gives $x = \dfrac{-2 \pm \sqrt{-4}}{2}$

Previously we dismissed solutions of this type, in which $b^2 - 4ac < 0$, as not being real. But now, because $\sqrt{-4} = 2i$, we see that the roots of this equation are the complex numbers $-1 + i$ and $-1 - i$.

Further, the roots are conjugate complex numbers.

We can show that if $b^2 - 4ac < 0$, the roots of the general quadratic equation $ax^2 + bx + c = 0$ are *always* conjugate complex numbers.

If $\qquad ax^2 + bx + c = 0 \quad$ and $\quad b^2 - 4ac < 0$

then $\qquad x = \dfrac{-b \pm \sqrt{(b^2 - 4ac)}}{2a} = \dfrac{-b}{2a} \pm \dfrac{i\sqrt{(4ac - b^2)}}{2a}$

$\therefore \; x = p \pm qi \;$ where $\; p = -b/2a \;$ and $\; q = \sqrt{(4ac - b^2)}/2a$

i.e. when $b^2 - 4ac < 0$, the roots of the equation $ax^2 + bx + c = 0$ are

$$p + qi \quad \text{and} \quad p - qi$$

and these are conjugate complex numbers.

> So, if one root of a quadratic equation with real coefficients is known to be complex, the other root must also be complex and the conjugate of the first.

When the quadratic equation $ax^2 + bx + c = 0$ has real roots α and β we know that

$$\alpha + \beta = \frac{-b}{a} \quad \text{and} \quad \alpha\beta = \frac{c}{a}$$

We now show that these relationships are valid also when the roots are complex.

If $\alpha = p + qi$ and $\beta = p - qi$ then

$$\alpha + \beta = 2p = 2\left(\frac{-b}{2a}\right) = \frac{-b}{a}$$

and $\quad \alpha\beta = (p + qi)(p - qi) = p^2 + q^2 = \dfrac{b^2}{4a^2} + \dfrac{(4ac - b^2)}{4a^2} = \dfrac{c}{a}$

Hence, for *any* quadratic equation $ax^2 + bx + c = 0$, the roots α and β satisfy the relationships

$$\alpha + \beta = -b/a \quad \text{and} \quad \alpha\beta = c/a$$

Examples 21b

1. One root of the equation $x^2 + px + q = 0$ is $2 - 3i$. Find the values of p and q

If one root is $2 - 3i$ the other must be $2 + 3i$

Then $\quad \alpha + \beta = 4 \quad$ and $\quad \alpha\beta = (2 - 3i)(2 + 3i) = 13$

Now any quadratic equation can be written in the form

$$x^2 - (\text{sum of roots})x + (\text{product of roots}) = 0$$

So the equation with roots $2 \pm 3i$ is

$$x^2 - 4x + 13 = 0$$

Therefore $p = -4$ and $q = 13$

2. Find the complex roots of the equation $2x^2 + 3x + 5 = 0$
   If these roots are α and β, confirm the relationships
   $$\alpha + \beta = -\frac{b}{a} \quad \text{and} \quad \alpha\beta = \frac{c}{a}$$

If $\qquad\qquad\qquad 2x^2 + 3x + 5 = 0$

then $\qquad\qquad\qquad x = \dfrac{-3 \pm \sqrt{(9 - 40)}}{4}$

$\Rightarrow \qquad\qquad \alpha = -\dfrac{3}{4} + \dfrac{\sqrt{31}}{4}i \qquad \beta = -\dfrac{3}{4} - \dfrac{\sqrt{31}}{4}i$

Hence $\qquad \alpha + \beta = \left(-\dfrac{3}{4} + \dfrac{\sqrt{31}}{4}i\right) + \left(-\dfrac{3}{4} - \dfrac{\sqrt{31}}{4}i\right)$

$$= -\frac{3}{2} = -\frac{b}{a}$$

and $$\alpha\beta = \left(-\frac{3}{4} + \frac{\sqrt{31}}{4}i\right)\left(-\frac{3}{4} - \frac{\sqrt{31}}{4}i\right)$$

$$= \frac{9}{16} - \frac{31}{16}i^2$$

$$= \frac{40}{16} = \frac{5}{2} = \frac{c}{a}$$

EXERCISE 21b

1. Solve the following equations.
   (a) $x^2 + x + 1 = 0$ (b) $2x^2 + 7x + 1 = 0$
   (c) $x^2 + 9 = 0$ (d) $x^2 + x + 3 = 0$
   (e) $x^4 - 1 = 0$ (f) $3x^2 + x + 3 = 0$

2. Form the equation whose roots are
   (a) $i, -i$ (b) $2 + i, 2 - i$
   (c) $1 - 3i, 1 + 3i$ (d) $1 + i, 1 - i, 2$

3. Without calculating a, b and c, evaluate $-b/a$ and c/a if one root of the equation $ax^2 + bx + c = 0$ is
   (a) $2 + i$ (b) $3 - 4i$ (c) i (d) $5i - 12$

4. Find the equation, one of whose roots is
   (a) $-1 - i$ (b) $-5 + i$ (c) $1 - 3i$ (d) $4 + i$
   Explain why this question cannot be answered if the given root is 2

THE ARGAND DIAGRAM

In the complex number $a + bi$, a and b are both real numbers so $a + bi$ can be represented by the ordered pair $\begin{pmatrix} a \\ b \end{pmatrix}$.

This suggests that a and b could be used as the coordinates of a point $A(a, b)$ in the xy plane.

THE HENLEY COLLEGE LIBRARY

Then the vector $\overrightarrow{OA}$ provides a visual representation of the complex number.

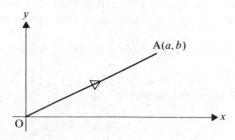

This idea was introduced by a French mathematician and his name, Argand, is given to the diagram which illustrates a complex number in this way. On an Argand diagram, the real part of a complex number is taken as the x-coordinate, and the coefficient of i is the y-coordinate. For this reason the x and y axes are often called the real and imaginary axes. It must be remembered however that the y-coordinate is the *real* number b.

A general complex number $x + y$i is represented by the vector $\overrightarrow{OP}$ where P is the point (x, y)

In an Argand diagram, the magnitude and direction of a line are used to represent a complex number in the same way that a section of line can be used to represent a vector quantity. The techniques and operations used in vector analysis can therefore be applied equally well to complex number analysis.

A COMPLEX NUMBER AS A VECTOR

On an Argand diagram a complex number such as $5 + 3$i can be represented by the vector $\overrightarrow{OA}$ where A is the point $(5, 3)$. However it can equally well be represented by any other vector with the same length and direction as $\overrightarrow{OA}$, e.g. by $\overrightarrow{BC}$ or $\overrightarrow{DE}$ in the diagram below.

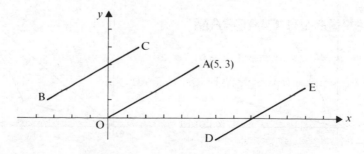

A complex number can be treated in this way only when it behaves as a *free* vector. If, on the other hand, $5 + 3i$ is regarded as a *position* vector, then *only* the vector $\overrightarrow{OA}$ represents $5 + 3i$. In this case the *actual point* $A(5, 3)$ is sometimes taken to represent $5 + 3i$

A vector representing a complex number is usually denoted by the symbol z, so

for a general complex number we have

$$z = x + yi$$

and for unique complex numbers we use z_1, z_2, etc.,

e.g. $z_1 = 5 + 3i, \quad z_2 = 7 - 4i$

When z is used on an Argand diagram, an arrow is needed to indicate the direction of the line representing the complex number.

For example, in the diagram below the line BC without an arrow could represent *either* $4 - 2i$ *or* $-4 + 2i$
Whereas the arrow shows that this line represents the complex number $4 - 2i$ only

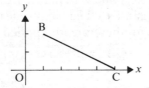
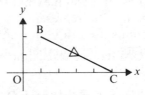

Graphical Addition and Subtraction of Complex Numbers

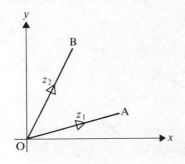
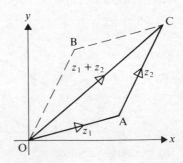

Consider two complex numbers, z_1 and z_2, represented on an Argand diagram by $\overrightarrow{OA}$ and $\overrightarrow{OB}$.

We know that the sum of the two *vectors* $\overrightarrow{OA}$ and $\overrightarrow{OB}$ is $\overrightarrow{OC}$ where AC is equal and parallel to OB. Therefore, as z_1 and z_2 behave as vectors, we have

> $z_1 + z_2$ is represented by the diagonal $\overrightarrow{OC}$ of the parallelogram OACB.

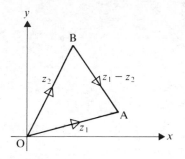 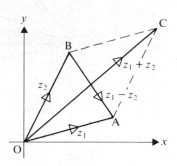

Now considering the vector triangle OAB we see that

$$\overrightarrow{BA} = \overrightarrow{BO} + \overrightarrow{OA}$$

But $\overrightarrow{OA}$ represents z_1 and $\overrightarrow{BO} = -\overrightarrow{OB}$ represents $-z_2$ therefore

> $z_1 - z_2$ is represented by the diagonal joining B to A in parallelogram OACB.

Note carefully the direction of the vector represented by this diagonal.

Hence the two diagonals of the parallelogram OACB represent the sum and the difference of z_1 and z_2

If $z_1 = x_1 + y_1 i$ and $z_2 = x_2 + y_2 i$ then

$$z_1 + z_2 = (x_1 + x_2) + (y_1 + y_2)i$$

Therefore C is the point $(\{x_1 + x_2\}, \{y_1 + y_2\})$,

e.g. if $z_1 = 3 + 5i$ and $z_2 = 7 - 2i$ then $z_1 + z_2$ can be represented on the Argand diagram by $\overrightarrow{OC}$ where C is the point $(10, 3)$.

Example 21c

If $z_1 = 5 - i$ and $z_2 = 3 + 4i$, represent on one Argand diagram the complex numbers $z_1, z_2, z_1 + z_2, z_1 - z_2$

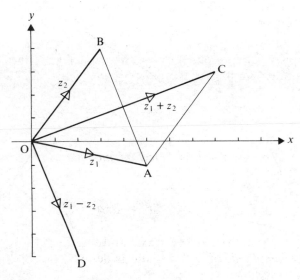

$\overrightarrow{OA}$ represents z_1 and $\overrightarrow{OB}$ represents z_2

$\overrightarrow{OC}$, the diagonal of the parallelogram OACB, represents $z_1 + z_2$

$z_1 - z_2$ is represented by the diagonal $\overrightarrow{BA}$ and also by $\overrightarrow{OD}$ which is equal and parallel to BA

EXERCISE 21c

1. Represent each complex number on an Argand diagram.

   (a) $5 + i$ (b) $-2 + 6i$ (c) $3 - 5i$ (d) $-4 - 4i$

   (e) $4 + 4i$ (f) -3 (g) $3i$ (h) $-7i$

2. If $z_1 = 3 - i$, $z_2 = 1 + 4i$, $z_3 = -4 + i$, $z_4 = -2 - 5i$, represent the following by lines on Argand diagrams, showing the direction of each line by an arrow.

   (a) $z_1 + z_2$ (b) $z_2 - z_3$ (c) $z_1 - z_3$ (d) $z_2 + z_4$

   (e) $z_4 - z_1$ (f) $z_3 - z_4$ (g) z_1

   (h) z_4 (i) $z_2 - z_1$ (j) $z_1 + z_3$

3. If $z_1 = x_1 + y_1 i$ and $z_2 = x_2 + y_2 i$ show on an Argand diagram the position of the points representing $\frac{1}{2}(z_1 + z_2)$ and $\frac{1}{3}(2z_1 + z_2)$

MODULUS AND ARGUMENT

Consider the point A(a, b), representing the complex number $a + b$i

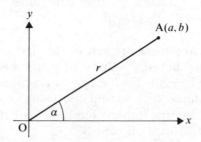

The length of the line OA is called the *modulus* of $a + b$i and is denoted by r,

i.e.
$$|a + b\mathrm{i}| = r = \sqrt{(a^2 + b^2)}$$

The angle between the positive x-axis and the line OA is called the *argument* or *amplitude* of $a + b$i and is denoted by α, where α is within the range $-\pi < \alpha \leqslant \pi$

i.e.
$$\arg(a + b\mathrm{i}) = \alpha = \arctan b/a$$

Even within this range there are usually two angles with the same tangent so, when giving the argument of a complex number in the form $\arctan b/a$, it is often necessary to include a diagram. Consider, for example, the complex numbers

$$4 + 3\mathrm{i}, \quad -4 - 3\mathrm{i}, \quad -4 + 3\mathrm{i} \quad \text{and} \quad 4 - 3\mathrm{i}$$

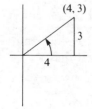
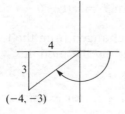
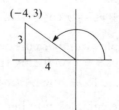
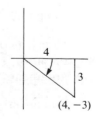

For $\quad 4 + 3$i, $\alpha = \arctan \frac{3}{4}$ and is positive and acute

$\Rightarrow \qquad \alpha = 0.644$ rad

For $-4 - 3$i, $\alpha = \arctan \frac{3}{4}$ and is negative and obtuse

$\Rightarrow \qquad \alpha = -2.498$ rad

For $-4 + 3i$, $\alpha = \arctan -\frac{3}{4}$ and is positive and obtuse

$\Rightarrow$ $\alpha = 2.498$ rad

For $4 - 3i$, $\alpha = \arctan -\frac{3}{4}$ and is negative and acute

$\Rightarrow$ $\alpha = -0.644$ rad

We see that the argument of each of the first two complex numbers is given by $\arctan \frac{3}{4}$ but the values of α are not the same: a similar observation can be made for the last two complex numbers.

This example shows that stating the value of $\arctan \alpha$ is ambiguous and that more information is required to define α uniquely, e.g. a diagram of the type used on page 385.

THE MODULUS/ARGUMENT FORM FOR A COMPLEX NUMBER

Consider a general complex number $x + yi$ represented on an Argand diagram by $\overrightarrow{OP}$ where P is the point (x, y) and OP is inclined at an angle θ to Ox.

From the diagram we see that $x = r \cos \theta$ and $y = r \sin \theta$

Therefore $x + yi = r \cos \theta + ir \sin \theta = r(\cos \theta + i \sin \theta)$

Hence a complex number can be changed from the form $x + yi$ into the form $r(\cos \theta + i \sin \theta)$ by finding the modulus, r, and the argument, θ

To convert the complex number $1 - i$, for example, we have

$r = |1 - i| = \sqrt{(1^2 + \{-1\}^2)} = \sqrt{2}$

$\theta = \arg(1 - i) = \arctan(-1) = -\frac{1}{4}\pi$

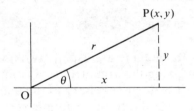

Hence, in modulus/argument form,

$1 - i = \sqrt{2}(\cos\{-\frac{1}{4}\pi\} + i \sin\{-\frac{1}{4}\pi\})$

Conversely, a complex number given in modulus/argument form can be changed directly into Cartesian form,

e.g. $4(\cos \frac{2}{3}\pi + i \sin \frac{2}{3}\pi) = 4(\{-\frac{1}{2}\} + i\{\frac{1}{2}\sqrt{3}\}) = -2 + 2i\sqrt{3}$

Note that the position where i is written varies according to the form of the complex term; the aim is always to make the term clear and unambiguous,

e.g. we write   6i,   i sin θ, $2i\sqrt{3}$, $3i \cos \frac{1}{2}\pi$,   etc.

Examples 21d _____

1. Express in the form $r(\cos \theta + i \sin \theta)$

    (a) $\sqrt{3} - i$ (b) -2 (c) $-5i$ (d) $-2 + 2i$

(a) For $\sqrt{3} - i$

$$r = \sqrt{(\{\sqrt{3}\}^2 + \{-1\}^2)} = 2$$

$$\tan \theta = -1/\sqrt{3} \quad \Rightarrow \quad \theta = -\tfrac{1}{6}\pi$$

$$\therefore \quad \sqrt{3} - i = 2(\cos \{-\tfrac{1}{6}\pi\} + \sin \{-\tfrac{1}{6}\pi\})$$

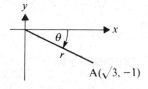

(b) For -2, $r = 2$ and $\theta = \pi$

$$(\theta \neq -\pi \text{ as } -\pi < \theta < \pi)$$

$$\therefore \qquad -2 = 2(\cos \pi + i \sin \pi)$$

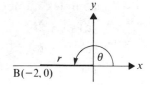

(c) For $-5i$, $r = 5$ and $\theta = -\tfrac{1}{2}\pi$

$$\therefore \qquad -5i = 5(\cos \{-\tfrac{1}{2}\pi\} + i \sin \{-\tfrac{1}{2}\pi\})$$

(d) For $-2 + 2i$

$$r = \sqrt{(\{-2\}^2 + 2^2)} = 2\sqrt{2}$$

$$\tan \theta = 2/-2 = -1 \quad \Rightarrow \quad \theta = \tfrac{3}{4}\pi$$

$$\therefore \quad -2 + 2i = 2\sqrt{2}(\cos \tfrac{3}{4}\pi + i \sin \tfrac{3}{4}\pi)$$

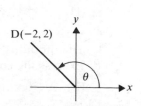

2. Find the modulus and argument of $\dfrac{7-i}{3-4i}$

First we express $\dfrac{7-i}{3-4i}$ in the form $a+bi$ (1, 1)

$$\frac{7-i}{3-4i}=\frac{(7-i)(3+4i)}{(3-4i)(3+4i)}=\frac{25+25i}{25}=1+i$$

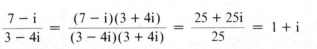

Then $|1+i|=\sqrt{2}$

and $\arg(1+i)$ is the positive acute angle $\arctan 1$, i.e. $\tfrac{1}{4}\pi$

$\therefore$ the modulus and argument of $\dfrac{7-i}{3-4i}$ are $\sqrt{2}$ and $\tfrac{1}{4}\pi$

EXERCISE 21d

1. Represent each of the following complex numbers by a line on an Argand diagram. Find the modulus and argument of each complex number.

(a) $3-2i$ (b) $-4+i$ (c) $-3-4i$ (d) $5+12i$

(e) $1-i$ (f) $-1+i$ (g) 6 (h) $-4i$

(i) $(1+i)^2$ (j) $i(1-i)$ (k) $i^2(1-i)$ (l) $(3+i)(4+i)$

(m) $\cos\tfrac{3}{4}\pi + i\sin\tfrac{3}{4}\pi$ (n) $2(\cos\tfrac{2}{3}\pi + i\sin\tfrac{2}{3}\pi)$

(p) $\cos(-\tfrac{5}{6}\pi) + i\sin(-\tfrac{5}{6}\pi)$ (q) $a+bi$

2. Express in the form $r(\cos\theta + i\sin\theta)$

(a) $1+i$ (b) $\sqrt{3}-i$ (c) $-3-4i$ (d) $-5+12i$

(e) $2-i$ (f) 6 (g) -3 (h) $4i$

(i) $-3-i\sqrt{3}$ (j) $24+7i$

3. The modulus, r, and argument, θ, of a complex number are given. Express the complex number in the form $x+yi$ when r and θ are

(a) $2, \tfrac{1}{6}\pi$ (b) $3, -\tfrac{1}{4}\pi$ (c) $1, \tfrac{2}{3}\pi$ (d) $3, 0$

(e) $4, \pi$ (f) $1, -\tfrac{3}{4}\pi$ (g) $2, \tfrac{1}{2}\pi$ (h) $2, -\tfrac{1}{2}\pi$

PRODUCTS AND QUOTIENTS

The product and quotient of two complex numbers can already be found algebraically when the numbers are given in the form $a + bi$

Now we are going to investigate what happens when we multiply or divide complex numbers in the form $r(\cos \theta + i \sin \theta)$

Taking $z_1 = r_1(\cos \theta_1 + i \sin \theta_1)$ and $z_2 = r_2(\cos \theta_2 + i \sin \theta_2)$ we have

$$
\begin{aligned}
z_1 z_2 &= r_1 r_2 (\cos \theta_1 + i \sin \theta_1)(\cos \theta_2 + i \sin \theta_2) \\
&= r_1 r_2 [\cos \theta_1 \cos \theta_2 + i \sin \theta_1 \cos \theta_2 + i \sin \theta_2 \cos \theta_1 + i^2 \sin \theta_1 \sin \theta_2] \\
&= r_1 r_2 [\cos \theta_1 \cos \theta_2 - \sin \theta_1 \sin \theta_2 + i(\sin \theta_1 \cos \theta_2 + \cos \theta_1 \sin \theta_2)] \\
&= r_1 r_2 [\cos (\theta_1 + \theta_2) + i \sin (\theta_1 + \theta_2)]
\end{aligned}
$$

Hence

$z_1 z_2$ gives a complex number with modulus $r_1 r_2$ and argument $\theta_1 + \theta_2$

i.e. $\qquad |z_1 z_2| = r_1 r_2 \quad \text{and} \quad \arg (z_1 z_2) = \theta_1 + \theta_2$

Similarly it can be shown that $\dfrac{z_1}{z_2} = \dfrac{r_1}{r_2} [\cos (\theta_1 - \theta_2) + i \sin (\theta_1 - \theta_2)]$

i.e. $\qquad \left| \dfrac{z_1}{z_2} \right| = \dfrac{r_1}{r_2} \quad \text{and} \quad \arg \left(\dfrac{z_1}{z_2} \right) = (\theta_1 - \theta_2)$

Note that when the argument of a product or a quotient is found in this way, the angle obtained may not lie between $-\pi$ and π. In such cases the corresponding argument within this range should be stated,

e.g. if $\theta_1 = \frac{2}{3}\pi$ and $\theta_2 = \frac{1}{2}\pi$

then $\qquad \theta_1 + \theta_2 = \frac{7}{6}\pi$

but $\qquad \arg (z_1 z_2) = -\frac{5}{6}\pi$

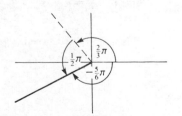

Example 21e

Write down the modulus and argument of $-\sqrt{3}+i$ and of $4+4i$.
Hence express in the form $r(\cos\theta + i\sin\theta)$ the complex numbers

$$(-\sqrt{3}+i)(4+4i) \quad\text{and}\quad \frac{(-\sqrt{3}+i)}{4+4i}$$

For $-\sqrt{3}+i$

$\quad r_1 = 2$ and $\theta_1 = \frac{5}{6}\pi$

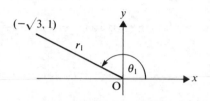

For $4+4i$

$\quad r_2 = 4\sqrt{2}$ and $\theta_2 = \frac{1}{4}\pi$

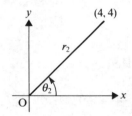

If $(-\sqrt{3}+i)(4+4i) = r_3(\cos\theta_3 + i\sin\theta_3)$

then $r_3 = r_1 r_2 = 8\sqrt{2}$ and $\theta_3 = \theta_1 + \theta_2 = \frac{13}{12}\pi$

As θ_3 is not between $-\pi$ and π
we refer to the diagram to find that
the required argument is $-\frac{11}{12}\pi$

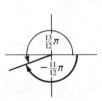

$\therefore \quad (-\sqrt{3}+i)(4+4i) = 8\sqrt{2}\left(\cos\left\{-\frac{11}{12}\pi\right\} + i\sin\left\{-\frac{11}{12}\pi\right\}\right)$

If $\dfrac{(-\sqrt{3}+i)}{(4+4i)} = r_4(\cos\theta_4 + i\sin\theta_4)$

then $r_4 = \dfrac{r_1}{r_2} = \dfrac{\sqrt{4}}{2}$ and $\theta_4 = \theta_1 - \theta_2 = \frac{7}{12}\pi$

this time θ_4 is between $-\pi$ and π so it is the required argument.

$\therefore \qquad \dfrac{(-\sqrt{3}+i)}{(4+4i)} = \dfrac{\sqrt{2}}{4}\left(\cos\frac{7}{12}\pi + i\sin\frac{7}{12}\pi\right)$

EXERCISE 21e

In each Question from 1 to 4

(a) find the modulus and argument of the given complex number, without first expressing it in the form $a + bi$ and illustrate your result on an Argand diagram.

(b) express the given complex number in the form $a + bi$, find the modulus and argument and check the results against the answers to part (a).

1. $(3 - \sqrt{3}i)(1 - i)$

2. $(1 + i)(1 - i)$

3. $\dfrac{(-2 - i\sqrt{3})}{(-2 + i\sqrt{3})}$

4. $\dfrac{2}{(1 + i)}$

5. Illustrate on an Argand diagram lines representing z, $1/z$, z^2 and $z - z^2$ when z is

(a) $2 + i$ (b) $\frac{1}{2} - \frac{1}{2}i$ (c) $3 + 4i$ (d) $\frac{1}{2}\sqrt{3} + \frac{1}{2}i$ (e) $5 - 12i$

MIXED EXERCISE 21

1. Find, in the form $a + bi$,

(a) $\dfrac{1}{7 + 5i}$ (b) $(2 - i)(4 + 3i)$ (c) $\dfrac{2 - i}{3 + i}$ (d) $(2 + 5i)^2$

2. If $z = 3 - i$ express $z + 1/z$ in the form $a + bi$ where a and b are real.

3. Find the square roots of z where $z = 12 + 5i$. Illustrate z and its square roots on an Argand diagram.

4. Find the modulus and argument of each of the following complex numbers.

(a) $-1 + i$ (b) $3 + 4i$ (c) $(-1 + i)(3 + 4i)$

Represent each of these numbers by points on an Argand diagram and label these points A, B and C. Find the area of triangle ABC.

5. If $z_1 = 1 + i$ and $z_2 = 7 - i$, find the modulus of

(a) $z_1 - z_2$ (b) $z_1 z_2$ (c) $\dfrac{z_1}{z_2}$ (d) $\dfrac{z_1 - z_2}{z_1 z_2}$

6. Given that $(2 + 3i)\lambda - 2\mu = 3 + 6i$ find the values of λ and μ.

7. Two complex numbers z_1 and z_2 are such that $z_1 + z_2 = 1$.
   If $z_1 = \dfrac{a}{1+i}$ and $z_2 = \dfrac{b}{1+2i}$ find a and b.

8. If $z = x + yi$ find the real and imaginary parts of $z - 1/z$.

9. One root of the equation $x^2 - \lambda x - \mu = 0$ is $2 - i$. Find λ and μ.

10. Find the modulus and argument of each root of the equation $y^2 + 4y + 8 = 0$

11. If α and β are the roots of the equation $3x^2 + x + 2 = 0$, find the value of

    (a) $\alpha + \beta$ (b) $\alpha\beta$ (c) $3\alpha^2 + \alpha + 2$

12. Given that z^* is the conjugate of z and $z = a + bi$ where a and b are real, find the possible values of z if $zz^* - 2iz = 7 - 4i$

CONSOLIDATION E

SUMMARY

VECTORS

A vector is a quantity with both magnitude and direction and can be represented by a line segment.

If lines representing several vectors are drawn 'head to tail' in order, then the line which completes a closed polygon represents the sum of the vectors (or the resultant vector).

A position vector has a fixed location in space.

The position vector of a point C which divides AB in the ratio $\lambda : \mu$ is given by $\dfrac{\lambda b + \mu a}{\lambda + \mu}$ where $\overrightarrow{OA} = \mathbf{a}$ and $\overrightarrow{OB} = \mathbf{b}$

Cartesian Unit Vectors

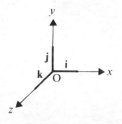

$\mathbf{i}$, $\mathbf{j}$ and $\mathbf{k}$ are unit vectors in the directions of Ox, Oy and Oz respectively,

Any vector can be given in the form $a\mathbf{i} + b\mathbf{j} + c\mathbf{k}$

$$(a_1\mathbf{i} + b_1\mathbf{j} + c_1\mathbf{k}) \pm (a_2\mathbf{i} + b_2\mathbf{j} + c_2\mathbf{k}) = (a_1 \pm a_2)\mathbf{i} + (b_1 \pm b_2)\mathbf{j} + (c_1 \pm c_2)\mathbf{k}$$

$$|a\mathbf{i} + b\mathbf{j} + c\mathbf{k}| = \sqrt{(a^2 + b^2 + c^2)}$$

For two vectors $\mathbf{v}_1 = a_1\mathbf{i} + b_1\mathbf{j} + c_1\mathbf{k}$ and $\mathbf{v}_2 = a_2\mathbf{i} + b_2\mathbf{j} + c_2\mathbf{k}$
    $\mathbf{v}_1$ and $\mathbf{v}_2$ are parallel if $\mathbf{v}_1 = \lambda\mathbf{v}_2$,
        i.e. $a_1 = \lambda a_2,\ b_1 = \lambda b_2,\ c_1 = \lambda c_2$
    $\mathbf{v}_1$ and $\mathbf{v}_2$ are equal if $a_1 = a_2,\ b_1 = b_2,\ c_1 = c_2$

The Vector Equation of a Line

For a line in the direction of the vector $\mathbf{d} = a\mathbf{i} + b\mathbf{j} + c\mathbf{k}$ and passing through a point with position vector $\mathbf{a} = x_1\mathbf{i} + y_1\mathbf{j} + z_1\mathbf{k}$
 a vector equation is $\mathbf{r} = \mathbf{a} + \lambda\mathbf{d}$

Two lines with equations $\mathbf{r}_1 = \mathbf{a}_1 + \lambda\mathbf{d}_1$ and $\mathbf{r}_2 = \mathbf{a}_2 + \mu\mathbf{d}_2$
 are parallel if $\mathbf{d}_1$ is a multiple of $\mathbf{d}_2$
 intersect if there are values of λ and μ for which $\mathbf{r}_1 = \mathbf{r}_2$
 are skew in all other cases.

The Scalar Product of Two Vectors

If θ is the angle between two vectors $\mathbf{a}$ and $\mathbf{b}$ then

$$\mathbf{a} \cdot \mathbf{b} = |\mathbf{a}||\mathbf{b}|\cos\theta$$

$\mathbf{a}$ and $\mathbf{b}$ are perpendicular $\Rightarrow \mathbf{a} \cdot \mathbf{b} = 0$

If $\mathbf{a} = x_1\mathbf{i} + y_1\mathbf{j} + z_1\mathbf{k}$ and $\mathbf{b} = x_2\mathbf{i} + y_2\mathbf{j} + z_2\mathbf{k}$ then
$$\mathbf{a} \cdot \mathbf{b} = x_1x_2 + y_1y_2 + z_1z_2$$

COMPLEX NUMBERS

A complex number is of the form $a + b\mathrm{i}$ where a and b are real and

$$\mathrm{i} = \sqrt{(-1)} \quad \text{and} \quad \sqrt{(-n^2)} = n\mathrm{i}$$

$(a + b\mathrm{i}) \pm (c + d\mathrm{i}) \equiv (a \pm c) + (b \pm d)\mathrm{i}$

$(a + b\mathrm{i})(c + d\mathrm{i}) \equiv (ac - bd) + (ad + bc)\mathrm{i}$

$\dfrac{(a + b\mathrm{i})}{(c + d\mathrm{i})} \equiv \dfrac{(a + b\mathrm{i})(c - d\mathrm{i})}{(c + d\mathrm{i})(c - d\mathrm{i})} \equiv \dfrac{(a + b\mathrm{i})(c - d\mathrm{i})}{c^2 + d^2}$

$x + y\mathrm{i} = a + b\mathrm{i} \quad \Rightarrow \quad x = a$ and $y = b$

$x + y\mathrm{i} = 0 \quad \Rightarrow \quad x = 0$ and $y = 0$

$|a + b\mathrm{i}| = \sqrt{(a^2 + b^2)}$

$\arg(a + b\mathrm{i}) = \alpha$ where $\tan\alpha = \dfrac{b}{a}$ and $-\pi < \alpha \leqslant \pi$

$x + y\mathrm{i} = r(\cos\theta + \mathrm{i}\sin\theta)$ where $r = |x + y\mathrm{i}|$ and $\theta = \arg(x + y\mathrm{i})$

If $z = a + b\mathrm{i}$, $\bar{z} = a - b\mathrm{i}$, where $\bar{z}$ and z are conjugate.

If a quadratic or cubic equation has any complex roots, they occur in conjugate pairs.

If $z_1 = r_1(\cos\theta_1 + i\sin\theta_1)$ and $z_2 = r_2(\cos\theta_2 + i\sin\theta_2)$

then $\quad |z_1 z_2| = r_1 r_2$ and $\arg z_1 z_2 = \theta_1 + \theta_2$

$$\left|\frac{z_1}{z_2}\right| = \frac{r_1}{r_2} \quad \text{and} \quad \arg\frac{z_1}{z_2} = \theta_1 - \theta_2$$

APPLICATIONS OF CALCULUS

Small Increments

If $y = f(x)$ and x increases by a small amount, δx, then

$$\delta y \approx \left(\frac{dy}{dx}\right)(\delta x)$$

Comparative Rates of Change

If a quantity p depends on a quantity q and the rate at which q increases with time t is known, then

$$\frac{dp}{dt} = \frac{dp}{dq} \times \frac{dq}{dt}$$

FINDING AREA AND VOLUME

Integration as a Process of Summation

$$\lim_{\delta x \to 0} \sum_{x=a}^{x=b} f(x)\,\delta x = \int_a^b f(x)\,dx$$

The area bounded by the graphs of $y = f(x)$, $y = g(x)$ and the lines, $x = a$ and $x = b$, can be found by summing the areas of vertical strips of width δx and taking the limiting value of this sum. This is equal to the definite integral between $x = a$ and $x = b$.

The area shown is given by

$$\text{Area} = \lim_{\delta x \to 0} \sum_{x=a}^{x=b} (y_1 - y_2)\,\delta x = \int_a^b (y_1 - y_2)\,dx$$

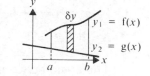

Similarly if the boundaries are $x = f(y)$, $x = g(y)$ and the lines, $y = a$ and $y = b$, the area is given by using horizontal strips, i.e.

$$\text{Area} = \lim_{\delta y \to 0} \sum_{y=a}^{y=b} (x_2 - x_1)\,\delta y = \int_a^b (x_1 - x_2)\,dy$$

The Trapezium Rule

An approximate value for the area shown in the diagram on the previous page can be found by taking strips of equal width, d, and using

$$\int_a^b f(x)\, dx \approx \tfrac{1}{2}d[y_0 + 2y_1 + \ldots + 2y_{n-2} + y_{n-1}]$$

Volume of Revolution

When the area bounded by the x-axis, the ordinates at a and b, and part of the curve $y = f(x)$ rotates completely about the x-axis, the volume generated is given by

$$\text{Volume} = \lim_{\delta x \to 0} \sum_{x=a}^{x=b} \pi y^2\, \delta x = \int_a^b \pi y^2\, dx$$

For rotation about the y-axis, $\quad \text{Volume} = \lim_{\delta y \to 0} \sum_{y=a}^{y=b} \pi x^2\, \delta y = \int_a^b \pi x^2\, dy$

MULTIPLE CHOICE EXERCISE E

TYPE I

1. The modulus of the vector $6\mathbf{i} - 2\mathbf{j} - 3\mathbf{k}$ is

   A $\sqrt{23}$ B 7   C 1   D 49   E $\sqrt{11}$

2. $\overrightarrow{OP} = 2\mathbf{i} - 6\mathbf{j}$ and $\overrightarrow{OQ} = 4\mathbf{i} - 3\mathbf{k}$. The vector of magnitude 7 in the direction of $\overrightarrow{PQ}$ is

   A $2\mathbf{i} + 6\mathbf{j} - 3\mathbf{k}$ D $7(2\mathbf{i} + 6\mathbf{j} - 3\mathbf{k})$
   B $7(-2\mathbf{i} - 6\mathbf{j} + 3\mathbf{k})$ E $7(6\mathbf{i} - 6\mathbf{j} - 3\mathbf{k})$
   C $-2\mathbf{i} - 6\mathbf{j} + 3\mathbf{k}$

3. If $\mathbf{a} = \mathbf{i} - \mathbf{j} + \mathbf{k}$ and $\mathbf{b} = \mathbf{i} + \mathbf{j} - \mathbf{k}$ then $\mathbf{a}.\mathbf{b}$ is

   A $2\mathbf{i}$ B -1 C $-2\mathbf{j} + 2\mathbf{k}$ D 2   E 3

4. The angle between the lines whose equations are $\mathbf{r}_1 = \mathbf{a}_1 + \lambda\mathbf{b}_1$ and $\mathbf{r}_2 = \mathbf{a}_2 + \mu\mathbf{b}_2$ is

   A $\arccos \dfrac{\mathbf{b}_1.\mathbf{b}_2}{b_1 b_2}$ C $\arccos \dfrac{\mathbf{a}_1.\mathbf{a}_2}{a_1 a_2}$ E $\arccos \lambda.\mu$

   B $\mathbf{b}_1.\mathbf{b}_2$ D $\mathbf{r}_1.\mathbf{r}_2$

5. The modulus of $12 - 5i$ is

   A 119   **B** 7   **C** 13   **D** $\sqrt{119}$ **E** $\sqrt{7}$

6. On an Argand diagram OP represents a complex number z. The conjugate of z is $\bar{z}$

   If P and Q are the points $(3, 5)$ and $(5, -3)$ then OQ represents

   A $-\bar{z}$ **B** $i\bar{z}$ **C** $-z$ **D** iz **E** $-iz$

7. $\dfrac{3 + 2i}{3 - 2i}$ is equal to

   A $\dfrac{5 + 12i}{13}$ **C** $\dfrac{5 + 6i}{13}$ **E** $\dfrac{13 + 12i}{5}$

   B $\dfrac{13 + 12i}{13}$ **D** $\dfrac{5 + 6i}{5}$

8. When $\sqrt{3} - i$ is divided by $-1 - i$ the modulus and argument of the quotient are respectively

   A $2\sqrt{2}, \frac{7}{12}\pi$ **C** $\sqrt{2}, \frac{7}{12}\pi$ **E** $\sqrt{2}, \frac{11}{12}\pi$

   B $\sqrt{2}, -\frac{11}{12}\pi$ **D** $2\sqrt{2}, -\frac{11}{12}\pi$

9. The equation $x^2 + 3x + 1 = 0$ has

   A no roots   **D** two real roots
   B one real and one complex root   **E** two complex roots
   C two imaginary roots

10. $\arg\left(\dfrac{1 - i}{1 + i}\right) =$

   A $\frac{1}{2}\pi$ **B** 0   **C** $-\frac{1}{4}\pi$ **D** $-\frac{1}{2}\pi$ **E** π

11.

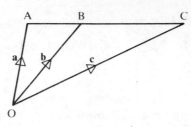

Given that $\overrightarrow{BC} = 2\overrightarrow{AB}$ then

A $a - 3b + 2c = 0$ D $2a + 3b - c = 0$
B $a + 3b - 2c = 0$ E $2a - 3b + c = 0$
C $a + b - 2c = 0$

12.

x	1	2	3
y	0	3	6

Given the information in the table, using the trapezium rule with 3 ordinates to find $\int_1^3 y \, dx$ gives

A   12   B   9/2   C   9   D   6   E   8

13. Given that $z_1 = 5 - 2i$ and $z_2 = 2 - i$, then if $\theta = \arg\left(\dfrac{z_1}{z_2}\right)$

A $\tan \theta = \frac{3}{4}$ C $\tan \theta = -\frac{3}{4}$ E $\tan \theta = -\frac{1}{12}$
B $\tan \theta = \frac{1}{12}$ D $\tan \theta = \frac{1}{8}$

14.

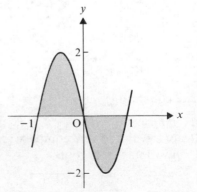

The shaded area is equal to

A $\displaystyle\int_{-1}^{1} y \, dx$ C $\displaystyle 2\int_{0}^{2} x \, dy$ E $\displaystyle\int_{-1}^{0} y \, dx - \int_{0}^{1} y \, dx$

B   zero   D $\displaystyle\int_{-1}^{0} y \, dx + \int_{0}^{1} y \, dx$

TYPE II

15. A line has equation $r = i + 2j + 3k + \lambda(4i - j + 7k)$.

    A The length of the line is $\sqrt{14}$

    B The line passes through the point $(1, 2, 3)$

    C The line is parallel to the line $r = \mu(i + 2j + 3k)$.

16. $-\frac{3}{4}\pi$ is an argument of

    A $1 - i$ **B** $\cos\frac{3}{4}\pi - i \sin\frac{3}{4}\pi$ **C** $-1 - i$

17.

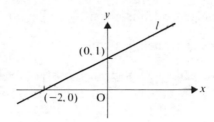

    The line l

    A is perpendicular to the vector $-i + 2j$

    B passes through the point with position vector $2i + j$

    C is parallel to the vector $i + 2j$

18. Given that $z_1 = 2 + 3i$ and $z_2 = 3 - 2i$,

    A $|z_2| = \sqrt{5}$

    B $|z_1| = \sqrt{13}$

    C z_1 and z_2 are conjugate complex numbers.

19.

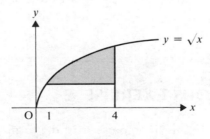

    The area shaded in the diagram is given by

    A $\displaystyle\int_1^4 y \, dx$ **B** $\displaystyle\int_1^4 (y - 1) \, dx$ **C** $\displaystyle\int_1^2 y \, dx$

20.

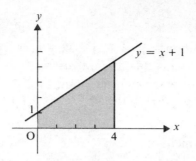

The shaded area is rotated through $360°$ about the x-axis.

A   The volume generated is given by $\pi \displaystyle\int_0^4 (x + 1)\,dx$

B   The volume generated is a cone.

C   The volume generated is given by $\pi \displaystyle\int_0^4 (x + 1)^2\,dx$.

TYPE III

21. $\mathbf{a.b} = 0 \quad \Rightarrow \quad \mathbf{a} = 0 \text{ or } \mathbf{b} = 0$

22.   Any complex number whose modulus is unity can be expressed as $\cos\theta + i\sin\theta$

23.   A complex number $a + bi$ is zero if $a = -b$

24.   Given that $y = \ln x^2$ and x increases by δx, then $\delta y \approx \left(\dfrac{1}{x^2}\right)(\delta x)$.

MISCELLANEOUS EXERCISE E

1.   Given that $z = 2 + 2i$, express z in the form $r(\cos\theta + i\sin\theta)$, where r is a positive real number and $-\pi < \theta \leqslant \pi$

On the same Argand diagram, display and label clearly the numbers
$$z, \ z^2 \ \text{and} \ 4/z$$

Find the values of $|z + z^2|$ and $\arg(z + 4/z)$.   (AEB)

2. (a) Find all possible values of the real numbers a and b which satisfy

$$2 + a\mathrm{i} = \frac{6 - 2\mathrm{i}}{b + \mathrm{i}}$$

(b) Given that $w = -\frac{1}{2} + \frac{1}{2}\mathrm{i}$, find the modulus and the argument of $\dfrac{1}{1 + w}$, giving the argument in radians between $-\pi$ and π

(AEB)

3. Given that $z = 1 + \mathrm{i}$, show that $z^3 = -2 + 2\mathrm{i}$. For this value of z, the real numbers p and q are such that

$$\frac{p}{1 + z} + \frac{q}{1 + z^3} = 2\mathrm{i}$$

Find the values of p and q. (JMB)

4. Given that $z_1 = -1 + \mathrm{i}\sqrt{3}$ and $z_2 = \sqrt{3} + \mathrm{i}$, find $\arg z_1$ and $\arg z_2$

Express z_1/z_2 in the form $a + \mathrm{i}b$, where a and b are real, and hence find $\arg(z_1/z_2)$

Verify that $\arg(z_1/z_2) = \arg z_1 - \arg z_2$ (U of L)

5. Find the values of the real numbers a and b so that

$$(a + b\mathrm{i})^2 = 16 - 30\mathrm{i}$$

Write down the two square roots of $16 - 30\mathrm{i}$ (AEB)

6. Find, in terms of π, the argument of the complex number

(a) $(1 + \mathrm{i})^2$

(b) $\dfrac{3 + \mathrm{i}}{1 + 2\mathrm{i}}$

(c) $(\cos \frac{1}{3}\pi + \mathrm{i} \sin \frac{1}{3}\pi)(\cos \frac{1}{4}\pi - \mathrm{i} \sin \frac{1}{4}\pi)$ (AEB)

7. Given that the real and imaginary parts of the complex number $z = x + \mathrm{i}y$ satisfy the equation

$$(2 - \mathrm{i})x - (1 + 3\mathrm{i})y - 7 = 0$$

find x and y

State the values of

(a) $|z|$ (b) $\arg z$ (U of L)

8. Given that p and q are real and that $1 + 2i$ is a root of the equation
$$z^2 + (p + 5i)z + q(2 - i) = 0$$
   determine   (a) the values of p and q
              (b) the other root of the equation.   (AEB)

9. Given that $z_1 = 1 + i\sqrt{3}$ and $z_2 = 4 + 3i$, calculate the moduli and arguments of $z_1 z_2$ and z_2/z_1, giving the arguments in degrees to one decimal place.

   Illustrate in one Argand diagram
   (a) $z_1 - z_2$ (b) $z_1 z_2$ (c) z_2/z_1 (U of L)

10. The points A and B have coordinates $(2, 3, -1)$ and $(5, -2, 2)$ respectively. Calculate the acute angle between AB and the line with equation
$$r = \begin{pmatrix} 2 \\ 3 \\ -1 \end{pmatrix} + t \begin{pmatrix} 1 \\ -2 \\ -2 \end{pmatrix}$$
    giving your answer correct to the nearest degree.   (C)
    $[r = 2i + 3j - k + t(i - 2j - 2k)]$

11. Referred to a fixed origin O, the lines l_1 and l_2 have equations
$$l_1 : r = (i + j + k) + \lambda(i - 7j + 2k)$$
$$l_2 : r = (4i - 4j + 9k) + \mu(i + j + 3k)$$
    where λ and μ are scalar parameters.

    Show that the lines l_1 and l_2 are perpendicular and that they intersect at the point P whose position vector is $2i - 6j + 3k$

    Determine, to the nearest half-degree, the acute angle between $\overrightarrow{OP}$ and l_2 (AEB)

12. The position vectors of the points A and B are given by
$$\overrightarrow{OA} = 3i - 2j + 2k \quad \overrightarrow{OB} = -i + j + 3k$$
    where O is the origin. Find a vector equation of the straight line passing through A and B. Given that this line is perpendicular to the vector $i + 2j + pk$, find the value of p.   (C)

13. With respect to a fixed origin O, the points L and M have position vectors $6\mathbf{i} + 3\mathbf{j} + 2\mathbf{k}$ and $2\mathbf{i} + 2\mathbf{j} + \mathbf{k}$ respectively.

    (a) Form the scalar product $\overrightarrow{OL}.\overrightarrow{OM}$ and hence find the cosine of angle LOM.

    (b) The point N is on the line LM produced such that angle MON is $90°$. Find an equation for the line LM in the form $\mathbf{r} = \mathbf{a} + \mathbf{b}t$ and hence calculate the position vector of N.

                                                             (AEB)

14. The points A and B have position vectors $4\mathbf{i} + \mathbf{j} - 7\mathbf{k}$ and $2\mathbf{i} + 6\mathbf{j} + 2\mathbf{k}$ respectively relative to the origin O. Show that the angle AOB is a right angle.

Find a vector equation for the median AM of the triangle OAB.

                                                            (U of L)p

15. Relative to an origin O, the points P and Q have position vectors
$$\mathbf{p} = 4\mathbf{i} - 3\mathbf{j} + 4\mathbf{k}, \quad \mathbf{q} = 6\mathbf{i} + \mathbf{j} - 2\mathbf{k} \quad \text{respectively.}$$

    (a) Prove that triangle OPQ is isosceles.

    (b) Evaluate $(\mathbf{q} - \mathbf{p}).\mathbf{q}$ and deduce the value of angle PQO, to the nearest $0.1°$.   (AEB)p

16. In $\triangle OPQ$, $\overrightarrow{OP} = 2\mathbf{p}$, $\overrightarrow{OQ} = 2\mathbf{q}$ and M and N are the midpoints of OP and OQ respectively.

    (a) By expressing the vectors $\overrightarrow{PQ}$ and $\overrightarrow{MN}$ in terms of $\mathbf{p}$ and $\mathbf{q}$, prove that
$$\overrightarrow{MN} = \tfrac{1}{2}\overrightarrow{PQ}$$

The lines PN and QM intersect at the point G.

    (b) Express, in terms of $\mathbf{p}$ and $\mathbf{q}$, the vectors $\overrightarrow{PN}$ and $\overrightarrow{QM}$,

    (c) Given that $\overrightarrow{GN} = \lambda\overrightarrow{PN}$ prove that
$$\overrightarrow{OG} = 2\lambda\mathbf{p} + (1 - \lambda)\mathbf{q}$$

    (d) Given that $\overrightarrow{GM} = \mu\overrightarrow{QM}$ find $\overrightarrow{OG}$ in terms of μ, $\mathbf{p}$ and $\mathbf{q}$.

    (e) Hence prove that
$$\overrightarrow{OG} = \tfrac{2}{3}(\mathbf{p} + \mathbf{q}) \qquad\qquad\qquad\text{(U of L)}$$

17. Referred to a fixed origin O, the points A and B have position vectors $a(5\mathbf{i} - \mathbf{j} - \mathbf{k})$ and $a(\mathbf{i} - 5\mathbf{j} + 7\mathbf{k})$ respectively, where a is a positive constant.

   (a) Find an equation of the line AB.

   (b) Show that the point C with position vector $a(4\mathbf{i} - 2\mathbf{j} + \mathbf{k})$ lies on AB.

   (c) Show that OC is perpendicular to AB.

   (d) Find the position vector of the point D, where $D \neq A$, on AB such that $|\overrightarrow{OD}| = |\overrightarrow{OA}|$.   (U of L)

18. O is the origin, and points A and B have position vectors relative to O given by

$$\overrightarrow{OA} = \mathbf{i} + 7\mathbf{j} \quad \text{and} \quad \overrightarrow{OB} = 5\mathbf{i} + 5\mathbf{j}$$

   Show that the vectors $\overrightarrow{OA}$ and $\overrightarrow{OB}$ have equal magnitudes.

   Points C and D have position vectors given by

$$\overrightarrow{OC} = 2\overrightarrow{OA} \quad \text{and} \quad \overrightarrow{OD} = \overrightarrow{OA} + \overrightarrow{OB}$$

   Express $\overrightarrow{OC}$ and $\overrightarrow{OD}$ in terms of $\mathbf{i}$ and $\mathbf{j}$, and draw a diagram showing the positions of A, B, C and D.

   Use a scalar product to calculate the angle between the vectors $\overrightarrow{OD}$ and $\overrightarrow{BC}$.   (O/C, SU & C)

19. The part of the curve $y = x^2 + 3$ between the lines $x = 0$ and $x = 1$ is rotated through $360°$ about the x-axis. Find the volume of the solid generated.   (O/C, SU & C)

20. The finite region bounded by the x-axis, the curve $y = e^{-x}$ and the lines $x = \pm a$ is denoted by R. Find, in terms of a, the volume of the solid generated when R is rotated through one revolution about the x-axis.   (C)p

21. The vertices A, B, C of a triangle have position vectors $\mathbf{a}$, $\mathbf{b}$, $\mathbf{c}$ respectively. The points D, E lie on BC, AC respectively and $AE:EC = CD:DB = 2:1$. The lines AD and BE intersect at the point F.

   Show that the position vector of F is

$$\tfrac{1}{7}\mathbf{a} + \tfrac{4}{7}\mathbf{b} + \tfrac{2}{7}\mathbf{c}$$ (WJEC)

22. Find the maximum and minimum points on the curve
$y = \cos x + \sqrt{3} \sin x$ in the interval $0 \leqslant x \leqslant 2\pi$ radians, carefully
distinguishing between them. Show that the curve passes through the
point $P(\frac{5}{6}\pi, 0)$ and find the other point Q in the interval $0 \leqslant x \leqslant 2\pi$
radians, where the curve crosses the x-axis. Sketch the curve in this
interval.

Calculate the volume generated when that part of the curve between
the origin and P is rotated through 2π radians about the x-axis.

(WJEC)

23. The points P, Q have position vectors $\mathbf{p}$, $\mathbf{q}$ relative to a fixed origin O.
Show that the point R which divides the line joining PQ in the ratio
$\lambda : 1 - \lambda$ has position vector

$$\mathbf{r} = \lambda \mathbf{q} + (1 - \lambda)\mathbf{p}$$

Show on a diagram the relative positions of P, Q, R when $\lambda = 2$

The vertices A, B, C of a triangle have position vectors $\mathbf{a}$, $\mathbf{b}$, $\mathbf{c}$
respectively. The points D, E lie on AB, AC respectively and are
such that $AD = 2DB$ and $2AE = EC$. Show that the point of
intersection G of BE and CD has position vector

$$\tfrac{2}{7}\mathbf{a} + \tfrac{4}{7}\mathbf{b} + \tfrac{1}{7}\mathbf{c}$$

The line AG meets BC at F. Find the ratios

(a) $BF : FC$ (b) $AG : GF$ (WJEC)

24. The region bounded by the curve $y = \dfrac{1}{2x + 1}$, the x-axis, the
y-axis and the line $x = 1$ is rotated through $360°$ about the x-axis.
Calculate the volume of the solid thus formed.   (O/C, SU & C)

25. The radius of a sphere is increasing at a rate of $4\,\text{cm/s}$. Obtain, as a
multiple of π, the rate of increase of the volume of the sphere when
the radius is $10\,\text{cm}$.   (C)

26. A straight metal bar, of square cross-section, is expanding due to
heating. After t seconds the bar has dimensions x cm by x cm by
$10x$ cm. Given that the area of the cross-section is increasing at
$0.024\,\text{cm}^2\,\text{s}^{-1}$ when $x = 6$, find the rate of increase of the side of
the cross-section at this instant. Find also the rate of increase of the
volume when $x = 6$ (JMB)

27.

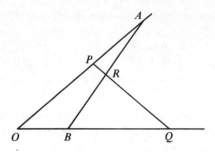

In the figure, $\overrightarrow{OA} = \mathbf{a}$, $\overrightarrow{OB} = \mathbf{b}$, $OP:PA = 3:2$ and $OQ:QB = 3:2$

(a) Write down $\overrightarrow{AB}$, $\overrightarrow{OP}$, $\overrightarrow{OQ}$ and $\overrightarrow{PQ}$, giving each answer in terms of $\mathbf{a}$, $\mathbf{b}$ or $\mathbf{a}$ and $\mathbf{b}$.

(b) Given that $AR = hAB$, where h is a scalar, show that
$$\overrightarrow{OR} = (1 - h)\mathbf{a} + h\mathbf{b}$$

(c) Given that $PR = kPQ$, where k is a scalar, write down an expression for $\overrightarrow{OR}$ in terms of $\mathbf{a}$, $\mathbf{b}$ and k.

(d) Using the two expressions for $\overrightarrow{OR}$ obtained in (b) and (c), calculate the values of h and k and hence write down the value of the ratio $PR:RQ$. (U of L)

28. The region bounded by the curve $y = (1 + \cos x)^{-1/2}$, the x-axis, and the lines $x = 0$ and $x = \frac{1}{2}\pi$, is denoted by R. Use the trapezium rule with ordinates at $x = 0$, $x = \frac{1}{4}\pi$ and $x = \frac{1}{2}\pi$ to estimate the area of R, giving two significant figures in your answer.
(C)

29. Use the trapezium rule with 5 ordinates (at 0, 0.5, 1, 1.5, 2) to estimate the value of the integral
$$\int_0^2 \ln(1 + x^3)\, dx$$
giving your answer to 3 significant figures. (O/C, SU & C)

30. The complex number ω has modulus 3 and argument $\frac{1}{4}\pi$.
Find the modulus and argument of

(a) ω^2 (b) $i\omega$

Mark in an Argand diagram the points P, Q and R representing the complex numbers ω, ω^2 and $i\omega$, respectively.

Find the area of quadrilateral $OPQR$, where O is the origin. (JMB)

31. The complex number $u = -10 + 9i$.

   (a) Show the complex number u on an Argand diagram.

   (b) Giving your answer to the nearest degree, calculate the argument of u.

   (c) Find the complex number v which satisfies the equation

$$uv = -11 + 28i$$

   (d) Verify that $|u + v| = 8\sqrt{2}$ (AEB)

32. The complex number z is such that

$$z - 2z^* = 3 + 6i$$

   where z^* denotes the complex conjugate of z. Find z in the form $a + ib$, where a and b are real.   (JMB)

33. Referred to an origin O, the points A, B and C have position vectors $-2i + 3j$, $i + 2j$ and $7i$ respectively.

   (a) Show that the points A, B and C lie on a straight line.

   (b) Find the position vector of the point D which lies on BC where $BD:DC = 2:1$

   (c) Find the position vector of the point E which is such that $OCEA$ is a parallelogram.   (U of L)p

34. Find the area of the region bounded by the curve with equation $y = (\sin 2x) + 1$ and the lines $x = 0$, $x = \frac{1}{3}\pi$ and $y = 0$. Leave your answer in terms of π.   (U of L)

35. Referred to a fixed origin O, the points A, B and C have position vectors

$$i - 2j + 2k, \quad 3i - k \quad \text{and} \quad -i + j + 4k$$

   respectively.

   Calculate the cosine of the angle BAC.

   Hence, or otherwise, find the area of the triangle ABC, giving your answer to three significant figures.   (AEB)

36. With respect to a fixed origin O, the points A and B have position vectors $(2\mathbf{i} + 3\mathbf{j} + 6\mathbf{k})$ and $(2\mathbf{i} + 4\mathbf{j} + 4\mathbf{k})$ respectively.

   (a) Calculate $|\overrightarrow{OA}|$, $|\overrightarrow{OB}|$ and, by using the scalar product $\overrightarrow{OA}.\overrightarrow{OB}$, calculate the value of the cosine of angle AOB.

   (b) The point C has position vectors $5\mathbf{i} + 12\mathbf{j} + 6\mathbf{k}$.
   Show that OC and AB are perpendicular. Show also that the line through O and C intersects the line through A and B and find the position vector of the point E where they intersect.

   (c) Given that $\overrightarrow{AE} = \lambda\overrightarrow{EB}$, find the value of λ and explain briefly why λ is negative.   (AEB)

37. A stream of width 25 m is measured for depth at intervals of 5.00 m across its width. The results are given in the following table.

Distance from one bank (m)	0	5.00	10.00	15.00	20.00	25.00
Depth (m)	1.12	2.38	3.98	3.64	2.54	0

By using the trapezium rule, estimate the cross-sectional area of the flow of water at this point of the stream.   (U of L)

38. The pressure p units and the volume v units of a gas are related by the law

$$pv^{1.4} = k$$

where k is a constant. Find an expression for $\dfrac{dp}{dv}$ in terms of p and v only.

Given that v undergoes a small increase of $c\%$, find by means of calculus an approximation for the percentage decrease in p.   (JMB)

39. The roots of the equation

$$z^2 + 4z + 29 = 0$$

are z_1 and z_2. Show that $|z_1| = |z_2|$ and calculate, in degrees, the argument of z_1 and the argument of z_2.

In an Argand diagram, O is the origin and z_1 and z_2 are represented by the points P and Q. Calculate the radius of the circle passing through the points O, P and Q.   (AEB)

40. Shade on a sketch the finite region R in the first quadrant bounded by the x-axis, the curve $y = \ln x$ and the line $x = 5$
By means of integration, calculate the area of R.

The region R is rotated completely about the x-axis to form a solid of revolution S.

x	1	2	3	4	5
$(\ln x)^2$	0	0.480	1.207	1.922	2.590

Use the given table of values and apply the trapezium rule to find an estimate of the volume of S, giving your answer to one decimal place.

(AEB)

41. Use the trapezium rule with 5 ordinates (at $x = 0, 0.25, 0.5, 0.75, 1$) to estimate the value of

$$\int_0^1 \frac{1}{1+x^2}\, dx$$

giving your answer to 2 significant figures.   (O/C, SU & C)

42. At time t seconds, where $t > 0$, the radius of an expanding spherical star is r km. Find the rate, in $km^3 s^{-1}$, at which the volume of the star is increasing when

$$r = 7 \times 10^5 \quad \text{and} \quad \frac{dr}{dt} = 20 \qquad \text{(U of L)}$$

43. Assuming the standard formulae for the volume $V(m^3)$ and surface area $S(m^2)$ of a sphere, show that $V^2 = kS^3$ where k is a constant. Give the value of this constant.

By taking logarithms and differentiating, show that

$$\frac{dV}{dS} = \frac{3V}{2S}$$

Deduce the approximate percentage increase in V if S increases by 1%.

(WJEC)

44. The equations of two lines, l_1 and l_2, are respectively
$$r_1 = 2\mathbf{i} + 3\mathbf{j} + 2\mathbf{k} + \lambda(\mathbf{i} - 2\mathbf{j} + s\mathbf{k}) \text{ and}$$
$$r_2 = 4\mathbf{i} - 7\mathbf{j} + 9\mathbf{k} + \mu(\mathbf{i} + 4\mathbf{j} - \mathbf{k})$$

(a) Show that l_1 and l_2 intersect and find the position vector of the point of intersection.

The coordinates of the points P and Q are $(4, -1, 6)$ and $(12, 2, 5)$ respectively,

(b) Verify that P lies on l_1

(c) Show that PQ is perpendicular to l_1

(d) If R is the reflection of Q in l_1, find the coordinates of R.

ANSWERS

Answers to questions taken from past examination papers are the sole responsibility of the authors and have not been approved by the Examining Boards.

CHAPTER 1

Exercise 1a – p. 6

1. (a)

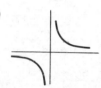

(b)

(c)

(d)

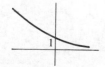

2. $16, 8, 4, 2, 1, \frac{1}{2}, \frac{1}{4}, \frac{1}{8}, \frac{1}{16}$
   as $x \to \infty$, $f(x) \to 0$ and
   as $x \to -\infty$, $f(x) \to \infty$

3. -2, as $x \to -2$ from below, $f(x) \to -\infty$
   as $x \to -2$ from above, $f(x) \to \infty$

4. (a)

(b)

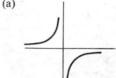

(c)

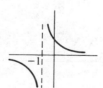

(d)

5. (a)

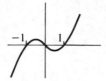

(b)

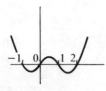

(c)

(d)

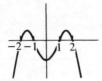

6. (a) 2.7   (b) 2.43   (c) 2.187
$y = 3(0.9)^x$

Exercise 1b – p. 9

1. (a) continuous, periodic
   (b) periodic, continuous for
      $x \neq \frac{1}{2}(2n+1)\pi$
   (c) continuous
   (d) continuous
   (e) periodic, continuous for
      $x \neq \frac{1}{2}(2n+1)\pi$
   (f) continuous
   (g) continuous
   (h) continuous for $x \neq 0$

2.

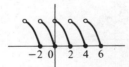

3.

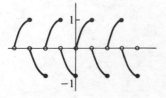

4.

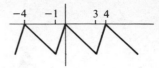

Exercise 1c – p. 11

1.

2.

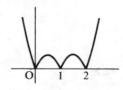

3.

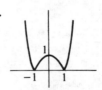

4.

5.

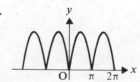

6.

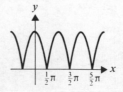

7.

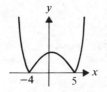

8.

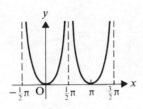

9.

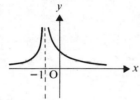

10.

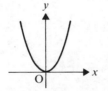

Exercise 1d – p. 13
1. $(-1, 1)$, $(0, 0)$, $(1, 1)$
2. $(2, 5)$, $[\frac{1}{2}(3 + \sqrt{41}), \frac{1}{2}(7 + 3\sqrt{41})]$
3. $[\frac{1}{2}(\sqrt{5} - 1), \frac{1}{2}(\sqrt{5} - 1)]$,
   $[\frac{1}{2}(\sqrt{5} + 1), \frac{1}{2}(\sqrt{5} + 1)]$
4. $(1, 3)$, $(1 + \sqrt{6}, \ 3 + 2\sqrt{6})$
5. $(1, 1)$, $(-1, 1)$
6. $(\frac{1}{2}\{-1 - \sqrt{13}\}, \frac{1}{2}\{5 + \sqrt{13}\})$,
   $(\frac{1}{2}\{\sqrt{13} - 1\}, \frac{1}{2}\{5 - \sqrt{13}\})$
7. $x = -\frac{1}{4}$
8. $x = 0$
9. $x = \frac{1}{2}$
10. $x = \frac{1}{2}(\sqrt{17} - 3)$, $x = 3$
11. $x = -1$, $x = -1 - \sqrt{2}$
12. $x = 0$, $x = -\frac{1}{2}(5 + \sqrt{57})$

CHAPTER 2

Exercise 2a – p. 18
1. $a = 2, b = -3$

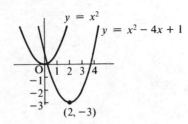

2. (a)

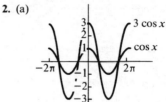

(b)

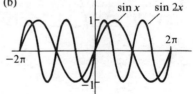

(c)

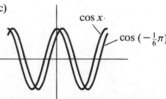

(d)

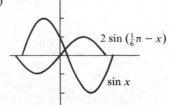

3.
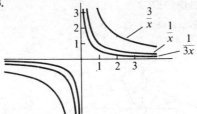

4.
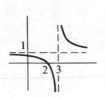

5.
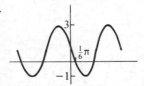

6.
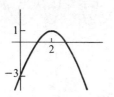

7.
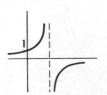

8.
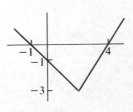

9.
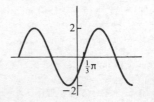

10.
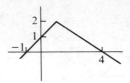

11.
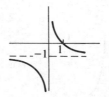

12.

13.

14.
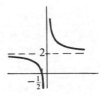

15.
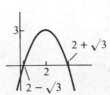

16.
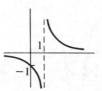

17.

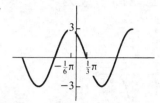

18.

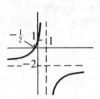

19.

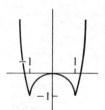

20.

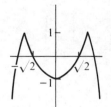

21.

22.

23.

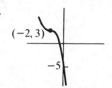

$(-2, 3)$

-5

24.

Exercise 2b – p. 22

1.

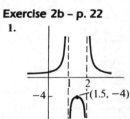

-4 $(1.5, -4)$

2.

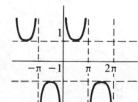

3.

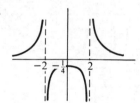

4.

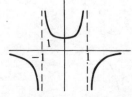

5.

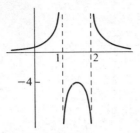

Mixed Exercise 2 – p. 23

1.

$y = 1$ and $x = 3$; (4, 0) and (0, $\frac{4}{3}$)

2. (a) f is undefined when $x = 1$

(b)

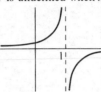

6.

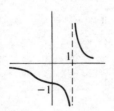

(c)

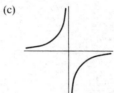

7.

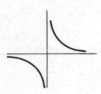

(d) $1 - 2f(x)$

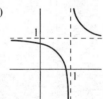

8.

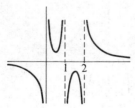

3. (a)

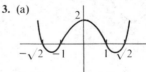

9.

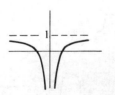

(b)

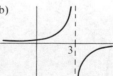

10.

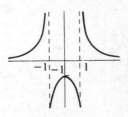

(c)

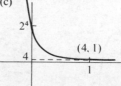

(d)

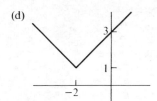

11. $-\frac{49}{8}$; $x = -2$ and $\frac{3}{2}$

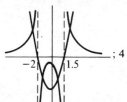

; 4

4.

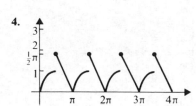

5.

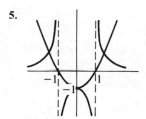

6.

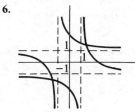

$g^{-1}(x) > -1, g^{-1}(x) < -1$

7. $x = -\frac{1}{2}$

8.

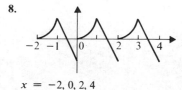

$x = -2, 0, 2, 4$

9. (a) odd with period π (b) odd
   (c) even   (d) even with period 2π

10.

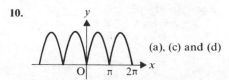

(a), (c) and (d)

CHAPTER 3

Exercise 3a – p. 26

1. $\dfrac{3}{2(x + 1)} - \dfrac{1}{2(x - 1)}$

2. $\dfrac{13}{6(x - 7)} - \dfrac{1}{6(x - 1)}$

3. $\dfrac{4}{5(x - 2)} - \dfrac{4}{5(x + 3)}$

4. $\dfrac{7}{9(2x - 1)} + \dfrac{28}{9(x + 4)}$

5. $\dfrac{1}{x - 2} - \dfrac{1}{x}$

6. $\dfrac{3}{x - 2} - \dfrac{1}{x - 1}$

7. $\dfrac{1}{2(x - 3)} - \dfrac{1}{2(x + 3)}$

8. $\dfrac{7}{3x} - \dfrac{1}{3(x + 1)}$

9. $\dfrac{9}{x} - \dfrac{18}{2x + 1}$

10. $\dfrac{2}{5(x - 1)} - \dfrac{1}{5(3x + 2)}$

Exercise 3b – p. 28

1. $\dfrac{9}{8(x - 5)} - \dfrac{1}{8(x + 3)}$

2. $\dfrac{4}{(x - 1)} - \dfrac{5}{(2x - 1)}$

3. $\dfrac{1}{5(x - 2)} + \dfrac{6}{5(4x - 3)}$

4. $\dfrac{5}{3(2x - 1)} - \dfrac{4}{3(x + 1)}$

5. $\dfrac{3}{x} - \dfrac{6}{2x - 1}$

6. $\dfrac{4}{9(x - 8)} - \dfrac{4}{9(x + 1)}$

7. $\dfrac{1}{2x - 3} + \dfrac{1}{2x + 3}$

8. $\dfrac{5}{x + 2} - \dfrac{1}{x}$

9. $\dfrac{1}{(x - 2)} + \dfrac{1}{2(x + 1)}$

10. $\dfrac{2}{(x - 2)} - \dfrac{1}{2x}$

Exercise 3c – p. 31

1. $\dfrac{1}{x - 1} - \dfrac{x + 1}{x^2 + 1}$

2. $\dfrac{1}{x} - \dfrac{x}{2x^2 + 1}$

3. $\dfrac{3}{2x} - \dfrac{x}{2(x^2 + 2)}$

4. $\dfrac{22}{19(x - 3)} + \dfrac{1 - 6x}{19(2x^2 + 1)}$

5. $\dfrac{3}{5(x + 2)} - \dfrac{3}{5(2x + 1)} + \dfrac{x - 1}{5(x^2 + 1)}$

6. $\dfrac{1}{x} - \dfrac{6x + 3}{2x^2 - 1} + \dfrac{2}{x - 1}$

7. $\dfrac{1}{x - 1} - \dfrac{1}{x - 2} + \dfrac{2}{(x - 2)^2}$

8. $\dfrac{2}{x} + \dfrac{1}{x^2} - \dfrac{3}{2x + 1}$

9. $\dfrac{3}{x} - \dfrac{9}{3x - 1} + \dfrac{9}{(3x - 1)^2}$

10. $1 + \dfrac{1}{2(x - 1)} - \dfrac{1}{2(x + 1)}$

11. $1 - \dfrac{7}{4(x + 3)} - \dfrac{1}{4(x - 1)}$

12. $x + \dfrac{2}{x - 1} - \dfrac{1}{x + 1}$

Exercise 3d – p. 33

1. $\log_{10} 1000 = 3$
2. $\log_2 16 = 4$
3. $\log_{10} 10\,000 = 4$

4. $\log_3 9 = 2$
5. $\log_4 16 = 2$
6. $\log_5 25 = 2$
7. $\log_{10} 0.01 = -2$
8. $\log_9 3 = \frac{1}{2}$
9. $\log_5 1 = 0$
10. $\log_4 2 = \frac{1}{2}$
11. $\log_{12} 1 = 0$
12. $\log_8 2 = \frac{1}{3}$
13. $\log_q p = 2$
14. $\log_x 2 = y$
15. $\log_p r = q$
16. $10^5 = 100\,000$
17. $4^3 = 64$
18. $10^1 = 10$
19. $2^2 = 4$
20. $2^5 = 32$
21. $10^3 = 1000$
22. $5^0 = 1$
23. $3^2 = 9$
24. $4^2 = 16$
25. $3^3 = 27$
26. $36^{1/2} = 6$
27. $a^0 = 1$
28. $x^z = y$
29. $a^b = 5$
30. $p^r = q$

Exercise 3e – p. 34

1. 2
2. 6
3. 6
4. 4
5. 2
6. 3
7. $\frac{1}{2}$
8. -2
9. -1
10. $\frac{1}{2}$
11. 0
12. 1
13. $\frac{1}{3}$
14. 0
15. $\frac{1}{3}$
16. 3

Exercise 3f – p. 36

1. $\log p + \log q$
2. $\log p + \log q + \log r$
3. $\log p - \log q$
4. $\log p + \log q - \log r$
5. $\log p - \log q - \log r$

6. $2 \log p + \log q$

7. $\log q - 2 \log r$

8. $\log p + \frac{1}{2} \log q$

9. $2 \log p + 3 \log q - \log r$

10. $\frac{1}{2} \log q - \frac{1}{2} \log r$

11. $n \log q$

12. $n \log p + m \log q$

13. $\log pq$

14. $\log p^2 q$

15. $\log q/r$

16. $\log q^3 p^4$

17. $\log p^n/q$

18. $\log pq^2/r^3$

Exercise 3g – p. 37

1. (a) $\dfrac{\log 8}{\log 3}$ (b) $\dfrac{1}{\log 5}$ (c) $\dfrac{\log 5}{2}$

2. (a) $\dfrac{\log_a y}{\log_a x}$ (b) $\dfrac{1}{\log_a x}$ (c) $\dfrac{\log_a 8}{\log_a y}$

3. (a) 1.19 (b) 2.86 (c) 2.46

Exercise 3h – p. 39

1. 1.63

2. 0.861

3. 1.16

4. 2.77

5. -0.104

6. 1.22

7. $-1, 1$

8. 2

9. 3

10. 2

11. 3

12. $\sqrt{3}$ or $1/\sqrt{3}$

13. $x = \frac{1}{2}, y = 1$

14. $x = 2, y = 4$

15. $x = 1, y = 0$

16. $x = 1, y = 0$

Mixed Exercise 3 – p. 39

1. (a) $\dfrac{1}{x-1} - \dfrac{1}{x+1}$

(b) $\dfrac{1}{x-2} - \dfrac{1}{x+1}$

(c) $\dfrac{1}{3(x-3)} - \dfrac{1}{3x}$

(d) $\dfrac{1}{x-1} - \dfrac{1}{x+3}$

(e) $\dfrac{1}{2(x-1)} - \dfrac{1}{2(x+1)}$

(f) $\dfrac{1}{2x-1} - \dfrac{1}{2x+1}$

2. (a) $\dfrac{2}{x-1} - \dfrac{2}{x} - \dfrac{2}{x^2}$

(b) $\dfrac{1}{x} - \dfrac{x}{x^2+1}$

(c) $1 + \dfrac{1}{2(x-1)} - \dfrac{1}{2(x+1)}$

(d) $\dfrac{1}{12(x-2)} - \dfrac{3}{4(x+2)} + \dfrac{2}{3(x+1)}$

(e) $\dfrac{4}{9(x-1)} + \dfrac{1}{3(x-1)^2} - \dfrac{8}{9(2x+1)}$

(f) $\dfrac{x+3}{2(x^2+1)} - \dfrac{1}{2(x+1)} - \dfrac{1}{(x+1)^2}$

3. $y = \dfrac{1}{x-1} - \dfrac{1}{x}, \dfrac{dy}{dx} = \dfrac{1}{x^2} - \dfrac{1}{(x-1)^2}$

4. $f(x) = \dfrac{1}{x-1} + \dfrac{1}{x+1}$.

$f'(x) = -\dfrac{1}{(x-1)^2} - \dfrac{1}{(x+1)^2}$

$= \dfrac{-2(x^2+1)}{(x-1)^2(x+1)^2}$

and there is no value of x for which $x^2 + 1 = 0$

5. (a) $x = 1.05$ (3 s.f.)

(b) $x = 1$

(c) $x = -2$

(d) $x = 2^{1/\sqrt{2}}$

(e) $x = 2^{11} (\sqrt[3]{3})$

(f) $x = 1.09$ (3 s.f.)

CHAPTER 4

Exercise 4a – p. 45

1. (a) $\dfrac{\sqrt{5}}{5}$ (b) $\frac{1}{5}$

(c) $\dfrac{|ma - b + c|}{\sqrt{(m^2+1)}}$ (d) $\dfrac{3\sqrt{2}}{2}$

(e) $\dfrac{|ax + by + c|}{\sqrt{(a^2+b^2)}}$ (f) $\dfrac{|c|}{\sqrt{(a^2+b^2)}}$

2. (a) opposite (b) same

(c) same

3. (a) $\frac{15}{16}$ (b) $\frac{5}{31}$

  (c) $\left| \dfrac{(a_1b_2 - a_2b_1)}{(a_1a_2 + b_2b_1)} \right|$

4. $\frac{3}{14}, \frac{9}{83}, \frac{3}{29}; \dfrac{9}{\sqrt{170}}, \dfrac{9}{\sqrt{41}}, \dfrac{3}{\sqrt{5}}$

5. $5y = x, 5x + y = 0$

6. (a) $(-1, 8)$ (b) $(-11, 10)$
   (c) $(-3, -14)$

7. $4X^2 + Y^2 - 14X + 12Y - 4XY + 11 = 0$

8. no

9. $\dfrac{|5X - 12Y + 3|}{13}, \dfrac{|3X + 4Y - 6|}{5}$;

   $14X + 112Y = 93$ and $64X - 8Y = 63$

10. $x + 2y = 7$

Exercise 4b – p. 48

1. $y^2 = 2y + 12x - 13$
2. $x + y = 4$
3. $8x^2 + 9y^2 - 20x = 28$
4. $11y = 3x$ and $99x + 27y + 130 = 0$
5. $x^2 + y^2 - 6x - 10y + 30 = 0$
6. $4x - 3y = 26$ and $4x - 3y + 24 = 0$
7. $x^2 + y^2 = 1$

Exercise 4c – p. 51

1. (a) $x^2 + y^2 - 2x - 4y = 4$
   (b) $x^2 + y^2 - 8y + 15 = 0$
   (c) $x^2 + y^2 + 6x + 14y + 54 = 0$
   (d) $x^2 + y^2 - 8x - 10y + 32 = 0$
2. (a) $(-4, 1); 5$
   (b) $(-\frac{1}{2}, -\frac{3}{2}); 3\sqrt{2}/2$
   (c) $(-3, 0); \sqrt{14}$
   (d) $(\frac{3}{4}, -\frac{1}{2}); \sqrt{5}/4$
   (e) $(0, 0); 2$
   (f) $(2, -3); 3$
   (g) $(1, 3); 3$
   (h) $(-1, \frac{1}{2}); \sqrt{69}/6$
3. (a) and (f)

Exercise 4d – p. 58

1. (a) yes   (b) yes   (c) no
   (d) yes
2. (a) $3y + 4x = 23$
   (b) $4y = x + 22$
   (c) $3y = 2x + 25$
3. $2x^2 + 2y^2 - 15x - 24y + 77 = 0$
4. $x^2 + y^2 - 8x + 6y = 0$
5. $x^2 + y^2 - 8x - 6y + 17 = 0$
6. $x^2 + y^2 - 4x - 14y - 116 = 0$
7. $x^2 + y^2 - 6y = 0$

8. $x^2 + y^2 + 4x + 4y + 4 = 0$
   or $196(x^2 + y^2) + 84(x + y) + 9 = 0$
9. $y = 0, 8y = 15x$
10. $x + 2y = 0$
11. $2x + y - 11 = 0, 2x + y + 9 = 0$
12. $(-\frac{4}{3}, \frac{2}{3}); 4\sqrt{5}/3$
13. (a), (b) and (c)
14. $9 \pm 6\sqrt{5}$
15. $2mac = a^2 - c^2$

CHAPTER 5

Note that the general solution of a trig equation can be given in different forms. An answer that is different to one given here can be checked by giving n values in both answers.

Exercise 5a – p. 65

1. $\frac{1}{3}\pi + 2n\pi, \frac{2}{3}\pi + 2n\pi$
2. $\pm\frac{1}{2}\pi + 2n\pi \ (= \frac{1}{2}\pi + n\pi)$
3. $-\frac{1}{3}\pi + n\pi$
4. $-14.5° + 360n°, -165.5° + 360n°$
5. $\pm\frac{2}{3}\pi + 2n\pi$
6. $\frac{1}{4}\pi + n\pi$
7. $2n\pi$
8. $\frac{1}{6}\pi + 2n\pi, \frac{5}{6}\pi + 2n\pi$
9. $\pm\frac{1}{6}\pi + 2n\pi, \pm\frac{5}{6}\pi + 2n\pi$
10. $n\pi$
11. $\pm41.4° + 360n°$
12. $-18° + 360n°, -162° + 360n°$
13. $45° + 180n°, -14° + 180n°$
14. $\pm\frac{1}{3}\pi + 2n\pi, \frac{1}{2}\pi + 2n\pi$
15. $n\pi, \pm\frac{1}{3}\pi + 2n\pi$
16. $\pm90° + 360n°, 14.5° + 360n°, 165.5° + 360n°$
17. $n\pi, \pm\frac{1}{6}\pi + 2n\pi$
18. $\pm51.8° + 360n°$
19. $360n°, \pm131.8° + 360n°$
20. $38.2° + 360n°, 141.8° + 360n°$
21. $\pm\frac{1}{2}\pi + 2n\pi$
22. $n\pi, \pm\frac{1}{3}\pi + 2n\pi$
23. $n\pi$
24. $n\pi$

Exercise 5b – p. 67

1. $\frac{1}{12}\pi + \frac{1}{3}n\pi$
2. $\pm\frac{1}{6}\pi + \frac{1}{2}n\pi$

3. $\frac{1}{2}\pi + 6n\pi,\ \frac{5}{2}\pi + 6n\pi$

4. $\pm\frac{1}{9}\pi + \frac{2}{3}n\pi$

5. $\pm\frac{1}{3}\pi + 4n\pi$

6. $-\frac{1}{8}\pi + \frac{1}{2}n\pi$

7. $\frac{1}{2}n\pi + \frac{3}{8}\pi$

8. $-\frac{8}{3}\pi + 8n\pi$

9. $\frac{1}{6}\pi + n\pi$

10. $\frac{1}{12}\pi + n\pi,\ n\pi - \frac{1}{4}\pi$

Exercise 5c – p. 68

1. $-165°, -105°, 15°, 75°$

2. $53.1°$

3. $-105°, -45°, 75°, 135°$

4. $60°$

5. π

6. $\frac{1}{9}\pi, \frac{5}{9}\pi, \frac{7}{9}\pi$

7. $\frac{5}{8}\pi, \frac{7}{8}\pi$

8. $\frac{1}{8}\pi, \frac{3}{8}\pi, \frac{5}{8}\pi, \frac{7}{8}\pi$

Exercise 5e – p. 72

1. 0

2. $\frac{1}{2}$

3. $\frac{1}{4}(\sqrt{6} - \sqrt{2})$

4. $-(2 + \sqrt{3})$

5. $\frac{1}{4}(\sqrt{6} - \sqrt{2})$

6. $\frac{1}{4}(\sqrt{6} + \sqrt{2})$

7. $\sin 3\theta$

8. 0

9. $\tan 3A$

10. $\tan \beta$

11. (a) $\frac{3}{5}$ (b) $-\frac{4}{5}$ (c) $-\frac{3}{4}$

12. (a) $1, 115°$ (b) $1, 30°$
    (c) $1, 310°$ (d) $1, 330°$

19. $67.5°, 247.5°$

20. $7.4°, 187.4°$

21. $37.9°, 217.9°$

22. $15°, 195°$

23. $\frac{1}{12}\pi + n\pi$

24. $\frac{7}{12}\pi + n\pi$

Exercise 5f – p. 75

1. $\frac{1}{2}$

2. $\dfrac{1}{\sqrt{2}}$

3. $\frac{1}{2}\sin 2\theta$

4. $\cos 8\theta$

5. $-\dfrac{1}{\sqrt{3}}$

6. $\tan 6\theta$

7. $\sqrt{2}\cos 3\theta$

8. $-\dfrac{1}{\sqrt{2}}$

9. $\tan(x + 45°)$

10. $\dfrac{1}{\sqrt{2}}$

11. (a) $\frac{24}{25}, -\frac{7}{25}$ (b) $\frac{336}{625}, \frac{527}{625}$
    (c) $\frac{120}{169}, -\frac{119}{169}$

12. (a) $-\frac{336}{527}$ (b) $\frac{527}{625}$ (c) $-\frac{336}{625}$
    (d) $\frac{164833}{390625}$

13. (a) $x(1 - y^2) = 2y$
    (b) $x = 2y^2 - 1$ (c) $x = 1 - \dfrac{2}{y^2}$
    (d) $2x^2y + 1 = y$

15. (a) $\frac{3}{2}\pi + 2n\pi, \frac{1}{6}\pi + 2n\pi, \frac{5}{6}\pi + 2n\pi$
    (b) $\pm\frac{1}{2}\pi + 2n\pi, \frac{7}{6}\pi + 2n\pi, \frac{11}{6}\pi + 2n\pi$
    (c) $\pm\frac{1}{3}\pi + 2n\pi$
    (d) $\pm 35.3° + 180n°$
    (e) $\frac{1}{2}\pi + n\pi, \frac{1}{4}\pi + n\pi$
    (f) $90° + 180n°, 23.6° + 360n°,$
        $156.4° + 360n°$

Exercise 5g – p. 79

1. (a) $\frac{24}{25}$ (b) $-\frac{7}{24}$ (c) $\frac{1}{2}, -2$
   (d) $\pm\dfrac{2}{\sqrt{5}}, \pm\dfrac{1}{\sqrt{5}}$

2. t^2

3. $\dfrac{1}{t}$

4. $\dfrac{(1 - t^2)}{2t^2}$

5. $\dfrac{1}{(3 + 6t - 5t^2)}$

8. $0, 67.4°$

9. $-153°, 130.4°$

10. $36.9°, 126.9°$

11. $-90°, 36.8°$

Mixed Exercise 5 – p. 80

1. $y = 1 - 2x^2$

4. $\frac{56}{65}, -\frac{16}{65}$

5. $x = 2y - 1$

6. $-155.7°, 24.3°, -114.3°, 65.7°$

7. $n\pi$

9. $\cot^2 x$

10. $90°, 270°$

11. $\frac{17}{24}\pi + n\pi, \frac{1}{24}\pi + n\pi$

12. $6\pi + 8n\pi, 6\pi$

CHAPTER 6

Exercise 6a – p. 87

1. $\frac{1}{2}(1-\sqrt{5}) < x < 0, x > \frac{1}{2}(1+\sqrt{5})$
2. $0 < x < 1$
3. $0 < x < 1$
4. $-1 < x < 0$
5. $-2 < x < -1$
6. $x > 0$
7. $-\frac{1}{2} < x < \frac{1}{2}$
8. $x > -\frac{1}{2}$
9. $-1 < x < \frac{1}{3}$
10. $x > 1$
11. $x < 1$
12. $x < -\frac{5}{4}, x > -\frac{1}{4}$
13. $x < 0, x > 2$
14. $x < 0, x > 2$
15. $0 < x < 1 + \sqrt{3}$
16. $-1 < x < 1$
17. $-3 < x < 3$ and $x > 5$
18. $1 < x < 2$ and $x < 0$
19. $x > 3$ and $1 < x < 2$
20. $2 < x < 8$
21. $0 < x < \frac{1}{4}\pi, \frac{3}{4}\pi < x < \frac{5}{4}\pi, \frac{7}{4}\pi < x < 2\pi$

Exercise 6b – p. 90

1. $-\frac{1}{7} \leqslant f(x) \leqslant 1$
2. $f(x) \geqslant 2$ and $f(x) \leqslant -2$
3. all values
4. $-\frac{1}{2} \leqslant f(x) \leqslant \frac{1}{2}$
5. $f(x) \geqslant 3 + \sqrt{8}$ and $f(x) \leqslant 3 - \sqrt{8}$
6. $f(x) \geqslant 2$ and $f(x) \leqslant -2$
7. $0 < k < \frac{4}{9}$
8. $-1 < p < 2$
10. $-\frac{1}{2} \leqslant f(x) \leqslant 1$
11.

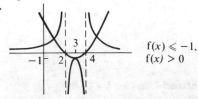

$f(x) \leqslant -1,$
$f(x) > 0$

12. (a) $f(x) < 0, f(x) > 0$
   (b) $f(x) \leqslant -\frac{1}{4}, f(x) > 0$
   (c) $f(x) \geqslant -\frac{1}{4}$
   (d) $f(x) \leqslant -4, f(x) > 0$

Mixed Exercise 6 – p. 91

1. $1 < x < 3$
2. $x > -1$

3. $x \geqslant 1$ and $x \leqslant -3$
4. $-2 < x \leqslant -1$ and $1 \leqslant x < 2$
   (Note: $x \neq 2$ or -2)
5. $\frac{2}{3} \leqslant f(x) \leqslant 2$
7.

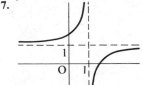

$g \equiv g^{-1}$

8. $x < -\frac{1}{2}$
9. $-1 < x < 0$ and $x > 2$
10. $(1, \frac{1}{2}), x < -1, 0 < x < 1$
11. (a) all values of x
   (b) $x > -\frac{1}{2}$

CONSOLIDATION A

Multiple Choice Exercise A – p. 96

1. C	9. B	17. F
2. A	10. E	18. T
3. D	11. E	19. F
4. B	12. B, C	20. T
5. E	13. B	21. F
6. B	14. B, C	22. F
7. C	15. A, B, C	23. F
8. D	16. A, C	

Miscellaneous Exercise A – p. 98

1. $A = \frac{5}{7}, B = -\frac{3}{7}$
2. (a) $\frac{3}{2}$, (b) $\frac{7}{4}$
3. $x < 1$
4. $0, 60°, 120°, 180°$
5. $x < -4$ and $-1 < x < 2$
6. $x = 6$
7. 0.774
8. (a)

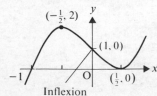

Inflexion

(b)

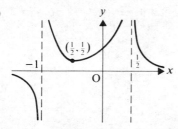

9. (a) $(3, 3)$ (b) $(\frac{14}{3}, \frac{1}{2})$

10. (a) $\dfrac{1}{2(x-1)} - \dfrac{1}{2(x+1)}$

(b) $\dfrac{1}{4(x+1)} - \dfrac{1}{4(x-1)} + \dfrac{1}{2(x-1)^2}$

(c) $\dfrac{1}{2(x-1)} - \dfrac{x+1}{2(x^2+1)}$

11. $x < -2\frac{1}{2}, x > -1$

12. (a)

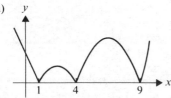

(b)

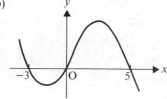

(c)

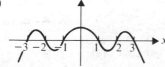

13. $n\pi, n\pi \pm \frac{1}{6}\pi$

14. $x = \frac{1}{6}(y^2 + 12)$

15. $\dfrac{3}{x} - \dfrac{3}{x+1}$

16. $(3, -2), 5; \frac{1}{5}|6 + k|; k = 19$ or -31

17. $\pm\frac{1}{4}\pi + n\pi; \pm\frac{1}{3}\pi + n\pi$

18. $x < -1$

19. $x < 1, x > 2\frac{1}{2}$

20. (a) $2^{(3+p+2q)}$,   (b) $3p + 2q$

21.

22. $\pm 66.4° + 360n°, \pm 104.5° + 360n°$

23. $2 < x < 4$

24. (b) $45°$ (c) $(6, -1); 10;$
$(x - 6)^2 + (y + 1)^2 = 100$

25. (a) 64 or $\frac{1}{64}$ (b) 21

26. $x > 2, x < -5$

27. $0°, 60°, 90°, 120°, 180°$

28. $\frac{4}{3}, y = -\frac{3}{4}(x - 2), 63.4°$

CHAPTER 7

Exercise 7a – p. 109

1. $(0, 0)$, point of inflexion

2. $(-1, -8)$, minimum; $(1, 8)$, maximum

3. $(-1, 2)$, maximum; $(1, -2)$, minimum

4. $(\frac{1}{2}, \frac{1}{4})$, maximum

5. $(-2, -4)$, maximum; $(2, 4)$, minimum

6. $(-1, 2)$ and $(1, 2)$, both minimum

7. $(-\sqrt{\frac{3}{2}}, 3\sqrt{\frac{3}{2}})$, maximum; $(0, 0)$, point of inflexion; $(\sqrt{\frac{3}{2}}, -3\sqrt{\frac{3}{2}})$, minimum

8. $(0, 0)$, minimum

9. $(1, \frac{3}{2}, \text{minimum})$

10. 54, maximum; -54, minimum

11. 2, minimum; -2, maximum

12. -3, minimum

13. -2, minimum

14. 16, maximum; -16, minimum

15. 1, inflexion; 0, minimum

16. (a) y decreases to zero, then increases, i.e. y has a minimum value at O.

(b) $\dfrac{dy}{dx}$ increases from $-$ve values through 0 at O to $+$ve value

(c) $\dfrac{dy}{dx} = 3x^3$

(d) $0; \dfrac{dy}{dx}$ has a stationary value at O.

17. (a)   (b)

(c)

— $f(x)$
-- $f'(x)$

In each case y has a stationary value
and $d^2y/dx^2 = 0$

Exercise 7b – p. 112

1. $3(x-1)(x-3)$

2. $\dfrac{3\sqrt{x}}{2} - \dfrac{3}{x}$

3. $(x-2)^4(6x+8)$

4. $(2x+3)^2(8x+3)$

5. $(x-1)^2(5x+3)$

6. $\dfrac{1}{2\sqrt{x}}(x-3)^3 + 3\sqrt{x}(x-3)^2$

   $= \dfrac{(x-3)^2(7x-3)}{2\sqrt{x}}$

7. $\dfrac{(x+5)^3(3x-17)}{(x-3)^2}$

8. $\dfrac{2}{(3x+2)^2}$

9. $\dfrac{(2x-7)^{1/2}(x+7)}{x^2}$

10. $\dfrac{x^2(7x-6)}{2\sqrt{(x-1)}}$

11. $3(x+3)^{-2}$

12. $2x(2x-3)(4x-3)$

Exercise 7c – p. 113

1. $\dfrac{2x(x-3)-(x-3)^2}{x^2} = \dfrac{(x-3)(x+3)}{x^2}$

2. $\dfrac{(x+3)(2x)-x^2}{(x+3)^2} = \dfrac{x(x+6)}{(x+3)^2}$

3. $\dfrac{-x^2-2x(4-x)}{x^4} = \dfrac{x-8}{x^3}$

4. $\dfrac{2x^3(x+1)-(x+1)^2(3x^2)}{x^6}$

   $= \dfrac{-(x+1)(x+3)}{x^4}$

5. $\dfrac{4(1-x)^3-12x(1-x)^2}{(1-x)^4} = \dfrac{4(1-4x)}{(1-x)^4}$

6. $\dfrac{(x-2)(4x)-2x^2}{(x-2)^2} = \dfrac{2x(x-4)}{(x-2)^2}$

7. $\dfrac{\frac{5}{3}x^{2/3}(3x-2)-3x^{5/3}}{(3x-2)^2} = \dfrac{2x^{2/3}(3x-5)}{3(3x-2)^2}$

8. $\dfrac{-3(1-2x)^2}{x^4}$

9. $\dfrac{(3x-2)(x+1)^{3/2}}{2x^2}$

Mixed Exercise 7 – p. 114

1. $\dfrac{3x+2}{2\sqrt{(x+1)}}$

2. $6x(x^2-8)^2$

3. $\dfrac{1-x^2}{(x^2+1)^2}$

4. $\dfrac{-4x^3}{3(2-x^4)^{2/3}}$

5. $\dfrac{2x}{(x^2+2)^2}$

6. $\frac{1}{2}x(5\sqrt{x}-8)$

7. $6x(x^2-2)^2$

8. $\dfrac{1-2x}{2\sqrt{(x-x^2)}}$

9. $\dfrac{\sqrt{x}+2}{2(\sqrt{x}+1)^2}$

10. $\dfrac{x(5x-8)}{2\sqrt{(x-2)}}$

11. $\dfrac{-(3x+4)}{2x^3\sqrt{(x+1)}}$

12. $6x^5(x^2+1)^2(2x^2+1)$

13. $\dfrac{x}{\sqrt{(x^2-8)}}$

14. $x^2(5x^2-18)$

15. $6x(x^2-6)^2$

16. $\dfrac{-(x^2+6)}{(x^2-6)^2}$

17. $-8x^3(x^4+3)^{-3}$

18. $\dfrac{(2-x)^2(2-7x)}{2\sqrt{x}}$

19. $\dfrac{2+5x}{2\sqrt{x}(2-x)^4}$

20. $(x-2)(3x-4)$

21. $30x^2(2x^3+4)^4$

CHAPTER 8

Exercise 8a – p. 118

1. (a) $2\sin 2\theta \cos\theta$
   (b) $2\cos 4\theta \cos\theta$
   (c) $2\cos 3\theta \sin\theta$
   (d) $-2\sin 4\theta \sin 3\theta$
   (e) $2\cos 4A \sin A$
   (f) $+2\sin 3A \sin 2A$
   (g) $2\sin(A+10°)\cos(A-10°)$
   (h) $2\sin(\theta+45°)\cos(\theta-45°)$

2. (a) $\sin 3\theta + \sin\theta$
   (b) $\cos 5\theta + \cos\theta$
   (c) $\sin 5\theta + \sin 3\theta$
   (d) $\cos 4\theta - \cos 2\theta$
   (e) $\cos 2\theta - \cos 6\theta$
   (f) $\frac{1}{2}\cos 5\theta + \frac{1}{2}\cos 3\theta$

3. (a) $\cos\theta$ (b) $-\sqrt{2}\cos x$
   (c) 0

4. (a) $\frac{1}{2}\sqrt{6}$ (b) $\frac{1}{2}\sqrt{6}$

5. (a) $\pm\frac{1}{2}\pi + 2n\pi$, $\pm\frac{1}{6}\pi + \frac{2}{3}n\pi$
   (b) $\pm\frac{1}{4}\pi + n\pi$, $n\pi$
   (c) $\frac{1}{2}n\pi$ (d) $\frac{1}{3}n\pi$, $\frac{1}{2}\pi + n\pi$
6. $\cos 2\theta(2\cos\theta + 1)$, $\pm\frac{1}{4}\pi + n\pi$,
   $\pm\frac{2}{3}\pi + 2n\pi$
7. $4\cos\theta\sin 4\theta\cos 2\theta$, $\pm\frac{1}{8}\pi + \frac{1}{2}n\pi$,
   $\pm\frac{1}{4}\pi + n\pi$
8. (a) $20°, 90°, 100°, 140°,$
   (b) $22\frac{1}{2}°, 90°, 112\frac{1}{2}°,$
   (c) $0°, 30°, 40°, 60°, 80°, 90°, 120°,$
       $150°, 160°, 180°,$
   (d) $30°, 45°, 135°, 150°,$
   (e) $0°, 60°, 105°, 120°, 165°, 180°,$
   (f) $10°, 130°$

CHAPTER 9

Exercise 9a – p. 124
1. (a) 7.39   (b) 0.368   (c) 4.48
   (d) 0.741
2. (a) $2e^x$ (b) $2x - e^x$ (c) e^x
3. $e^2 - 2$
4. $2 + 2e$
5. 1
6. 0

Exercise 9b – p. 126
1. (a) 1.24   (b) 0.183   (c) 2.85
   (d) 0.693
2. (a) $\ln x - \ln(x - 1)$
   (b) $\ln 5 + 2\ln x$
   (c) $\ln(x + 2) + \ln(x - 2)$
   (d) $\ln\sin x - \ln\cos x$
   (e) $2\ln\sin x$
   (f) $\frac{1}{2}\ln(x + 1) - \frac{1}{2}\ln(x - 1)$
3. (a) $\ln\dfrac{x}{(1 - x)^2}$ (b) $\ln\dfrac{e}{x}$
   (c) $\ln(\sin x\cos x) = \ln(\frac{1}{2}\sin 2x)$
   (d) $\ln x^2\sqrt{(x - 1)}$
4. (a) $\frac{2}{3}\ln x$ (b) $5\ln x$
5. (a) 2.10   (b) 0   (c) 1.05
   (d) 0
6. (a) $x = 1$ (b) $x = 3$
   (c) $x = 0$
7. They are the reflection of each other in
   the line $y = x$; $f = g^{-1}$

Exercise 9c – p. 129
1. $\dfrac{3}{x}$ 2. $\dfrac{1}{x}$ 3. $-\dfrac{2}{x}$
4. $-\dfrac{1}{2x}$ 5. $-\dfrac{5}{x}$ 6. $\dfrac{1}{2x}$

7. $-\dfrac{3}{2x}$ 8. $\dfrac{5}{2x}$
9. $(1, -1)$
10. $(2^{1/3}, \{2 - 2\ln 2\})$
11. $(4, \{\ln 4 - 2\})$
12.   13.

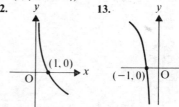

14.   15.

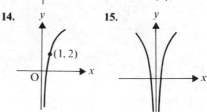

Exercise 9d – p. 131
(pr $\equiv$ product; f of f $\equiv$ function of a function)
1. (a) pr   (b) $u = e^x, v = x^2 + 1$
2. (a) f of f   (b) $u = x^2 + 1, y = e^u$
3. (a) pr   (b) $u = x, v = \ln x$
4. (a) f of f   (b) $u = x + 1, y = e^{u/2}$
5. (a) pr   (b) $u = e^x, v = \ln x$
6. (a) f of f   (b) $u = 3 - x^2, y = \ln u$
7. (a) f of f   (b) $u = \ln x, y = u^2$
8. (a) f of f   (b) $u = -2x, y = e^u$
9. (a) f of f   (b) $u = \ln x, y = \dfrac{1}{u}$
10. $fg(x) = e^{2x}$; $gf(x) = e^{x^2}$
11. (a) $\left(\dfrac{1}{x}\right)^2$ (b) $\ln x^2$
    (c) $\ln\left(\dfrac{1}{x}\right)$ (d) $(\ln x)^2$
    (e) $\ln\left(\dfrac{1}{x}\right)^2$ (f) $\left(\dfrac{1}{x}\right)^2$

Exercise 9e – p. 133
1. $e^x(x + 1)$
2. $x(2\ln x + 1)$
3. $e^x(x^3 + 3x^2 - 2)$
4. $12x\ln(x - 2) + \dfrac{6x^2}{(x - 2)}$
5. xe^x
6. $x\ln x + \dfrac{(x^2 + 4)}{2x}$

7. $\dfrac{4 + 3x}{2\sqrt{(2 + x)}}$

8. $\frac{1}{2}\ln(x - 5) + \dfrac{x}{2(x - 5)}$

9. $(x^2 + 2x - 2)e^x$

10. $\dfrac{1 - x}{e^x}$

11. $\dfrac{e^x(x - 2)}{x^3}$

12. $\dfrac{1 - 3\ln x}{x^4}$

13. $\dfrac{x\ln x - 2(x + 1)}{2x\sqrt{(x + 1)}(\ln x)^2}$

14. $\dfrac{e^x(x^2 - 2x - 1)}{(x^2 - 1)^2}$

15. $\dfrac{-2}{(e^x - e^{-x})^2}$

16. $4e^{4x}$

17. $\dfrac{2x}{x^2 - 1}$

18. $2xe^{x^2}$

19. $-6e^{(1 - x)}$ or $-6e(e^{-x})$

20. $2xe^{(x^2 + 1)}$

21. $\dfrac{1}{2(x + 2)}$

22. $\dfrac{2\ln x}{x}$

23. $\dfrac{-1}{x(\ln x)^2}$

24. $\frac{1}{2}\sqrt{(e^x)}$

Mixed Exercise 9 – p. 133

1. $1 + \ln x$

2. $\frac{8}{3}(4x - 1)^{-1/3}$

3. $\dfrac{e^x(x - 2)}{(x - 1)^2}$

4. $\dfrac{-(x^3 + 4)}{2x^3\sqrt{(1 + x^3)}}$

5. $\dfrac{(x - 1)\ln(x - 1) - x\ln x}{x(x - 1)\{\ln(x - 1)\}^2}$

6. $3(\ln 10)10^{3x}$

7. $\dfrac{2x}{(1 + x^2)^2}$

8. $\dfrac{2}{x^2 e^{2/x}}$

9. $\dfrac{-e^x}{1 - e^x}$ or $\dfrac{e^x}{e^x - 1}$

10. $3x^2 e^{3x}(x + 1)$

11. $\dfrac{4}{5(2x - 1)^2} - \dfrac{6}{5(x - 3)^2}$
$= \dfrac{2(3 - 2x^2)}{(2x - 1)^2(x - 3)^2}$

12. $\dfrac{e^{x/2}(x - 10)}{2x^6}$

13. $\dfrac{2}{x} - \dfrac{1}{x + 3} - \dfrac{2x}{x^2 - 1}$

14. $\dfrac{5x + 9}{x(x + 3)}$

15. $\dfrac{4}{x}(\ln x)^3$

16. $\dfrac{(x + 3)^2(x^2 - 6x + 6)}{(x^2 + 2)^2}$

17. $\dfrac{e^x - 1}{2\sqrt{(e^x - x)}}$

18. $\dfrac{8x}{x^2 + 1}$

19. $\dfrac{dy}{dx} = \dfrac{4}{(1 - 2x)^2}; \dfrac{d^2y}{dx^2} = \dfrac{16}{(1 - 2x)^3}$

20. $\dfrac{dy}{dx} = \dfrac{1}{x(x + 1)}; \dfrac{d^2y}{dx^2} = \dfrac{-(2x + 1)}{x^2(x + 1)^2}$

21. $\dfrac{dy}{dx} = \dfrac{-4e^x}{(e^x - 4)^2}; \dfrac{d^2y}{dx^2} = \dfrac{4e^x(e^x + 4)}{(e^x - 4)^3}$

CHAPTER 10

Exercise 10a – p. 137

1. $\dfrac{1}{2r} - \dfrac{1}{2(r + 2)}; \dfrac{n(3n + 5)}{4(n + 1)(n + 2)}$

2. $\dfrac{1}{r + 1} - \dfrac{1}{r + 2}; \dfrac{n - 2}{4(n + 2)}$

3. $\dfrac{1}{r} - \dfrac{1}{r + 1}; \dfrac{n + 1}{n(2n + 1)}$

4. $\dfrac{1}{16(2r - 1)} + \dfrac{1}{8(2r + 1)} - \dfrac{3}{16(2r + 3)};$
$\dfrac{n(n + 1)}{2(2n + 1)(2n + 3)}$

5. $\frac{1}{2}n(n + 1)(2n^2 + 2n + 1)$

6. $\dfrac{n(n + 3)}{4(n + 1)(n + 2)}$

7. $\dfrac{n(n + 2)}{(n + 1)^2}$

8. $\frac{1}{4}n^2(n + 1)^2$

9. $\dfrac{\cos 2\theta - \cos(2n + 2)\theta}{2\sin\theta}$

Exercise 10b – p. 140

1. $\frac{1}{3}n(n + 1)(n + 2)$

2. $\frac{n}{4}(n + 1)(n + 2)(n + 3)$

3. $\frac{1}{12}n(n + 1)(45n^2 + 37n + 2)$

4. $42\,075$

5. $\frac{1}{6}n(2n + 5)(n - 1)$

6. $-n(2n + 1)$

7. $\frac{1}{3}n(4n^2 + 6n - 1)$

Exercise 10c – p. 142

1. $\frac{3}{4}$

2. 3024

3. $\dfrac{ab^2(1 - b^{n-1})}{(1 - b)}$

4. 650

5. $6n(n + 1)$

6. $\frac{1}{4}n(n + 1)(n^2 + n + 2)$

7. $\frac{3}{4}$

8. $210 \ln 3$

9. $(1 - e^n)/(1 - e)$

10. $\frac{1}{6}n(n + 1)(3n^2 + 13n + 11)$

11. $\frac{1}{2}$

12. 2

13. $\frac{2}{3}n(n + 1)(2n + 1)$

14. $-\ln n$

Exercise 10d – p. 147

1. $1 - x - \dfrac{x^2}{2} - \dfrac{x^3}{3}, -\frac{1}{2} < x < \frac{1}{2}$

2. $\frac{1}{3} - \dfrac{x}{9} + \dfrac{x^2}{27} - \dfrac{x^3}{81}, -3 < x < 3$

3. $1 - \dfrac{x}{4} + \dfrac{3x^2}{32} - \dfrac{5x^3}{128}, -2 < x < 2$

4. $1 + 2x + 3x^2 + 4x^3, -1 < x < 1$

5. $1 - \frac{1}{2}x + \frac{3}{8}x^2 - \frac{5}{16}x^3, -1 < x < 1$

6. $1 + \frac{1}{2}x - \frac{5}{8}x^2 - \frac{3}{16}x^3, -1 < x < 1$

7. $-2 - 3x - 3x^2 - 3x^3, -1 < x < 1$

8. $2 + 2x + \frac{21}{4}x^2 + \frac{27}{8}x^3, -\frac{1}{3} < x < \frac{1}{3}$

9. $\frac{1}{2} - \frac{3}{4}x + \frac{13}{8}x^2 - \frac{51}{16}x^3, -\frac{1}{2} < x < \frac{1}{2}$

10. $1 + x + \frac{1}{2}x^2 + \frac{1}{2}x^3, -1 < x < 1$

11. $1 - \frac{1}{9}x^2, -3 < x < 3$

12. $x - x^2 + x^3, -1 < x < 1$

13. $1 - 3p^{-1} + 6p^{-2} - 10p^{-3} + 15p^{-4}, |p| < 1$

14. 1.732

15. $3.162\,28$

16. $1 + 2x + 2x^2$

18. $1 - 3x + \frac{7}{2}x^2$

19. $1 + 2x + 5x^2$

20. $\frac{3}{2} + \frac{15}{4}x$

23. $\frac{1}{729}[27 + 27x + 18x^2 + 20x^3]$

Mixed Exercise 10 – p. 149

1. $1 + 18x + 144x^2 + \ldots + 2^9x^9$; all x

2. $1 - 2x + 4x^2 - \ldots; -\frac{1}{2} < x < \frac{1}{2}$

3. $x + x^2 + x^3 + \ldots; -1 < x < 1$

4. $1 + 3x + 6x^2 + \ldots; -\frac{1}{2} < x < \frac{1}{2}$

5. $\dfrac{1}{3(1 + x)} + \dfrac{2}{3(1 - 2x)}; 1 + x + 3x^2$

6. 8

7. $1, 7, 19, 37; 3n^2 - 3n + 1$

9. (a) $2n^2(n + 1)^2$ (b) $\ln\left(\dfrac{1}{n}\right)$

10. (a) $1 + \frac{3}{2}x + \frac{27}{8}x^2; |x| < \frac{1}{3}$

   (b) $1.343\,75$ with $x = \frac{1}{6}$

11. (a) 1 (b) $-16x^3$

12. (a) $1\frac{1}{2}$ (b) 1

CHAPTER 11

Exercise 11a – p. 160

1. (a) $\left(\dfrac{1}{y}\right) = a(x) \div b$

  (b) $(y^2 - x) = b(y) - a$

  (c) $(y) = a(e^x/y) + b$

  (d) $(x) = (y^2 + y) - k; m = 1$

  (e) $(\ln y) = (n + 2)(\ln x) + \ln a$

  (f) $(x) = a(\ln y) - k$

  There are other alternatives.

2. $a = 2, b = -4$ (exactly)

3. $a = 6, b = 4$

4. $a = 0.5, b = -2$

5. $a = 30, b = 2$

6. $a = 2, b = \frac{1}{2}$

7. $a = 3, b = -2$ (or $a = -2, b = 3$)

8. $k = 4, n = 0.07$

9. $k = 5500, n = 1.5$

10. The 'incorrect' values are -0.26 and -0.40; the estimated correct values are -0.50 and -1.00.

CONSOLIDATION B

Multiple Choice Exercise B – p. 166

1. C	8. D	15. B
2. D	9. B	16. T
3. D	10. B, C	17. F
4. C	11. B, C	18. F
5. B	12. A	19. T
6. A	13. A	20. F
7. D	14. C	21. F

Miscellaneous Exercise B – p. 169

1. $1 - x + 2x^2 - \frac{14}{3}x^3$; $-\frac{1}{3} < x < \frac{1}{3}$
2. -1
3. $2 \sin 2x \cos x$;
$x = -\frac{1}{2}\pi, -\frac{1}{3}\pi, 0, \frac{1}{3}\pi, \frac{1}{2}\pi, \pi$;
$x = \frac{1}{2}n\pi, 2n\pi \pm \frac{1}{3}\pi$
4. $1 - 4x - 8x^2 - 32x^3$, $-\frac{1}{8} < x < \frac{1}{8}$
6. $\dfrac{2}{(1 + 2x)} - \dfrac{1}{1 + x}$, $1\frac{11}{27}$
7. $\frac{1}{4}\pi + \frac{1}{2}n\pi$, $\frac{1}{2}\pi + n\pi$
8. $\frac{3}{2}$, $-\frac{27}{8}$
9. (a) $12x(1 + x^2)^5$ (b) $\dfrac{2x}{(1 + x^2)^2}$

   (c) $2x \ln (1 + x^2) + \dfrac{2x^3}{1 + x^2}$
10. $\dfrac{-1}{x - 2} + \dfrac{4}{x + 6}$; $x = -\frac{2}{3}$; $\frac{1}{8}$ max, $\frac{9}{8}$ min
11. $4, 2x^2, -12x^3$
12.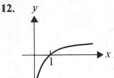

    0.14, 7.39,
    2 (the line $y = 3 - x$ cuts $y = |\ln x|$
    in 2 places)

13.

$\lg x$	1	1.30	1.60	1.88	2
$\lg y$	2.13	2.52	2.92	3.27	3.43

   $m = 1.3, c = 0.83, y = 6.8x^{1.3}$
14. $\frac{1}{2}\pi + n\pi$, $\frac{1}{12}\pi + n\pi$, $\frac{5}{12}\pi + n\pi$
15. $\frac{1}{2}$
16. $\dfrac{dy}{dx} = e^{-2x} - 2xe^{-2x}$;

    $\dfrac{d^2y}{dx^2} = 4xe^{-2x} - 4e^{-2x}$; $\frac{1}{2}$, max.
17. (a) $\ln R = n \ln V + \ln K$
    (b) $n = 1.8, k = 0.67$
18. $\cos 3\theta + \cos \theta$; $\frac{1}{3}(2n + 1)\pi$

19. 0; $\dfrac{x^2}{(1 + 2x)(1 + x)^2}$ which is +ve for
    $x > 0$;
    Since f(x) is increasing for all $x > 0$
    and since f(x) = 0 when $x = 0$,
    $\log_e (1 + 2x) - \dfrac{2x}{1 + x} > 0$ for $x > 0$,
    i.e. $\log_e (1 + 2x) > \dfrac{2x}{1 + x}$
20. $33\,000$, -1.4
21. 3.4
22. (a) $1 - x + x^2 - x^3$, $1 + 2x + 4x^2 + 8x^3$,
    $1 + 4x + 12x^2 + 32x^3$

    (b) $\dfrac{3}{1 + x} + \dfrac{1}{1 - 2x} + \dfrac{2}{(1 - 2x)^2}$
    (c) $6 + 7x + 31x^2 + 69x^3$
    (d) $-\frac{1}{2} < x < \frac{1}{2}$
23. (a) $f^{-1}: x \to \frac{1}{2}(3 + e^x)$, $x \in \mathbb{R}$

    (b) $g: x \to \dfrac{2}{2x - 3}$, $x \in \mathbb{R}$, $x > \frac{3}{2}$

    (c)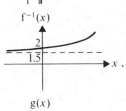
    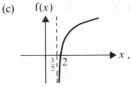
    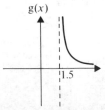

    (d) 2.78

CHAPTER 12

Exercise 12a – p. 178

1. (a) $2, 30°$ (b) $\sqrt{10}, 71.6°$
   (c) $5, 36.9°$
2. $\sqrt{2} \cos (2\theta + \frac{1}{4}\pi)$
3. $\sqrt{29} \sin (3\theta + 21.8°)$
4. $-2 \sin (\theta - \frac{1}{6}\pi)$; max 2 at $\theta = 300°$,
   min -2 at $\theta = 120°$

5. $25 \cos (\theta + 73.7^\circ)$; max 28 at
$\theta = 286.3^\circ$, min -22 at $\theta = 106.3^\circ$

6. $\sqrt{2}, -\sqrt{2}; -\sqrt{2}$ max, $\sqrt{2}$ min

7. $-\sqrt{\frac{2}{3}}$ max, $\sqrt{\frac{2}{3}}$ min; $\infty, -\infty$

8. (a) $45^\circ + 360n^\circ$
(b) $118.1^\circ + 360n^\circ$, $-36.9^\circ + 360n^\circ$
(c) $360n^\circ$, $360n^\circ - 143.2^\circ$
(d) $360n^\circ$, $360n^\circ - 53.1^\circ$

Exercise 12b – p. 181

1. $\frac{2}{7}n\pi$

2. $\frac{1}{5}n\pi$

3. $\frac{1}{2}(4n - 1)\pi$, $\frac{1}{10}(4n + 1)\pi$

4. $\frac{1}{18}(2n + 1)\pi$

5. $2n\pi$, $\frac{1}{7}(2n + 1)\pi$

6. $\frac{1}{2}n\pi$, $\frac{1}{3}n\pi$

7. $\frac{2}{3}n\pi - \frac{1}{6}\pi$, $\frac{2}{5}n\pi$

8. $\frac{1}{6}(2n + 1)\pi$

9. $0, \frac{1}{5}\pi, \frac{2}{5}\pi, \frac{1}{2}\pi, \frac{3}{5}\pi, \frac{4}{5}\pi, \pi$

10. $\frac{1}{10}\pi, \frac{3}{10}\pi, \frac{1}{2}\pi, \frac{7}{10}\pi, \frac{9}{10}\pi$

11. $0, \frac{1}{9}\pi, \frac{2}{9}\pi, \frac{1}{3}\pi, \frac{5}{9}\pi, \frac{7}{9}\pi, \frac{4}{5}\pi, \pi$

12. $0, \frac{2}{11}\pi, \frac{4}{11}\pi, \frac{6}{11}\pi, \frac{8}{11}\pi, \frac{10}{11}\pi$

13. $-140^\circ, -60^\circ, -20^\circ, 100^\circ$

14. $-115^\circ, -25^\circ, 65^\circ, 155^\circ$

15. $-150^\circ, -110^\circ, 10^\circ, 130^\circ$

16. $-110^\circ, -30^\circ, 10^\circ, 130^\circ$

17. $-150^\circ, -60^\circ, 30^\circ, 120^\circ$

18. $-155^\circ, -140^\circ, -65^\circ, 25^\circ, 40^\circ, 115^\circ$

Exercise 12c – p. 184

1. $\frac{1}{3}\pi$ **2.** $-\frac{1}{2}\pi$ **3.** $\frac{1}{2}\pi$

4. $-\frac{1}{3}\pi$ **5.** $\frac{2}{3}\pi$ **6.** $-\frac{1}{4}\pi$

7.

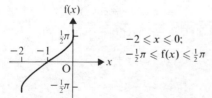

$-2 \leqslant x \leqslant 0$;
$-\frac{1}{2}\pi \leqslant f(x) \leqslant \frac{1}{2}\pi$

8.

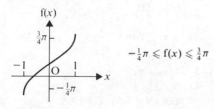

$-\frac{1}{4}\pi \leqslant f(x) \leqslant \frac{3}{4}\pi$

Exercise 12d – p. 186

1. $\frac{1}{2}$ (b) $\dfrac{\theta^2}{1 - 2\theta^2}$ (c) $\dfrac{\theta}{2 - \theta^2}$

(d) 2   (e) 1   (f) 1

3. (a) $\frac{1}{2}$ (b) $\frac{2}{3}$ (c) $\sqrt{\frac{3}{2}}$ (d) 1

Exercise 12e – p. 189

1. $\frac{1}{6}\pi(1 + 4n)$

2. $\frac{1}{8}\pi(1 + 4n)$, $\frac{1}{4}\pi(n - 1)$

3. $2n\pi$, $\pm\frac{2}{3}\pi + 2n\pi$

4. $2n\pi + \tan^{-1}\frac{12}{5}$

5. $n\pi$

6. $\pm\frac{1}{4}\pi + n\pi$, $\pm\frac{2}{3}\pi + 2n\pi$

7. $\pm\frac{1}{2}\pi + 2n\pi$

8. $\tan^{-1}\frac{3}{4} + \frac{1}{2}\pi + 2n\pi$

9. $\pm\frac{1}{2}\pi + 2n\pi$

10. $2n\pi + \sin^{-1}(\sqrt{2} - 1)$,
$(2n + 1)\pi + \sin^{-1}(\sqrt{2} - 1)$

11. $0, 30^\circ, 150^\circ, 180^\circ$

12. $53.1^\circ, 180^\circ$

13. $0, 180^\circ$

14. $201.5^\circ, 338.5^\circ$

15. $26.6^\circ, 135^\circ, 206.6^\circ, 315^\circ$

16. $34.6^\circ, 325.4^\circ$

17. $120^\circ, 300^\circ$

18. $0, 72^\circ, 90^\circ, 144^\circ, 180^\circ, 216^\circ, 270^\circ, 288^\circ$

19. $20^\circ, 100^\circ, 140^\circ, 220^\circ, 260^\circ, 340^\circ$

Mixed Exercise 12 – p. 189

1. $\dfrac{3\sqrt{3} - 4\sqrt{2}}{2(3 + \sqrt{6})}$ (b) $\frac{1}{6}(1 - 2\sqrt{6})$

2. $\pm 70.5^\circ$

3. $\pm\frac{1}{4}\pi + n\pi$, $\frac{1}{6}\pi + 2n\pi$, $\frac{5}{6}\pi + 2n\pi$

5. $n\pi$, $\pm\frac{1}{3}\pi + 2n\pi$,

7. $\pm\frac{1}{6}\pi + \frac{2}{3}n\pi$, $\frac{1}{12}\pi + n\pi$, $\frac{5}{12}\pi + n\pi$

8. $\sqrt{2} - 1$

9. $1 - 2y^2 = x$

10. $x = 60^\circ, 300^\circ, y = 120^\circ, 240^\circ$

12. $\frac{1}{4}\pi + 2n\pi$, $\frac{3}{4}\pi + 2n\pi$

13. $5 \sin (\theta - \alpha)$ where $\tan \alpha = \frac{3}{4}$;
1 min, $-\frac{7}{3}$ max, $\pm\infty$

14. $\sqrt{2} \sin (2\theta - \frac{1}{4}\pi)$; $\frac{3}{8}\pi$

15. $257.6^\circ, 349.8^\circ$

16. $y = \frac{1}{2}x, |x| \leqslant 2$

17. $0, 36^\circ, 108^\circ, 180^\circ, 252^\circ, 324^\circ, 360^\circ$

18. $ax = a^2 - 2y^2$

19. $49.1^\circ, 229.1^\circ$

21. $\pm\frac{1}{6}\pi + \frac{2}{3}n\pi$, $n\pi$

22. $\alpha = 90° - \theta$; $135°, 315°, 22\frac{1}{2}°, 112\frac{1}{2}°$,
$202\frac{1}{2}°, 292\frac{1}{2}°$

23. $-\frac{1}{4}$

24. $\sqrt{2}\cos(x - \frac{1}{4}\pi)$; $\frac{1}{4}\pi$

25. $0, \frac{2}{5}\pi, \frac{4}{5}\pi, \frac{6}{5}\pi, \frac{8}{5}\pi, 2\pi$

26. $24°, 96°, 168°, 80°$

28. $\frac{1}{14}(1 + 2n)\pi$

29. $5\cos(x + \alpha)$ where $\tan\alpha = \frac{4}{3}$;
$4 + 2\sec(x + \alpha)$

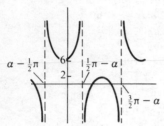

CHAPTER 13

Exercise 13a – p. 194

2. (a) $\cos x + \sin x$　　　(b) $\cos\theta$
(c) $-3\sin\theta$　　　(d) $5\cos\theta$
(e) $3\cos\theta - 2\sin\theta$
(f) $4\cos x + 6\sin x$

3. (a) -1　　(b) 1　　(c) -1
(d) 1　　(e) $2(\pi - 1)$　　(f) 4

4. (a) $\frac{1}{6}\pi$　　(b) $\frac{1}{6}\pi$　　(c) $\frac{1}{4}\pi$
(d) π

5. (a) $(\frac{1}{3}\pi, \sqrt{3} - \frac{1}{3}\pi)$, max;
$(\frac{5}{3}\pi, -\sqrt{3} - \frac{5}{3}\pi)$, min
(b) $(\frac{1}{6}\pi, \frac{1}{6}\pi + \sqrt{3})$, max;
$(\frac{5}{6}\pi, \frac{5}{6}\pi - \sqrt{3})$, min

6. $y + \theta = 3 + \frac{1}{2}\pi$

7. $2\pi y + x = 2\pi^3 - \pi$

8. $(0, 1)$

Exercise 13b – p. 197

1. $4\cos 4x$

2. $2\sin(\pi - 2x)$ or $2\sin 2x$

3. $\frac{1}{2}\cos(\frac{1}{2}x + \pi)$ or $-\frac{1}{2}\cos\frac{1}{2}x$

4. $\dfrac{x\cos x - \sin x}{x^2}$

5. $-\dfrac{(\cos x + \sin x)}{e^x}$

6. $\dfrac{\cos x}{2\sqrt{(\sin x)}}$

7. $2\sin x\cos x$ or $\sin 2x$

8. $\cos^2 x - \sin^2 x$ or $\cos 2x$

9. $\cos x\, e^{\sin x}$

10. $-\tan x$

11. $e^x(\cos x - \sin x)$

12. $x^2\cos x + 2x\sin x$

13. $2x\cos x^2$

14. $-\sin x\, e^{\cos x}$

15. $3\cot x$

16. $\dfrac{\sin x}{\cos^2 x} = \sec x\tan x$

17. $\sec^2 x$

18. $\dfrac{-\cos x}{\sin^2 x} = -\operatorname{cosec} x\cot x$

19. $\dfrac{-1}{\sin^2 x} = -\operatorname{cosec}^2 x$

Exercise 13c – p. 199

3. $2x\arcsin x + \dfrac{x^2}{\sqrt{(1 - x^2)}}$

4. $\dfrac{2\arctan x}{1 + x^2}$

5. $\dfrac{3}{\sqrt{(1 - 9x^2)}}$

6. $\dfrac{e^{\arctan x}}{1 + x^2}$

7. $\dfrac{1}{(\arcsin x)\sqrt{(1 - x^2)}}$

8. $\dfrac{-3x^2}{\sqrt{(1 - x^6)}}$

Mixed Exercise 13 – p. 200

1. (a) $-4\cos 4\theta$　　(b) $1 + \sin\theta$
(c) $3\sin^2\theta\cos\theta + 3\cos 3\theta$

2. (a) $3x^2 + e^x$　　(b) $2e^{(2x + 3)}$
(c) $e^x(\sin x + \cos x)$

3. $-\dfrac{3}{x}$　　(b) $-\dfrac{2}{x}$　　(c) $\dfrac{1}{2x}$

4. (a) $3\cos x + e^{-x}$　　(b) $\dfrac{1}{2x} + \frac{1}{2}\sin x$
(c) $4x^3 + 4e^x - \dfrac{1}{x}$
(d) $-\frac{1}{2}(e^{-x} + x^{-3/2}) - \dfrac{1}{x}$

5. $1 + \dfrac{1}{x} + \ln x$

6. $3\sin 6x$

7. $\frac{8}{3}(4x - 1)^{-1/3}$

8. $9 - 18\sqrt{x} + 8x$

9. $(x^4 + 4x^3 + 3)/(x + 1)^4$

10. $\cos^2 x(4\cos^2 x - 3)$

11. $-1/\sin x \cos x$ or $-2\csc 2x$

12. $2x\sin x + x^2\cos x$

13. $e^x(x-2)/(x-1)^2$

14. $2\cos x/(1-\sin x)^2$

15. $x(5x-4)/2\sqrt{(x-1)}$

16. $-2(1-x)^2(2x+1)$

17. $\dfrac{3}{2(x+3)} - \dfrac{x}{x^2+2}$

18. $e^{\arcsin x}/\sqrt{(1-x^2)}$

19. $-3(\arccos x)^2/\sqrt{(1-x^2)}$

20. $e^x/(1+e^{2x})$

21. $a^x(1+x\ln a)$

22. $\dfrac{2x}{x^2+1}$

23. (a) $x = \ln 3$ (b) $x = 1\ (not\ -1)$
     (c) $x = \frac{1}{4}$

24. (a) 1 (b) $y - x = 1 - \frac{1}{2}\pi$
     (c) $y + x = 1 + \frac{1}{2}\pi$

25. (a) $1 + e$ (b) $y = x(1+e)$
     (c) $y(1+e) + x = (1+e)^2 + 1$

26. (a) 2 (b) $y = 2x + 1$
     (c) $2y + x = 2$

27. (a) -1 (b) $x + y = 3$
     (c) $x - y + 1 = 0$

28. (a) $(\frac{1}{2}\pi, 0)$, min; $(\frac{3}{2}\pi, 2)$, max
     (b) $(\frac{1}{6}\pi, \{\frac{1}{12}\pi + \frac{1}{2}\sqrt{3}\})$, max;
      $(\frac{5}{6}\pi, \{\frac{5}{12}\pi - \frac{1}{2}\sqrt{3}\})$, min
     (c) $(\ln 3, \{3 - 3\ln 3\})$, min; only one
      turning point

29. (a) $(\frac{1}{6}\pi, \{\frac{1}{2}\pi - \sqrt{3}\})$ (b) $(1, -1)$

CHAPTER 14

Exercise 14a – p. 206

1. $2x + 2y\dfrac{dy}{dx} = 0$

2. $2x + y + (x+2y)\dfrac{dy}{dx} = 0$

3. $2x + x\dfrac{dy}{dx} + y = 2y\dfrac{dy}{dx}$

4. $-\dfrac{1}{x^2} - \dfrac{1}{y^2}\dfrac{dy}{dx} = e^y\dfrac{dy}{dx}$

5. $-\dfrac{2}{x^3} - \dfrac{2}{y^3}\dfrac{dy}{dx} = 0$

6. $\dfrac{x}{2} - \dfrac{2y}{9}\dfrac{dy}{dx} = 0$

7. $\cos x + \cos y\dfrac{dy}{dx} = 0$

8. $\cos x\cos y - \sin x\sin y\dfrac{dy}{dx} = 0$

9. $e^y + xe^y\dfrac{dy}{dx} = 1$

10. $(1+x)\dfrac{dy}{dx} = 2x - 1 - y$

11. $\dfrac{dy}{dx} = \pm\dfrac{1}{\sqrt{(2x+1)}}$

12. $\dfrac{d^2y}{dx^2} = \pm\dfrac{x}{\sqrt{(2-x^2)^3}}$

13. $\pm\frac{1}{4}\sqrt{2}$

15. (a) $xx_1 - 3yy_1 = 2(y + y_1)$
     (b) $x(2x_1 + y_1) + y(2y_1 + x_1) = 6$

17. $3x + 12y - 7 = 0$

Exercise 14b – p. 209

1. $\dfrac{dy}{dx} = \dfrac{x-y}{x\ln x}$

2. $\dfrac{dy}{dx}\left(\dfrac{1}{y+1} - \ln x\right) = \dfrac{y}{x}$

3. $\dfrac{1}{y}\dfrac{dy}{dx} = \ln(x+x^2) + \dfrac{1+2x}{1+x}$

4. $\dfrac{-(x^2+8)}{(x+2)^2(x-4)^2}$

5. $\dfrac{2x(3-x)}{(x-1)^2(x-3)^2}$

6. $-(1-x)^4(7x^2 - 2x + 10)$

7. $2x\sqrt{\dfrac{(x-3)}{(x+1)}}$

8. $\dfrac{x}{(x+2)^2(x^2-1)}\left\{\dfrac{1}{x} - \dfrac{2}{x+2} - \dfrac{2x}{x^2-1}\right\}$
     $= \dfrac{-(3x^3 + 2x^2 - x + 2)}{(x+2)^3(x^2-1)^2}$

9. $-\dfrac{y}{2}\left\{\dfrac{9x^2 - 4x + 12}{(x^2+4)(3x-2)}\right\}$
     $= \dfrac{-(9x^2 - 4x + 12)}{2\sqrt{(x^2+4)^3}(3x-2)^3}$

Exercise 14c – p. 216

1. (a) $\dfrac{1}{4t}$ (b) $-\tan\theta$ (c) $-\dfrac{4}{t^2}$

2. $\dfrac{dy}{dx} = 1 - t^2;\ \frac{3}{4}$

3. (a) $\frac{3}{2}t$ (b) $\frac{3}{2}\sqrt{x}$
(c) $x = t^2 \Rightarrow t = \sqrt{x}$

4. (a) $x = 2y^2$; $\dfrac{dy}{dx} = \dfrac{1}{4y} = \dfrac{1}{4t}$

(b) $x^2 + y^2 = 1$; $\dfrac{dy}{dx} = -\dfrac{x}{y} = -\tan\theta$

(c) $xy = 4$; $\dfrac{dy}{dx} = -\dfrac{4}{x^2} = -\dfrac{4}{t^2}$

5. $(-\frac{1}{3}\sqrt{3}, \frac{2}{9}\sqrt{3})$, max; $(\frac{1}{3}\sqrt{3}, -\frac{2}{9}\sqrt{3})$, min
6. $(2n\pi + \frac{1}{2}\pi, 1)$
7. $2x + y + 2 = 0$
8. $ty + t^4 = t^3x + 1$
9. $ty = 2x + 2t^2$
10. $y = x$; $(-\frac{1}{2}\sqrt{2}, -\frac{1}{2}\sqrt{2})$
11. $2y + tx = 8t + t^3$; $(8 + t^2, 0)$,
$(0, \frac{1}{2}t\{8 + t^2\})$; $A = |\frac{1}{4}t(8 + t^2)^2|$

12. (a) $\dfrac{\cos t}{e^t}$ (b) $y = \sin(\ln x)$

(c) $\dfrac{\cos(\ln x)}{x}$

13. $t^2y + x = 2t$; $(2t, 0)$, $\left(0, \dfrac{2}{t}\right)$; $A = 2$

Mixed Exercise 14 – p. 217

1. (a) $4y^3\dfrac{dy}{dx}$ (b) $y^2 + 2xy\dfrac{dy}{dx}$

(c) $-\dfrac{1}{y^2}\dfrac{dy}{dx}$ (d) $\ln y + \dfrac{x}{y}\dfrac{dy}{dx}$

(e) $\cos y\dfrac{dy}{dx}$ (f) $e^y\dfrac{dy}{dx}$

(g) $\dfrac{dy}{dx}\cos x - y\sin x$

(h) $(\cos y - y\sin y)\dfrac{dy}{dx}$

2. $\dfrac{x}{2y}$

3. $-\dfrac{y^2}{x^2}$

4. $-\dfrac{2y}{3x}$

5. $-\dfrac{x^3(3x^3 + 6x^2 - 2x - 8)}{(1 + x)^4}$

6. $-\dfrac{(x - 1)^3(2x^2 - 3x - 15)}{x^6(x + 3)^2}$

7. $-\dfrac{y(y + 1)(3x + 2)}{x(x + 1)(y + 2)}$

8. $3t/2$

9. $\dfrac{t}{t + 1}$

10. $-\frac{3}{2}\cos\theta$

11. $-\dfrac{1}{t^2}$

12. $\dfrac{2y + 7}{2y - 2x + 3}$

13. $2t - t^2$

CHAPTER 15

Exercise 15a – p. 221
1. touch at $(1, 2)$
2. $(1, 1)$ and $(-5, 4)$
3. no intersection
4. $(0, 2)$ and $(\frac{8}{5}, \frac{14}{5})$
5. $(1, 0)$ and $(2, 0)$
6. $(0, 0)$, $(-3, 0)$ and $(-2, 0)$
7. touch at $(1, 0)$ and $(2, 0)$
8. touch and cross at $(-3, 0)$; $(-2, 0)$
9. touch at $(0, 0)$
10. $k = -\frac{1}{8}$
11. $k = \frac{1}{4}$
12. $k = \pm 4\sqrt{5}$
13. $(\frac{1}{2}, 2)$
14. $(\frac{1}{4}\sqrt{7}, \frac{7}{4})$
15. touch at $(-1, -2)$; cut at $(1, 2)$ and $(4, \frac{1}{2})$
16. (a) $\pm 2\sqrt{5}$ (b) $-2\sqrt{5} < \lambda < 2\sqrt{5}$
(c) $\lambda < -2\sqrt{5}, \lambda > 2\sqrt{5}$
17.

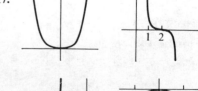

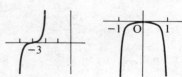

18. (a) (i) $y = 0, y = 16(1 - x)$
(ii) $64y + 32x - 1 = 0$
(b) (i) $y = 0, y = 36(1 - x)$
(ii) $2y + x = \pm 6\sqrt{2}$
(c) (i) $3y = 2(x - 1), y = 0$
(ii) $8y + 4x + 3 = 0$

Exercise 15b – p. 226
1. $3x^2 - y^2 - 60x + 256 = 0$; hyperbola
2. $16y = x^2$; parabola
3. $3x^2 + 4y^2 = 48$; ellipse
4. (a) parabola   (b) circle
   (c) rectangular hyperbola
   (d) hyperbola   (e) parabola
   (f) ellipse
5.

$(3, 0), (-3, 0)$ and $(0, 2), (0, -2)$

6.

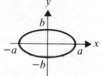

7. (a) $(\pm 4, 0), (0, \pm 2); (\pm 4, 0)$
   (b) they touch at $(\pm 4, 0)$

Exercise 15c – p. 231
1. (a)

   (b)

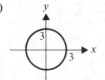

   (c)

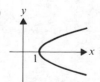

   (d)

(e)

(f)

(g)

(h)

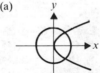

$(2, -1)$

2. (a)

(b)

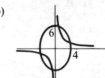

(c)

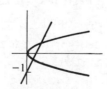

3.

$(2\sqrt{(2\sqrt{5} - 2)}, 2\sqrt{5} - 2),$
$(-2\sqrt{(2\sqrt{5} - 2)}, 2\sqrt{5} - 2)$

4. (a) cuts   (b) misses
   (c) touches   (d) cuts
5. (a) $y^2 = 8x$ (b) $x - 2y + 8 = 0$
   (c) $2x - y(p + q) + 4pq = 0$
   (d) $(18, 12)$ and $(2, -4)$ (e) 2
   (f) $(2, 4)$ and $(2, -4)$ (g) $(8, 8)$
6. (a)

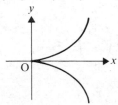

   (b) $\dfrac{dy}{dx} = \dfrac{3}{2}t$

   $\dfrac{dy}{dx} \to 0$ as $t \to 0$ from both

   positive and negative values.
   The curve forms a cusp at the
   origin; the x-axis is a tangent to the
   curve.

CONSOLIDATION C

Multiple Choice Exercise C – p. 236
1. C	8. C	15. C
2. A	9. D	16. T
3. A	10. C	17. F
4. D	11. B, C	18. F
5. D	12. A, B	19. F
6. A	13. A, D	20. T
7. C	14. B, D	21. F

Miscellaneous Exercise C – p. 239
1. (a) $\dfrac{1}{(2x + 1)^2}$ (b) $\dfrac{2x}{\sqrt{(1 - x^4)}}$
2. (a) $4x \cos(2x^2)$ (b) $2^x \ln 2$
3. (a) $\dfrac{1 + \cos x + \sin x}{(1 + \cos x)^2}$

   (b) $\dfrac{1}{1 - x^2}$
6. (a) $-\dfrac{1}{4}$ (b) 4
7. $2n\pi + \dfrac{1}{2}\pi, 2n\pi - \dfrac{1}{6}\pi$
8. $\dfrac{1}{4}\pi < x < \dfrac{2}{3}\pi$ and $\dfrac{3}{4}\pi < x < \pi$
9. (a) $80(4x - 1)^{19}$ (b) $\dfrac{1}{2\sqrt{x}(1 + x)}$
10. $\left(\dfrac{25}{4}a, -\dfrac{9}{4}a\right)$

11. $e^x\left\{\dfrac{2 \cos 2x}{1 + \sin 2x} + \ln(1 + \sin 2x)\right\}$

12. $x = 0, \pi, 2\pi;$ $\left(\dfrac{1}{3}\pi, \dfrac{3}{4}\sqrt{3}\right), (\pi, 0),$
    $\left(\dfrac{5}{3}\pi, -\dfrac{3}{4}\sqrt{3}\right)$

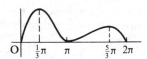

13. $\theta = 0, \dfrac{1}{10}\pi, \dfrac{3}{10}\pi, \dfrac{1}{3}\pi$
14. $x + y + 3 = 0$
15. $R = 5, \tan \alpha = \dfrac{4}{3}$
    (a) 2.1 rad, 2.9 rad
    (b) 0.3 rad, 0.9 rad, 2.4 rad, 4.5 rad
    A is $(0, 1)$; B is $(-2.2, 5)$;
    C is $\left(0.9, \dfrac{5}{6}\right)$
16. 2
17. $R = 4.47; \alpha = 1.11$ rad
18. $\theta = -\dfrac{1}{4}\pi, \dfrac{5}{4}\pi$ (or $\dfrac{7}{4}\pi$ etc);
    $x + y + 2 = 0, x - y = 4$
19. $3x - y + (y - x)\dfrac{dy}{dx} = 4;$ $(1, 3), (3, 1)$
20. (a) $e^t(\sin t + \cos t) = \sqrt{2}e^t \sin\left(t + \dfrac{1}{4}\pi\right);$
    1 m/s
    (b) $\dfrac{3}{4}\pi, \dfrac{7}{4}\pi, \dfrac{11}{4}\pi, \dfrac{15}{4}\pi$
    (c) $-e^\pi$
    (d)

22. $115°, 318°$
23. (a) $\dfrac{dx}{dt} = -e^{-t}(\sin t + \cos t);$ 1 m/s
    (b) $\dfrac{3}{4}\pi, \dfrac{7}{4}\pi$
    (c) $2 e^{-t} \sin t$
24. $82.4°$
25. (a) $5 \cos\left(\theta + \arctan \dfrac{4}{3}\right)$
    (b) 5
26. (a) $x \cos x + \sin x$
    (b) $2 \tan x \sec^2 x$
27. $\dfrac{dy}{dx} = \dfrac{6x - y}{x - 2y}$

28.

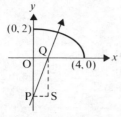

$S(3\cos\theta, -6\sin\theta)$;

$r = 6\sqrt{5}, \cos\alpha = \dfrac{1}{\sqrt{5}}, \sin\alpha = \dfrac{2}{\sqrt{5}};$

$6\sqrt{5}, \left(\dfrac{3}{\sqrt{5}}, -\dfrac{12}{\sqrt{5}}\right)$

CHAPTER 16

All indefinite integrals in this chapter require the term $+K$.

Exercise 16a – p. 247

1. $\frac{1}{4}e^{4x}$
2. $-4e^{-x}$
3. $\frac{1}{3}e^{(3x-2)}$
4. $-\frac{2}{5}e^{(1-5x)}$
5. $-3e^{-2x}$
6. $5e^{(x-3)}$
7. $2e^{(x/2+2)}$
8. $\dfrac{2^x}{\ln 2}$
9. $\dfrac{4^{(2+x)}}{\ln 4}$
10. $\frac{1}{2}e^{2x} - \dfrac{1}{2e^{2x}}$
11. $\dfrac{a^{(1-2x)}}{-2\ln a}$
12. $\dfrac{2^x}{\ln 2} + \frac{1}{3}x^3$
13. $\frac{1}{2}\{e^4 - 1\}$
14. $2\{e^2 - 1\}$
15. $1 - \dfrac{1}{e}$
16. $1 - e^2$

Exercise 16b – p. 250
Answers can be expressed in other forms also.

1. (a) $\frac{1}{2}\ln|x| + K$
   (b) $\frac{1}{2}\ln\{A|x|\}$ or $\ln(Ax^2)$

2. (a) $4\ln|x| + K$
   (b) $4\ln\{A|x|\}$ or $\ln(Ax^4)$
3. (a) $\frac{1}{3}\ln|3x + 1| + K$
   (b) $\frac{1}{3}\ln A|3x + 1|$ or $\ln A|\sqrt[3]{(3x + 1)}|$
4. (a) $-\frac{1}{2}\ln|1 - 2x| + K$
   (b) $-\frac{3}{2}\ln A|1 - 2x|$
   or $\ln A|(1 - 2x)^{-3/2}|$
5. (a) $2\ln|2 + 3x| + K$
   (b) $\ln A(2 + 3x)^2$
6. (a) $-\frac{3}{2}\ln|4 - 2x| + K$
   (b) $\ln A|(4 - 2x)^{-3/2}|$
   or $\ln \dfrac{A}{|(4x - 2x)^{3/2}|}$
   (or each of these results using $\ln|2 - x|$)
7. (a) $-4\ln|1 - x| + K$
   (b) $\ln \dfrac{A}{(1 - x)^4}$
8. (a) $-\frac{5}{7}\ln|6 - 7x| + K$
   (b) $\ln \dfrac{A}{|(6 - 7x)^{5/7}|}$
9. $3\ln 2$
10. $\frac{1}{2}\ln 2 = \ln\sqrt{2}$
11. $2\ln 2 = \ln 4$
12. $\ln 2$

Exercise 16c – p. 251
1. $-\frac{1}{2}\cos 2x$
2. $\frac{1}{7}\sin 7x$
3. $\frac{1}{4}\tan 4x$
4. $-\cos\left(\frac{1}{4}\pi + x\right)$
5. $\frac{3}{4}\sin\left(4x - \frac{1}{2}\pi\right)$
6. $\frac{1}{2}\tan\left(\frac{1}{3}\pi + 2x\right)$
7. $-\frac{1}{4}\cot 4x$
8. $-\frac{2}{3}\cos(3x - \alpha)$
9. $-10\sin\left(\alpha - \frac{1}{2}x\right)$
10. $\frac{5}{4}\sec 4x$
11. $\frac{1}{3}\sin 3x - \sin x$
12. $\frac{1}{2}\tan 2x + \frac{1}{4}\cot 4x$
13. $\frac{1}{3}$
14. $-\frac{1}{4}$
15. 0
16. $\frac{1}{2}$

Exercise 16d – p. 253
1. $\arcsin x$
2. $\arctan\left(\frac{1}{2}x\right)$

3. $\arcsin(\frac{1}{2}x)$
4. $3\arctan x$
5. $\frac{5}{3}\arctan(\frac{1}{3}x)$
6. $\arcsin(\frac{1}{3}x)$
7. $\frac{1}{4}\arctan(\frac{1}{4}x)$
8. $5\arcsin(x/\sqrt{2})$

Exercise 16e – p. 253
1. $\frac{1}{2}\cos(\frac{1}{2}\pi - 2x)$
2. $\frac{1}{4}e^{(4x-1)}$
3. $\frac{1}{7}\tan 7x$
4. $\frac{1}{2}\ln|2x-3|$
5. $\sqrt{(2x-3)}$
6. $-1/\{3(3x-2)\}$
7. $5^x/\ln 5$
8. $-2\operatorname{cosec}\frac{1}{2}x$
9. $\frac{1}{9}(3x-5)^3$
10. $\frac{1}{4}e^{(4x-5)}$
11. $\frac{1}{6}(4x-5)^{3/2}$
12. $-\frac{1}{3}\cot 3x$
13. $-\frac{3}{2}\ln|1-x|$
14. $10^{(x+1)}/\ln 10$
15. $\frac{1}{3}\sin(3x-\frac{1}{3}\pi)$
16. $\frac{1}{4}\tan^2 2x + \frac{1}{2}\ln|\cos x|$
17. $\frac{2}{3}\sqrt{2}$
18. $2e(e-1)$
19. $-\frac{1}{2}$
20. $\frac{1}{6}\pi$

Exercise 16f – p. 258
1. e^{x^4}
2. $-e^{\cos x}$
3. $e^{\tan x}$
4. e^{x^2+x}
5. $e^{(1-\cot x)}$
6. $e^{(x+\sin x)}$
7. $e^{(1+x^2)}$
8. $e^{(x^3-2)}$
9. $\frac{1}{10}(x^2-3)^5$
10. $-\frac{1}{3}(1-x^2)^{3/2}$
11. $\frac{1}{6}(\sin 2x + 3)^3$
12. $-\frac{1}{6}(1-x^3)^2$
13. $\frac{2}{3}(1+e^x)^{3/2}$
14. $\frac{1}{5}\sin^5 x$
15. $\frac{1}{4}\tan^4 x$
16. $\dfrac{1}{3(n+1)}(1+x^{n+1})^3$

17. $-\frac{1}{3}\cot^3 x$
18. $\frac{4}{9}(1+x^{3/2})^{3/2}$
19. $\frac{1}{12}(x^4+4)^3$
20. $-\frac{1}{4}(1-e^x)^4$
21. $\frac{2}{3}(1-\cos\theta)^{3/2}$
22. $\frac{1}{3}(x^2+2x+3)^{3/2}$
23. $\frac{1}{2}e^{(x^2+1)}$
24. $\frac{1}{2}(1+\tan x)^2$

Exercise 16g – p. 259
1. $\frac{1}{2}(e-1)$
2. $\frac{1}{5}$
3. $\frac{1}{2}(\ln 2)^2$
4. $7\frac{5}{15}$
5. $e-1$
6. 9
7. $\frac{1}{2}(e^3-1)$
8. $\frac{13}{24}$
9. $\frac{1}{3}(\ln 3)^3$
10. $\frac{7}{3}$

Exercise 16h – p. 264
1. $x\sin x + \cos x$
2. $e^x(x^2 - 2x + 2)$
3. $\frac{1}{16}x^4(4\ln|3x|-1)$
4. $-e^{-x}(x+1)$
5. $3(\sin x - x\cos x)$
6. $\frac{1}{5}e^x(\sin 2x - 2\cos 2x)$
7. $\frac{1}{5}e^{2x}(\sin x + 2\cos x)$
8. $\frac{1}{32}e^{4x}(8x^2 - 4x + 1)$
9. $-\frac{1}{2}e^{-x}(\cos x + \sin x)$
10. $x(\ln|2x|-1)$
11. xe^x
12. $\frac{1}{72}(8x-1)(x+1)^8$
13. $\sin(x+\frac{1}{6}\pi) - x\cos(x+\frac{1}{6}\pi)$
14. $\dfrac{1}{n^2}(\cos nx + nx\sin nx)$
15. $\dfrac{x^{n+1}}{(n+1)^2}[(n+1)\ln|x|-1]$
16. $\frac{3}{4}(2x\sin 2x + \cos 2x)$
17. $\frac{1}{5}e^x(\sin 2x - 2\cos 2x)$
18. $(2-x^2)\cos x + 2x\sin x$
19. $\dfrac{e^{ax}}{a^2+b^2}(a\sin bx - b\cos bx)$
20. $\frac{1}{3}\sin\theta(3\cos^2\theta + 2\sin^2\theta)$
21. $\frac{1}{2}e^{x^2-2x+4}$

22. $(x^2 + 1)e^x$
23. $-\frac{1}{4}(4 + \cos x)^4$
24. $e^{\sin x}$
25. $\frac{2}{15}\sqrt{(1 + x^5)^3}$
26. $\frac{1}{5}(e^x + 2)^5$
27. $\frac{1}{4}e^{2x-1}(2x - 1)$
28. $-\frac{1}{20}(1 - x^2)^{10}$
29. $\frac{1}{6}\sin^6 x$

Exercise 16i – p. 266
1. 1
2. $\frac{32}{3}\ln 2 - \frac{7}{4}$
3. e
4. $-\frac{1}{2}(e^\pi + 1)$
5. $\frac{16}{15}$
6. $\frac{1}{4}\pi^2 - 2$
7. $\frac{1}{2}\pi - 1$
8. $e - 2$
9. $\frac{1}{12}\pi(4\sqrt{3} - 3) - \ln\sqrt{2}$

Mixed Exercise 16 – p. 267
1. $\frac{1}{4}e^{2x}(2x^2 - 2x + 1)$
2. e^{x^2}
3. $\frac{1}{6}(3\tan x - 4)^2$
4. $\frac{1}{4}(x + 1)^2\{2\ln(x + 1) - 1\}$
5. $\frac{1}{4}\tan^4 x$
6. $(x^2 - 2)\sin x + 2x\cos x$
7. $-e^{\cos x}$
8. $\frac{1}{288}(2x + 3)^8(16x - 3)$
9. $-\frac{1}{2}e^{(1-x)^2}$
10. $\frac{1}{4}e^{(2x-1)}(2x - 1)$
11. $\frac{1}{6}\sin^6 x$
12. $-\frac{1}{4}(4 + \cos x)^4$
13. $\frac{1}{2}e^{(x^2 - 2x + 3)}$
14. $-\frac{1}{30}(1 - x^3)^{10}$
15. $\frac{1}{3}(e^9 - e^3)$
16. $\frac{1}{4}$
17. $\ln\frac{1}{2}$
18. 1
19. $\frac{1}{9}(e^3 - 1)$
20. $\frac{1}{8}(\pi - 2)$
21. $\frac{1}{2}(\ln 2)^2$
22. $\frac{1}{3}\ln 2^8 - \frac{7}{9}$

CHAPTER 17

All indefinite integrals in this chapter require the addition of a constant of integration.

Exercise 17a – p. 271
1. $\ln(4 + \sin x)$
2. $\frac{1}{3}\ln|3e^x - 1|$
3. $\dfrac{1}{4(1 - x^2)^2}$
4. $\dfrac{1}{2\cos^2 x}$
5. $\frac{1}{4}\ln(1 + x^4)$
6. $\ln|x^2 + 3x - 4|$
7. $\frac{2}{3}\sqrt{(2 + x^3)}$
8. $\dfrac{-1}{\sin x - 2}$
9. $\ln|\ln x|$
10. $\dfrac{-1}{5\sin^5 x}$
11. $\dfrac{1}{3\cot^3 x}$
12. $-2\sqrt{(1 - e^x)}$
13. $\frac{1}{6}\ln|3x^2 - 6x + 1|$
14. $\dfrac{-1}{(n - 1)\sin^{n-1} x}$ $(n \neq 1)$
15. $\dfrac{1}{(n - 1)\cos^{(n-1)} x}$ $(n \neq 1)$
16. $\ln|4 + \sec x|$
17. $\dfrac{1}{2(1 - \tan x)^2}$
18. $\dfrac{1}{(3 + \cos x)}$
19. $\ln|\sec x + \tan x|$
20. $-\ln|\text{cosec } x + \cot x|$
21. $\ln 3$
22. $\ln\sqrt{2}$
23. $\frac{1}{18} - \frac{1}{128}$
24. $\dfrac{e - 1}{2(e + 1)}$
25. 0
26. $\dfrac{1}{\ln 4}$

Exercise 17b – p. 274
1. $2\ln\left|\dfrac{x}{x + 1}\right|$

2. $\ln\left|\dfrac{x-2}{x+2}\right|$

3. $\frac{1}{2}\ln|x^2-1|$

4. $\frac{1}{2}\ln\left|\dfrac{(x+2)^3}{x}\right|$

5. $\ln\dfrac{(x-3)^2}{|x-2|}$

6. $\frac{1}{2}\ln\dfrac{|x^2-1|}{x^2}$

7. $x-\ln|x+1|$

8. $x+4\ln|x|$

9. $x-4\ln|x+4|$

10. $\ln\dfrac{|1-x|}{x^4}$

11. $x-\frac{1}{2}\ln\left|\dfrac{x+1}{x-1}\right|$

12. $x+\ln\dfrac{|x+1|}{(x+2)^4}$

13. $\frac{1}{2}\ln|x^2-1|$

14. $\dfrac{-1}{x^2-1}$

15. $\ln\left|\dfrac{x-1}{x+1}\right|$

16. $\ln|x^2-5x+6|$

17. $\ln\dfrac{(x-3)^6}{(x-2)^4}$

18. $\ln\left|\dfrac{(x-3)^3}{x-2}\right|$

19. $4+\ln 5$

20. $\ln\frac{1}{6}$

21. $\frac{1}{2}\ln\frac{12}{5}$

22. $\ln\frac{5}{3}$

23. $\frac{5}{36}$

24. $1-\frac{3}{2}\ln\frac{7}{5}$

Exercise 17c – p. 275

1. $\dfrac{2}{x-2}-\dfrac{2}{x-1}$; $\dfrac{-2}{(x-2)^2}+\dfrac{2}{(x-1)^2}$;

$\dfrac{4}{(x-2)^3}-\dfrac{4}{(x-1)^3}$;

2. $\dfrac{\frac{9}{5}}{x-3}-\dfrac{\frac{3}{5}}{2x-1}$;

$\dfrac{6}{5(2x-1)^2}-\dfrac{9}{5(x-3)^2}$;

$\dfrac{-24}{5(2x-1)^3}+\dfrac{18}{5(x-3)^3}$

3. $\dfrac{\frac{2}{3}}{x-4}+\dfrac{\frac{1}{3}}{x+2}$;

$\dfrac{-1}{3(x+2)^2}-\dfrac{2}{3(x-4)^2}$;

$\dfrac{2}{3(x+2)^3}+\dfrac{4}{3(x-4)^3}$

4. $\dfrac{1}{x-3}-\dfrac{1}{x+2}$; $\dfrac{1}{(x+2)^2}-\dfrac{1}{(x-3)^2}$;

$\dfrac{-2}{(x+2)^3}+\dfrac{2}{(x-3)^3}$;

5. $\dfrac{3}{2x+3}-\dfrac{1}{x+1}$;

$\dfrac{-6}{(2x+3)^2}+\dfrac{1}{(x+1)^2}$;

$\dfrac{24}{(2x+3)^3}-\dfrac{2}{(x+1)^3}$

6. $\dfrac{\frac{3}{2}}{x-1}-\dfrac{\frac{9}{2}}{3x-1}$;

$\dfrac{27}{2(3x-1)^2}-\dfrac{3}{2(x-1)^2}$;

$\dfrac{-81}{(3x-1)^3}+\dfrac{3}{(x-1)^3}$

Exercise 17d – p. 279

1. $\frac{1}{4}(2x-\sin 2x)$

2. $\sin x-\frac{1}{3}\sin^3 x$

3. $-\frac{1}{15}\cos x(15-10\cos^2 x+3\cos^4 x)$

4. $\tan x-x$

5. $\frac{1}{32}\{12x-8\sin 2x+\sin 4x\}$

6. $\frac{1}{2}\tan^2 x-\ln|\sec x|$

7. $\frac{1}{32}\{12x+8\sin 2x+\sin 4x\}$

8. $\frac{1}{3}\cos x(\cos^2 x-3)$

9. $\frac{1}{15}\sin^3\theta(5-3\sin^2\theta)$

10. $(\sin^{11}\theta)(\frac{1}{11}-\frac{1}{13}\sin^2\theta)$

11. $(\sin^{(n+1)}\theta)\left(\dfrac{1}{n+1}-\dfrac{1}{n+3}\sin^2\theta\right)$

$(n\neq -1 \text{ or } -3)$

12. $\frac{1}{32}(4\theta-\sin 4\theta)$

13. $\frac{1}{15}(10\sin 3\theta-3\sin 5\theta-15\sin\theta)$

14. $\frac{1}{15}\tan^3\theta(3\tan^2\theta+5)$

15. $-\frac{1}{7}(\cos 7t+7\cos t)$

16. $\frac{1}{21}(3\sin 7t+7\sin 3t)$

17. $\frac{1}{16}(2\cos 4t-\cos 8t)$

18. $\frac{1}{8}(2\sin 2t-\sin 4t)$

19. $\left(\dfrac{-1}{n+m}\right)\cos(n+m)t$

$-\left(\dfrac{1}{n-m}\right)\cos(n-m)t$

$(n^2 \neq m^2)$

20. $\dfrac{1}{2(n+m)}\sin(n+m)t$

$+\dfrac{1}{2(n-m)}\sin(n-m)t$

$(n^2 \neq m^2)$

21. $\frac{1}{12}$

22. $\frac{13}{15}-\frac{1}{4}\pi$

23. $\frac{1}{5}(2\sqrt{3}-3)$

24. $\frac{1}{16}\sqrt{3}$

CHAPTER 18

All indefinite integrals require the term $+K$.

Exercise 18a – p. 285

1. $\frac{1}{21}(x+3)^6(3x+2)$

2. $\frac{1}{2}\arctan\dfrac{x}{2}$

3. $-\frac{2}{3}(x+6)\sqrt{(3-x)}$

4. $\frac{2}{15}(3x-2)(x+1)^{3/2}$

5. $\dfrac{1-5x}{10(x-3)^5}$

6. $\ln|x+\sqrt{(1+x^2)}|$

7. $\frac{4}{135}(9x+8)(3x-4)^{3/2}$

8. $\frac{3}{10}\arctan\frac{2}{5}x$

9. $\ln|x+2+\sqrt{(x^2+4x+3)}|$

10. $-\frac{1}{36}(8x+1)(1-x)^8$

11. $\frac{1}{2}\arcsin 2x$

12. $\dfrac{5+4x}{12(4-x)^4}$

Exercise 18b – p. 286

1. $\frac{1}{2}e^{2x+3}$

2. $\frac{1}{6}(2x^2-5)^{3/2}$

3. $\frac{1}{12}(6x-\sin 6x)$

4. $-\frac{1}{2}e^{-x^2}$

5. $-\frac{1}{8}(\cos 4\theta + 2\cos 2\theta)$

6. $\frac{1}{110}(10u-7)(u+7)^{10}$

7. $-\dfrac{1}{12(x^3+9)^4}$

8. $\frac{1}{2}\ln|1-\cos 2y|$ $\quad(y \neq n\pi)$

9. $\frac{1}{2}\ln|2x+7|$

10. $\arcsin u$

11. $-\frac{2}{9}(1+\cos 3x)^{3/2}$

12. $\frac{1}{16}(\sin 4x - 4x\cos 4x)$

13. $\frac{1}{2}\ln|x^2+4x-5|$

14. $\frac{1}{2}\ln|x^2+4x+5|$

15. $-\frac{1}{4}(x^2+4x-5)^{-2}$

16. $-(9-y^2)^{3/2}$

17. $\frac{1}{13}e^{2x}(2\cos 3x + 3\sin 3x)$

18. $x(\ln|5x|-1)$

19. $\frac{1}{6}\sin 2x(3-\sin^2 2x)$

20. $-e^{\cot x}$

21. $-2\sqrt{(7+\cos y)}$

22. $e^x(x^2-2x+2)$

23. $\frac{1}{2}\ln|x^2-4|$

24. $x+\ln\left|\dfrac{x-2}{x+2}\right|$

25. $\frac{1}{4}\ln\left|\dfrac{x-2}{x+2}\right|$

26. $\frac{1}{30}(3\sin 5x + 5\sin 3x)$

27. $-\frac{1}{30}\cos 2\theta(15 - 10\cos^2 2\theta + 3\cos^4 2\theta)$

28. $\frac{1}{15}\cos^3 u(3\cos^2 u - 5)$

29. $\tan\theta - \theta$

30. $\arcsin x + 2\sqrt{(1-x^2)}$

31. $\frac{1}{27}(9y^2\sin 3y + 6y\cos 3y - 2\sin 3y)$

32. $-\ln|1-\tan x|$

33. $\frac{1}{3}(7+x^2)^{3/2}$

34. $-\frac{1}{5}\cos(5\theta - \frac{1}{4}\pi)$

35. $\sin\theta\{\ln|\sin\theta|-1\}$

36. $e^{\tan u}$

37. $\dfrac{2x-1}{10(3-x)^6}$

38. $\frac{1}{3}\tan^3 x$

Exercise 18c – p. 289

1. $y^2 = A - 2\cos x$

2. $\dfrac{1}{y}-\dfrac{1}{x} = A$

3. $2y^3 = 3(x^2+4y+A)$

4. $x = A\sec y$

5. $(A-x)y = 1$

6. $y = \ln\dfrac{A}{\sqrt{(1-x^2)}}$

7. $y = A(x-3)$

8. $x+A = 4\ln|\sin y|$

9. $u^2 = v^2 + 4v + A$

10. $16y^3 = 12x^4 \ln|x| - 3x^4 + A$

11. $y^2 + 2(x + 1)e^{-x} = A$

12. $\sin x = A - e^{-y}$

13. $2r^2 = 2\theta - \sin 2\theta + A$

14. $u + 2 = A(v + 1)$

15. $y^2 = A + (\ln|x|)^2$

16. $y^2 = Ax(x + 2)$

17. $4v^3 = 3(2 + t)^4 + A$

18. $1 + y^2 = Ax^2$

19. $Ar = e^{\tan\theta}$

20. $y^2 = A - \operatorname{cosec}^2 x$

21. $v^2 + A = 2u - 2\ln|u|$

22. $e^{-x} = e^{1-y} + A$

23. $A - \dfrac{1}{y} = 2\ln|\tan x|$

24. $y - 1 = A(y + 1)(x^2 + 1)$

8. $\ln\sqrt{(x^2 + 1)} - \arctan x$

9. $-2\sqrt{(\cos x)}$

10. $\sqrt{2}$

11. $\frac{256}{15}$

12. $\frac{1}{2} - \ln\sqrt{2}$

13. $9 - \frac{16}{3}\sqrt{2}$

14. $\ln\frac{9}{4}$

15. $\ln\left\{\dfrac{\ln 3}{\ln 2}\right\}$

16. $\ln\frac{1}{6}$

17. $3\frac{1}{2} + 6\ln\frac{2}{3}$

18. $x^3y = y - 1$

19. $y = \tan\{\frac{1}{2}(x^2 - 4)\}$

20. $y^2 = 2x$

Exercise 18d – p. 291

1. $s\dfrac{ds}{dt} = k$

2. $\dfrac{dh}{dt} = k\ln|H - h|$

3. $\dfrac{dp}{dt} kp(s - p)$

Exercise 18e – p. 294

1. $y^3 = x^3 + 3x - 13$

2. $e^t(5 - 2\sqrt{s}) = 1$

3. $3(y^2 - 1) = 8(x^2 - 1)$

4. $y = x^2 - x$

5. $y = e^x - 2$

6. $y = 5 - \dfrac{3}{x}$

7. $y = \pm\sqrt{(4 + e^{-3} - e)}$
$= \pm 1.154$

8. $(y + 1)^2(x + 1) = 2(x - 1)$

9. $2y = x^2 + 6x$

10. $4y^2 = (y + 1)^2(x^2 + 1)$

Mixed Exercise 18 – p. 295

1. $\frac{1}{10}(1 + x^2)^5$

2. $-\frac{1}{9}e^{-3x}(3x + 1)$

3. $\frac{1}{10}(\sin 5x + 5\sin x)$

4. $x + \ln|x + 2|$

5. $\dfrac{-1}{3(x^3 + 1)}$

6. $\ln\left|\dfrac{x - 4}{x - 1}\right|$

7. $\ln\left|\dfrac{x}{2x + 1}\right|$

CONSOLIDATION D

Multiple Choice Exercise D – p. 298

1. B	**6. B**	**11. F**
2. B	**7. A, C**	**12. F**
3. D	**8. B**	**13. T**
4. D	**9. B, C**	**14. F**
5. B	**10. A, D**	**15. F**

Miscellaneous Exercise D – p. 300

1. (a) $\frac{1}{4}e^{2x}(6x + 5)$ (b) $\frac{1}{64}(\pi + 2)$

2. (a) $\frac{1}{12}(6x + \sin 6x) + K$
(b) $\frac{1}{2}\arctan(\frac{1}{2}e^x) + K$
(c) $\frac{1}{4}e^{2x}(2x - 1) + K$

3. $A = -1, B = 1; \ln 2 - \frac{1}{2}$

4. $\arctan e - \frac{1}{4}\pi = 0.43$ to 2 d.p.

5. $\frac{1}{20}$

6. $2, 3, -1; y - 3 = -\frac{2}{5}(x - 1)$

7. $2\ln A\left|\dfrac{8 - x}{4 - x}\right|; t = 2\ln\left|\dfrac{8 - x}{3(4 - x)}\right|;$
$2\ln\frac{5}{3}$ minutes, i.e. 61 minutes

8. $\dfrac{\frac{1}{2}}{x + 1} - \dfrac{\frac{1}{2}}{x + 3}; y = \sqrt{\left(\dfrac{8(x - 1)}{x + 3}\right)}$

9. (a) $\cos 3x - 3x\sin 3x$
(b) $\frac{1}{3}x\sin 3x + \frac{1}{9}\cos 3x + k$

10. $\frac{1}{2}(e^2 + 3)$

11. $\ln 18 + \frac{1}{8}\pi$

12. (a) $\frac{8}{105}$ (b) $\sqrt{3} - 1 - \frac{1}{12}\pi$

13. $y = 1 - \dfrac{2}{x}$

14. (a) $-\frac{2}{9}$ (b) $\frac{2}{3}$

15. $s = \frac{1}{2}\arctan(\frac{1}{2}t) - \frac{1}{8}\pi$

16. 3

17. $y = \dfrac{10}{20x + 1}$

18. $\dfrac{4}{2x - 1} - \dfrac{1}{x - 1}$

19. $\dfrac{dx}{dt} = k(p - x)^2$ (b) $x = \dfrac{kp^2 t}{1 + kpt}$

20. (a) $\sec x (\sec^2 x + \tan^2 x)$
    (b) $\sec x$;
    $\tan y - y = \frac{1}{4}\sec 2x \tan 2x$
    $\qquad + \frac{1}{4}\ln(\sec 2x + \tan 2x) + K$

21. $y = x^2 - 5; x^2 > 5$

22. $\frac{1}{4}\pi$

23. (b) $\dfrac{dp}{dt} = kp(1 - p)$

    (d) $2t = \ln\left|\dfrac{p}{p - 1}\right|$

24. 0.034 i.e. $(\frac{11}{6}\sqrt{3} - \pi)$

CHAPTER 19

Exercise 19a – p. 312

1. (a) $\overrightarrow{AC}$ (b) $\overrightarrow{BD}$
   (c) $\overrightarrow{AD}$ (d) $\overrightarrow{DB}$

2. $\mathbf{a}, \mathbf{b} - \mathbf{a}, 2\mathbf{b} - 2\mathbf{a}$

5. $\mathbf{b} - \mathbf{a}, \mathbf{a} + \mathbf{b}, \mathbf{a} - 3\mathbf{b}, 2\mathbf{b}, 2\mathbf{a} - 2\mathbf{b}$

6. $\overrightarrow{DE} = \overrightarrow{CH} = \overrightarrow{BG} = \mathbf{c}$,
   $\overrightarrow{DC} = \overrightarrow{EH} = \overrightarrow{FG} = \mathbf{a}$,
   $\overrightarrow{FE} = \overrightarrow{GH} = \overrightarrow{BC} = \mathbf{b}$

8. $\mathbf{c} - \mathbf{a}, \mathbf{b} - \mathbf{a}, \mathbf{b} - \mathbf{c}$

9. $\mathbf{b} + \mathbf{c} - \mathbf{a}, \mathbf{b} - \mathbf{c} - \mathbf{a}, \mathbf{a} + \mathbf{b} + \mathbf{c}$,
   $\mathbf{a} + \mathbf{b} - \mathbf{c}$

10. (a) $90°$ (b) $19.1°$ (c) $40.9°$

Exercise 19b – p. 316

1. (a) $\frac{1}{2}(\mathbf{a} + \mathbf{c})$ (b) $\frac{1}{2}(\mathbf{b} + \mathbf{d})$

2. $\frac{1}{4}(3\mathbf{a} + \mathbf{d})$

3. $\frac{1}{7}(3\mathbf{b} + 4\mathbf{c})$

4. (a) $\frac{1}{4}(\mathbf{a} + 2\mathbf{b} + \mathbf{c})$ (b) $\frac{1}{4}(\mathbf{b} + 2\mathbf{c} + \mathbf{d})$

5. $\mathbf{PQ} = \frac{1}{2}(\mathbf{c} - \mathbf{a}), \mathbf{AC} = (\mathbf{c} - \mathbf{a}) = 2\mathbf{PQ}$

6. $\frac{3}{4}(\mathbf{d} - \mathbf{a})$

7. $\frac{1}{2}(\mathbf{c} + \mathbf{d} - 2\mathbf{a})$

8. (a) $\frac{1}{3}(\mathbf{a} + \mathbf{b} + \mathbf{c})$ (b) $\frac{1}{3}(\mathbf{a} + \mathbf{b} + \mathbf{c})$

Exercise 19c – p. 323

1. (a) $3\mathbf{i} + 6\mathbf{j} + 4\mathbf{k}$ (b) $\mathbf{i} - 2\mathbf{j} - 7\mathbf{k}$
   (c) $\mathbf{i} - 3\mathbf{k}$ (d) $-4\mathbf{k}$

2. (a) $(5, -7, 2)$ (b) $(1, 4, 0)$
   (c) $(0, 1, -1)$ (d) $(2, 0, -3)$

3. (a) $\sqrt{21}$ (b) 5   (c) 3   (d) 7

4. (a) 6   (b) 7   (c) $\sqrt{206}$ (d) 9

5. (a) $\frac{3}{5}\mathbf{i} + \frac{4}{5}\mathbf{j}$
   (b) $\frac{1}{9}\mathbf{i} + \frac{8}{9}\mathbf{j} - \frac{4}{9}\mathbf{k}$
   (c) $\frac{2}{3}\mathbf{i} - \frac{2}{3}\mathbf{j} + \frac{1}{3}\mathbf{k}$
   (d) $\frac{6}{7}\mathbf{i} - \frac{3}{7}\mathbf{j} - \frac{2}{7}\mathbf{k}$

6. (a) $3\mathbf{i} + 4\mathbf{k}$ (b) $2\mathbf{i} + 2\mathbf{j} + 2\mathbf{k}$
   (c) $2\mathbf{i} + 3\mathbf{j} + 3\mathbf{k}$ (d) $-6\mathbf{i} + 12\mathbf{j} - 8\mathbf{k}$

8. (b), (d) and (e)

9. (c)

10. $\lambda = \frac{1}{2}$

11. (b), (e) and (f)

12. (a) neither   (b) parallel
    (c) equal   (d) neither

13. (a) $-2\mathbf{i} - 3\mathbf{j} + 7\mathbf{k}$ (b) $-\mathbf{j} + 3\mathbf{k}$
    (c) $-2\mathbf{i} - 3\mathbf{j} + 4\mathbf{k}$

14. $\sqrt{62}, \sqrt{10}, 2\sqrt{6}$

15. $\sqrt{5}$

16. $y^2 = 4x$; parabola

17. $\overrightarrow{AB} = -\mathbf{i} + 3\mathbf{j}, \overrightarrow{BD} = -\mathbf{j} - 4\mathbf{k}$
    $\overrightarrow{CD} = -2\mathbf{i} + 2\mathbf{j} - 6\mathbf{k}$,
    $\overrightarrow{AD} = -\mathbf{i} + 2\mathbf{j} - 4\mathbf{k}$

Exercise 19d – p. 329

1. $\sqrt{13}; 3$

2. $\frac{1}{3}(4\mathbf{i} - \mathbf{j} + 4\mathbf{k})$

3. (a) $\frac{2}{5}\mathbf{i} + \frac{1}{5}\mathbf{j} + \frac{7}{5}\mathbf{k}$ (b) $-2\mathbf{i} + \mathbf{j} + 5\mathbf{k}$

4. (a) no   (b) no   (c) no

8. (a) $5\mathbf{i} + 6\mathbf{j} + \mathbf{k}$ (b) $7\mathbf{i} + 6\mathbf{j} + 8\mathbf{k}$

Exercise 19e – p. 333

1. (a) $\mathbf{i} + \mathbf{j} + \mathbf{k}$
   (b) $3\mathbf{i} + 5\mathbf{k}$
   (c) $2\mathbf{i} + 3\mathbf{j} + 4\mathbf{k}$

2. (a) $\mathbf{r} = \mathbf{i} - 3\mathbf{j} + 2\mathbf{k} + \lambda(5\mathbf{i} + 4\mathbf{j} - \mathbf{k})$;
   (b) $\mathbf{r} = 2\mathbf{i} + \mathbf{j} + \lambda(3\mathbf{j} - \mathbf{k})$;
   (c) $\mathbf{r} = \lambda(\mathbf{i} - \mathbf{j} - \mathbf{k})$

3. (a) no   (b) yes   (c) no

4. (a) $(\frac{5}{2}, \frac{7}{2}, \frac{9}{2})$ (b) $(3, 4, 5)$
   (c) There are alternative answers
   $\mathbf{r} = 2\mathbf{i} + 3\mathbf{j} + 4\mathbf{k} + \lambda(2\mathbf{i} + 2\mathbf{j} + 6\mathbf{k})$
   $\mathbf{r} = 2\mathbf{i} + 3\mathbf{j} + 4\mathbf{k} + \lambda(\mathbf{i} + \mathbf{j} + 5\mathbf{k})$
   $\mathbf{r} = \mathbf{i} + 2\mathbf{j} - \mathbf{k} + \lambda(3\mathbf{i} + 3\mathbf{j} + 11\mathbf{k})$

5. (a) $\mathbf{r} = 3\mathbf{i} + \mathbf{j} - 4\mathbf{k} + \lambda(\mathbf{i} - 3\mathbf{j} - 6\mathbf{k})$;
   $(\frac{7}{3}, 3, 0); (0, 10, 14); (\frac{10}{3}, 0, -6)$
   (b) $\mathbf{r} = \mathbf{i} + \mathbf{j} + 7\mathbf{k} + \lambda(2\mathbf{i} + 3\mathbf{j} - 6\mathbf{k})$;
   $(\frac{10}{3}, \frac{9}{2}, 0); (0, -\frac{1}{2}, 10); (\frac{1}{3}, 0, 9)$

6. $\mathbf{r} = 5\mathbf{i} - 2\mathbf{j} - 4\mathbf{k} + \lambda(3\mathbf{i} + 4\mathbf{j} + 5\mathbf{k})$;
$(\frac{13}{2}, 0, -\frac{3}{2})$

7. $\mathbf{r} = 2\mathbf{i} - \mathbf{j} - \mathbf{k} + \lambda(6\mathbf{i} + 3\mathbf{j} + 4\mathbf{k})$

8. (a) $\mathbf{r} = \mathbf{i} + 2\mathbf{j} + 4\mathbf{k} + \lambda(\mathbf{i} + 2\mathbf{j} - 2\mathbf{k})$
(b) $(3, 6, 0)$

Exercise 19f – p. 336

1. parallel
2. intersecting; $\mathbf{r} = \mathbf{i} + 2\mathbf{j}$
3. skew
4. $a = -3$; $\mathbf{r} = -\mathbf{i} + 3\mathbf{j} + 4\mathbf{k}$
5. $t = 6$; $(1, 2, 2)$
6. (a) $p = -2$ (b) $p = 0$

Exercise 19g – p. 342

1. (a) 30
(b) 0; $\mathbf{a}$ and $\mathbf{b}$ are perpendicular
(c) -1
2. (a) $7, \frac{1}{3}\sqrt{7}$ (b) $14, \sqrt{(\frac{7}{19})}$
(c) $3, \frac{3}{58}\sqrt{58}$ (d) $1, \frac{1}{5}$
3. (a) $a^2 - \mathbf{a}\cdot\mathbf{b}$ (b) $a^2 + b^2 - 2\mathbf{a}\cdot\mathbf{b}$
(c) $b^2 - a^2$ (d) $2\mathbf{b}\cdot\mathbf{c}$
4. (a) 0   (b) $-b^2$ (c) a^2
(d) $2a^2 - b^2$
5. 4
6. (a) -4 (b) 15   (c) 11
(d) -12 (e) 29
7. $-\sqrt{\frac{7}{34}}$; $\arccos\sqrt{\frac{7}{10}}$
11. (a) $\arccos\frac{19}{21}$
(b) 0, i.e. parallel lines

Mixed Exercise 19 – p. 344

1. $\overrightarrow{DE} = -\mathbf{a}$, $\overrightarrow{BC} = \mathbf{c} - \mathbf{a}$,
$\overrightarrow{CE} = \mathbf{e} - \mathbf{c}$, $\overrightarrow{CD} = \mathbf{a} + \mathbf{e} - \mathbf{c}$,
$\overrightarrow{AD} = \mathbf{a} + \mathbf{e}$

3.

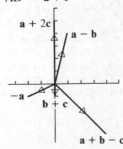

5. $\overrightarrow{CG} = \overrightarrow{CE} + \overrightarrow{EG}$
$\overrightarrow{CG} = \frac{1}{2}\overrightarrow{CA} + \dfrac{1}{1+\lambda}(\frac{1}{2}\overrightarrow{AC} + \overrightarrow{CB})$
$\overrightarrow{CG} = \overrightarrow{CD} + \overrightarrow{DG}$
$\overrightarrow{CG} = \frac{1}{2}\overrightarrow{CB} + \dfrac{1}{1+\mu}(\frac{1}{2}\overrightarrow{BC} + \overrightarrow{CA})$

6. (a) $\mathbf{r} = \mathbf{i} - 2\mathbf{j} + 2\mathbf{k} + \lambda(-2\mathbf{i} + 2\mathbf{j} - 6\mathbf{k})$
$= \mathbf{i} - 2\mathbf{j} + 2\mathbf{k} + \lambda(\mathbf{i} - \mathbf{j} + 3\mathbf{k})$
(b) $\mathbf{r} = 8\mathbf{i} + \mathbf{j} + 4\mathbf{k} + \lambda(7\mathbf{i} + 3\mathbf{j} + 2\mathbf{k})$
7. (a) $-\mathbf{j} + 3\mathbf{k}$ (b) $-\frac{17}{7}\mathbf{j} + \frac{12}{7}\mathbf{k}$
8. (a) $17\mathbf{i} + 2\mathbf{j} + 12\mathbf{k}$
(b) $-3\mathbf{i} + 2\mathbf{j} - 10\mathbf{k}$
9. AB and CD, BC and AD
10. (a) 17   (b) -3
11. (a) perpendicular; intersect at $2\mathbf{i} - \mathbf{j}$
(b) neither parallel nor perpendicular;
intersect at $3\mathbf{i} - \mathbf{j} - \mathbf{k}$
12. 81°

CHAPTER 20

Exercise 20a – p. 349

1. (a) 10.003 333; 10.003 332
(b) 2.083 333; 2.080 084
(c) 3.979 167; 3.979 057
2. (a) 0.857   (b) 0.515   (c) 0.719
3. $\left\{\dfrac{x}{1+x} + \ln(1+x)\right\}\delta x$; 0.75
4. $(\sec^2 x)\delta x$; $1 + \frac{1}{16}\pi$
5. 2.0125
6. $\frac{1}{8}\sqrt{2}$; $\frac{1}{8}a\sqrt{2}$

Exercise 20b – p. 352

1. 0.099 cm/s
2. 30 cm³/s
3. 8 m²/s
4. Decreasing at 0.126 cm/s
5. -2
6. $4a$ cm/minute

Exercise 20c – p. 357

1. (a) $\frac{8}{3}$ (b) $\frac{16}{3}$

2. 9

3. 36

4. $\frac{4}{3}$

5. 1

6. $e - 2$

7. $\frac{1}{6}$

8. $\frac{1}{2}\sqrt{3} - \frac{1}{6}\pi$

9. (a) 1   (b) $2 - \ln 3$

10. $8(\ln 2)$

11. $\frac{1}{4}\pi a^2$

12. (a)

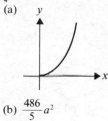

    (b) $\dfrac{486}{5}a^2$

13. (a) $\frac{4}{3}$ (b) $\frac{1}{3}$

14. $4\sqrt{3}$

Exercise 20d – p. 361

1. (a) 24   (b) 22

2. (a) 0.806   (b) 0.705

3. (a) 1.462   (b) 1.625

4. (a) 1.243   (b) 1.282

5.

24	0.806
22	0.705
21.33	0.667

6. 0.34 m

Exercise 20e – p. 368

1. $\frac{512}{15}\pi$

2. $\frac{1}{2}\pi(e^6 - 1)$

3. $\frac{1}{2}\pi$

4. $\frac{64}{5}\pi$

5. 2π

6. 8π

7. 8π

8. $\frac{3}{5}\pi(\sqrt[3]{32} - 1)$

9. $\frac{1}{2}\pi(e^2 - 1)$

10. $\frac{16}{15}\pi$

11. $\frac{16}{15}\pi$

12. $\frac{1}{2}\pi^2$

13. $\frac{5}{2}\pi$

14. $\frac{3}{10}\pi$

Mixed Exercise 20 – p. 369

1. $0.15 \text{ cm}^3/\text{s}$

2. $e^4 - 1$

3. $\frac{16}{3}$; 8π

4. (a) $4\sqrt{3}$

    (b) (i) $\dfrac{48\pi\sqrt{3}}{5}$ (ii) $\dfrac{9\pi}{2}$

5. (a) $e - \dfrac{1}{e}$ (b) $\dfrac{1}{2}\left(e^2 - \dfrac{1}{e^2}\right)$

6. (a) 4.0208   (b) 10.955

7. $\frac{32}{3}$ (b) 8π

8. 2; $\frac{1}{2}\pi^2$

9. (a) 4 sq units   (b) $-4\frac{1}{2}$

10. (a) $27\frac{1}{2}$ (b) $27\frac{1}{2}$

    (c) the value in (a) is exact because the area is that of a trapezium

CHAPTER 21

Exercise 21a – p. 377

1. $-i, i, i, -i, 1, i$

2. (a) $10 + 4i$ (b) $7 + 2i$

    (c) $6 - 2i$ (d) $(a + c) + (b + d)i$

3. (a) $-4 + 6i$ (b) $1 - 4i$

    (c) $-2 + 16i$ (d) $(a - c) + (b - d)i$

4. (a) $10 - 5i$ (b) $39 + 23i$

    (c) $11 - 7i$ (d) 25

    (e) $3 - 4i$ (f) $-2 + 2i$

    (g) $-4 + 3i$ (h) $x^2 + y^2$

    (i) $-3 + i$ (j) $(a^2 - b^2) + 2abi$

5. (a) $1 + i$ (b) $\frac{9}{25} + \frac{13}{25}i$

    (c) $\frac{4}{17} + \frac{16}{7}i$ (d) i (e) $-i$

    (f) $\dfrac{x^2 - y^2}{x^2 + y^2} + \dfrac{2xy}{x^2 + y^2}i$

    (g) $1 - 3i$ (h) $-3 - 2i$

6. (a) $x = 9, y = -7$

    (b) $x = -\frac{3}{2}, y = \frac{7}{2}$

    (c) $x = \frac{7}{2}, y = \frac{1}{2}$

    (d) $x = 2, y = 0$

    (e) $x = 13, y = 0$

    (f) $x = 15, y = 8$

    (g) $x = 11, y = 3$

    (h) $x = 2, y = 1$

    or $x = -2, y = -1$

7. (a) $7, -1$ (b) $-2, -2$

    (c) $25, 0$ (d) $\frac{10}{17}, \frac{11}{17}$

    (e) $\frac{9}{5}, -\frac{4}{5}$ (f) $0, \dfrac{-2y}{x^2 + y^2}$

8. (a) $\pm(2-i)$ (b) $\pm(5-2i)$
     (c) $\pm(1+i)$ (d) $\pm(4+i)$
     (e) $\pm(1+5i)$

Exercise 21b – p. 380

1. (a) $-\frac{1}{2}\pm\frac{1}{2}i\sqrt{3}$ (b) $-\frac{7}{4}\pm\frac{1}{4}\sqrt{41}$
     (c) $\pm 3i$ (d) $-\frac{1}{2}\pm\frac{1}{2}i\sqrt{11}$
     (e) $\pm 1, \pm i$ (f) $-\frac{1}{6}\pm\frac{1}{6}i\sqrt{35}$

2. (a) $x^2 + 1 = 0$
     (b) $x^2 - 4x + 5 = 0$
     (c) $x^2 - 2x + 10 = 0$
     (d) $x^3 - 4x^2 + 6x - 4 = 0$

3. (a) 4, 5   (b) 6, 25   (c) 0, 1
     (d) -24, 169

4. (a) $x^2 + 2x + 2 = 0$
     (b) $x^2 + 10x + 26 = 0$
     (c) $x^2 - 2x + 10 = 0$
     (d) $x^2 - 8x + 17 = 0$

When one root is real, so is the other and there is no special relationship between them.

Exercise 21d – p. 388

Arguments are given in terms of π when exact and in degrees to 3 s.f. otherwise.

1. (a) $\sqrt{13}, -33.7°$ (b) $\sqrt{17}, 166°$
     (c) $5, -127°$ (d) $13, 67.4°$
     (e) $\sqrt{2}, -\frac{1}{4}\pi$ (f) $\sqrt{2}, \frac{3}{4}\pi$
     (g) $6, 0$ (h) $4, -\frac{1}{2}\pi$
     (i) $2, \frac{1}{2}\pi$ (j) $\sqrt{2}, \frac{1}{4}\pi$
     (h) $\sqrt{2}, \frac{3}{4}\pi$ (l) $\sqrt{170}, 32.5°$
     (m) $1, \frac{3}{4}\pi$ (n) $2, \frac{2}{3}\pi$
     (p) $1, -\frac{5}{6}\pi$
     (q) $\sqrt{(a^2 + b^2)}$, $\arctan \frac{a}{b}$ (ambiguous
     until $a + b$ are known)

2. (a) $\sqrt{2}(\cos\frac{1}{4}\pi + i\sin\frac{1}{4}\pi)$
     (b) $2\{\cos(-\frac{1}{6}\pi) + i\sin(-\frac{1}{6}\pi)\}$
     (c) $5\{\cos(-127°) + i\sin(-127°)\}$
     (d) $13\{\cos(-113°) + i\sin(-113°)\}$
     (e) $\sqrt{5}\{\cos(-26.6°) + i\sin(-26.6°)\}$
     (f) $6\{\cos 0 + i\sin 0\}$
     (g) $3\{\cos \pi + i\sin \pi\}$
     (h) $4\{\cos\frac{1}{2}\pi + i\sin\frac{1}{2}\pi\}$
     (i) $2\sqrt{3}\{\cos(-\frac{5}{6}\pi) + i\sin(-\frac{5}{6}\pi)\}$
     (j) $25\{\cos(16.3°) + i\sin(16.3°)\}$

3. (a) $\sqrt{3} + i$ (b) $\frac{3}{2}\sqrt{2} - \frac{3}{2}i\sqrt{2}$
     (c) $-\frac{1}{2} + \frac{1}{2}i\sqrt{3}$ (d) $3 + 0i$
     (e) $-4 + 0i$ (f) $-\frac{1}{2}\sqrt{2} - \frac{1}{2}i\sqrt{2}$
     (g) $0 + 2i$ (h) $0 - 2i$

Exercise 21e – p. 391

1. (a) $(2\sqrt{3})(\sqrt{2}) = 2\sqrt{6}$;
     $-\frac{1}{6}\pi + (-\frac{1}{4}\pi) = -\frac{5}{12}\pi$
     (b) $(3 - \sqrt{3}) - (3 + \sqrt{3})i$

2. (a) $(\sqrt{2})(\sqrt{2}) = 2$; $\frac{1}{4}\pi + (-\frac{1}{4}\pi) = 0$
     (b) 2

3. (a) $(\sqrt{7}) \div (\sqrt{7}) = 1$;
     $(-139°) - (139°) = -278° \Rightarrow 172°$
     (b) $\frac{1}{7} + \frac{4}{7}i\sqrt{3}$

4. (a) $2 \div (\sqrt{2}) = \sqrt{2}$; $0 - (\frac{1}{4}\pi) = -\frac{1}{4}\pi$
     (b) $1 - i$

5. (a) $\dfrac{1}{z} = \dfrac{2-i}{5}$; $z^2 = 3 + 4i$

     (b) $\dfrac{1}{z} = 1 + i$; $z^2 = -\frac{1}{2}i$

     (c) $\dfrac{1}{z} = \dfrac{3-4i}{25}$; $z^2 = -7 + 24i$

     (d) $\dfrac{1}{z} = \frac{1}{2}\sqrt{3} - \frac{1}{2}i$; $z^2 = \frac{1}{2}(1 + i\sqrt{3})$

     (e) $\dfrac{1}{z} = \dfrac{5 + 12i}{13}$; $z^2 = -119 - 120i$

Mixed Exercise 21 – p. 391

1. (a) $\dfrac{7 - 5i}{74}$ (b) $11 + 2i$
     (c) $\frac{1}{2}(1 + i)$ (d) $-21 + 20i$

2. $\frac{3}{10}(11 - 3i)$

3. $\pm\frac{1}{2}\sqrt{2}(5 + i)$

4. (a) $\sqrt{2}, \frac{3}{4}\pi$ (b) 5, 53°
     (c) $5\sqrt{2}, -172°$, area $= 8.9$

5. (a) $2\sqrt{10}$ (b) 10   (c) 1
     (d) $\frac{1}{5}\sqrt{10}$

6. $\lambda = 2, \mu = \frac{1}{2}$

7. $a = 4, b = -5$

8. $\dfrac{x(x^2 + y^2 - 1)}{x^2 + y^2}, \dfrac{y(x^2 + y^2 + 1)}{x^2 + y^2}$

9. $\lambda = 4, \mu = -5$

10. $2\sqrt{2}, \frac{3}{4}\pi$; $2\sqrt{2}, -\frac{3}{4}\pi$

11. (a) $-\frac{1}{3}$ (b) $\frac{2}{3}$ (c) 0

12. $2 + i, 2 - 3i$

CONSOLIDATION E

Multiple Choice Exercise E – p. 396

1. B	9. D	17. A
2. A	10. D	18. B
3. B	11. E	19. B
4. A	12. D	20. C
5. C	13. B	21. F
6. E	14. E	22. T
7. A	15. B	23. F
8. C	16. B, C	24. F

Miscellaneous Exercise E – p. 400

1. $2\sqrt{2}\,(\cos\frac{1}{4}\pi + i\sin\frac{1}{4}\pi)$; $\sqrt{104}$;
   0.32 rad

2. (a)

a	−2	−4
b	2	1

   (b) $\sqrt{2}$; $-\frac{1}{4}\pi$

3. $p = -2, q = -4$

4. $\arg z_1 = \frac{2}{3}\pi$, $\arg z_2 = \frac{1}{6}\pi$; i, $\frac{1}{2}\pi$

5. (a)

a	5	−5
b	−3	3

   ; $\pm (5 - 3i)$

6. (a) $\frac{1}{2}\pi$ (b) $-\frac{1}{4}\pi$ (c) $\frac{1}{12}\pi$

7. $x = 3, y = -1$;
   (a) $\sqrt{10}$ (b) $-\arctan\frac{1}{3}$

8. (a) $-1, 7$ (b) $-7i$

9. 10, 96.9°; $\frac{5}{2}$, −23.1°

10. 69°

11. 77.5°

12. $\mathbf{r} = -\mathbf{i} + \mathbf{j} + 3\mathbf{k} + \lambda(4\mathbf{i} - 3\mathbf{j} - \mathbf{k})$; −2

13. (a) 20, $\frac{20}{21}$
    (b) $\mathbf{r} = (2\mathbf{i} + 2\mathbf{j} + \mathbf{k}) + t(4\mathbf{i} + \mathbf{j} + \mathbf{k})$;
    $-\frac{14}{11}\mathbf{i} + \frac{13}{11}\mathbf{j} + \frac{2}{11}\mathbf{k}$

14. $\mathbf{r} = \mathbf{i} + 3\mathbf{j} + \mathbf{k} + t(3\mathbf{i} - 2\mathbf{j} - 8\mathbf{k})$;

15. (b) 28, 54.2°

16. (b) $\overrightarrow{PN} = \mathbf{q} - 2\mathbf{p}$, $\overrightarrow{QM} = \mathbf{p} - 2\mathbf{q}$
    (d) $\overrightarrow{OG} = 2\mu\mathbf{q} + (1 - \mu)\mathbf{p}$

17. (a) $\mathbf{r} = a(5\mathbf{i} - \mathbf{j} - \mathbf{k}) + \lambda(4\mathbf{i} + 4\mathbf{j} - 8\mathbf{k})$
    (d) $a(3\mathbf{i} - 3\mathbf{j} + 3\mathbf{k})$

18. $\overrightarrow{OC} = 2\mathbf{i} + 14\mathbf{j}$, $\overrightarrow{OD} = 6\mathbf{i} + 12\mathbf{j}$; 45°

19. $\frac{56}{5}\pi$

20. $\dfrac{\pi(e^{4a} - 1)}{2e^{2a}}$

22. $(\frac{1}{3}\pi, 2)$, max; $(\frac{4}{3}\pi, -2)$, min; $\frac{11}{6}\pi$;
    $\frac{1}{6}\pi(10\pi + 3\sqrt{3})$

23. (a) $1:4$ (b) $5:2$

24. $\frac{1}{3}\pi$

25. 1600π cm^3/s

26. 0.002 cm/s; 2.16 cm^3/s

27. (a) $\mathbf{b} - \mathbf{a}, \frac{3}{5}\mathbf{a}, 3\mathbf{b}, 3\mathbf{b} - \frac{3}{5}\mathbf{a}$
    (c) $\overrightarrow{OR} = \frac{3}{5}(1 - k)\mathbf{a} + 3k\mathbf{b}$
    (d) $h = \frac{1}{2}, k = \frac{1}{6}$; PR:RQ $= 1:5$

28. 1.3

29. 1.69

30. (a) $9, \frac{1}{2}\pi$
    (b) $3, \frac{3}{4}\pi$;   , $\dfrac{27\sqrt{2}}{2}$

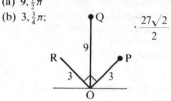

31. (b) 138°
    (c) $v = 2 - i$

32. $-3 + 2i$

33. (b) $5i + \frac{2}{3}\mathbf{j}$ (c) $5i + 3\mathbf{j}$

34. $\frac{3}{4} + \frac{1}{3}\pi$

35. $\frac{4}{17}$; 8.26

36. (a) $7; 6; \frac{20}{21}$ (b) $2\mathbf{i} + \frac{25}{4}\mathbf{j} + \frac{12}{5}\mathbf{k}$
    (c) $\lambda = -\frac{9}{4}$; E is on AB produced

37. 65.5 m^2

38. $\dfrac{dp}{dv} = -1.4\dfrac{p}{v}$; 1.4$c$%

39. 112°, −112° (to nearest degree); $\frac{29}{4}$

40. $5\ln 5 - 4$; 15.4

41. 0.78

42. 1.2315×10^{14} km^3/s

43. $k = \frac{1}{36}\pi$; 1.5%

44. (a) $5\mathbf{i} - 3\mathbf{j} + 8\mathbf{k}$ (b) $(-4, -4, 7)$
    (d) $(-4, -4, 7)$

INDEX

NOTES